The chemistry of
the carbonyl group
Volume 2

THE CHEMISTRY OF FUNCTIONAL GROUPS

A series of advanced treatises under the general editorship of
Professor Saul Patai

The chemistry of alkenes (published in 2 volumes)
The chemistry of the carbonyl group (published in 2 volumes)
The chemistry of the ether linkage (published)
The chemistry of the amino group (published)
The chemistry of the nitro and nitroso groups (published in 2 parts)
The chemistry of the carboxylic acids and esters (published)
The chemistry of the carbon-nitrogen double bond (published)

The chemistry of
the carbonyl group
Volume 2

Edited by

JACOB ZABICKY

The Weizmann Institute of Science, Rehovoth, Israel

1970

INTERSCIENCE PUBLISHERS

a division of John Wiley & Sons

LONDON – NEW YORK – SYDNEY – TORONTO

Library of Congress Catalogue Card No. 66–18177
ISBN 0 471 98051 X

Made and Printed in Great Britain by
William Clowes and Sons Limited, London and Beccles

To Z. Shifmanovich

Foreword

Two main purposes were to be served by publishing a second volume to *The Chemistry of the Carbonyl Group*, namely, to include chapters which failed to materialize in the first volume, and to widen the scope of the original plan.

Thus, a chapter on 'equilibrium additions', dealing with a group of fundamental reactions of carbonyl compounds, appears now. On the other hand, the commitment to write a chapter on the syntheses and applications of isotopically labelled carbonyl compounds remained, as before, unfulfilled.

The topic 'oxidations of carbonyl compounds' was restricted in the previous volume to the oxidation of aldehydes by the transition metals. In this volume the behaviour of many other oxidants is considered, and ketones are also included as substrates. A chapter is dedicated to 'enolization', one of the most widely studied organic rearrangements, which undoubtedly deserves special attention in a treatise on the carbonyl group.

In the first volume a chapter dealt with the effects of ultra-violet and visible radiation on carbonyl compounds. The effects of other types of radiation are considered here in two chapters, one on 'radiation chemistry' and one on 'mass spectroscopy'. The latter also affords an extension to the coverage of the analytical aspects of the carbonyl group presented earlier.

In most volumes in the series *The Chemistry of Functional Groups*, certain classes of compounds related to the functional group under consideration are assigned separate chapters. The chemical behaviour and properties of the 'oxocarbons' are treated in this volume.

Owing to technical reasons this volume appeared one year later than scheduled. I would like, therefore, to express my obligation to the contributing authors for their forbearance of such a delay.

I wish to thank Professor Saul Patai for his advice and also my wife for her encouragement and patience.

The Weizmann Institute of Science
Rehovoth, 1970   JACOB ZABICKY

The Chemistry of the Functional Groups
Preface to the series

The series 'The Chemistry of the Functional Groups' is planned to cover in each volume all aspects of the chemistry of one of the important functional groups in organic chemistry. The emphasis is laid on the functional group treated and on the effects which it exerts on the chemical and physical properties, primarily in the immediate vicinity of the group in question, and secondarily on the behaviour of the whole molecule. For instance, the volume *The Chemistry of the Ether Linkage* deals with reactions in which the C—O—C group is involved, as well as with the effects of the C—O—C group on the reactions of alkyl or aryl groups connected to the ether oxygen. It is the purpose of the volume to give a complete coverage of all properties and reactions of ethers in as far as these depend on the presence of the ether group, but the primary subject matter is not the whole molecule, but the C—O—C functional group.

A further restriction in the treatment of the various functional groups in these volumes is that material included in easily and generally available secondary or tertiary sources, such as Chemical Reviews, Quarterly Reviews, Organic Reactions, various 'Advances' and 'Progress' series as well as textbooks (i.e. in books which are usually found in the chemical libraries of universities and research institutes) should not, as a rule, be repeated in detail, unless it is necessary for the balanced treatment of the subject. Therefore each of the authors is asked *not* to give an encyclopaedic coverage of his subject, but to concentrate on the most important recent developments and mainly on material that has not been adequately covered by reviews or other secondary sources by the time of writing of the chapter, and to address himself to a reader who is assumed to be at a fairly advanced post-graduate level.

With these restrictions, it is realized that no plan can be devised for a volume that would give a *complete* coverage of the subject with *no* overlap between chapters, while at the same time preserving the readability of the text. The Editor set himself the goal of attaining *reasonable* coverage with *moderate* overlap, with a minimum of cross-references between the chapters of each volume. In this manner, sufficient freedom is given to each author to produce readable quasi-monographic chapters.

The general plan of each volume includes the following main sections:

(a) An introductory chapter dealing with the general and theoretical aspects of the group.

(b) One or more chapters dealing with the formation of the functional group in question, either from groups present in the molecule, or by introducing the new group directly or indirectly.

(c) Chapters describing the characterization and characteristics of the functional groups, i.e., a chapter dealing with qualitative and quantitative methods of determination including chemical and physical methods, ultraviolet, infrared, nuclear magnetic resonance, and mass spectra; a chapter dealing with activating and directive effects exerted by the group and/or a chapter on the basicity, acidity or complex-forming ability of the group (if applicable).

(d) Chapters on the reactions, transformations and rearrangements which the functional group can undergo, either alone or in conjunction with other reagents.

(e) Special topics which do not fit any of the above sections, such as photochemistry, radiation chemistry, biochemical formations and reactions. Depending on the nature of each functional group treated, these special topics may include short monographs on related functional groups on which no separate volume is planned (e.g. a chapter on 'Thioketones' is included in the volume *The Chemistry of the Carbonyl Group*, and a chapter on 'Ketenes' is included in the volume *The Chemistry of Alkenes*). In other cases, certain compounds, though containing only the functional group of the title, may have special features so as to be best treated in a separate chapter, as e.g., 'Polyethers' in *The Chemistry of The Ether Linkage*, or 'Tetraaminoethylenes' in *The Chemistry of the Amino Group*.

This plan entails that the breadth, depth and thought-provoking

1*

nature of each chapter will differ with the views and inclinations of the author and the presentation will necessarily be somewhat uneven. Moreover, a serious problem is caused by authors who deliver their manuscript late or not at all. In order to overcome this problem at least to some extent, it was decided to publish certain volumes in several parts, without giving consideration to the originally planned logical order of the chapters. If after the appearance of the originally planned parts of a volume it is found that either owing to non-delivery of chapters, or to new developments in the subject, sufficient material has accumulated for publication of an additional part, this will be done as soon as possible.

The overall plan of the volumes in the series 'The Chemistry of the Functional Groups' includes the titles listed below:

The Chemistry of the Alkenes (published in two volumes)
The Chemistry of the Carbonyl Group (published in two volumes)
The Chemistry of the Ether Linkage (published)
The Chemistry of the Amino Group (published)
The Chemistry of the Nitro and Nitroso Group (published in two parts)
The Chemistry of Carboxylic Acids and Esters (published)
The Chemistry of the Carbon–Nitrogen Double Bond (published)
The Chemistry of the Cyano Group (in press)
The Chemistry of the Amides (in press)
The Chemistry of the Carbon–Halogen Bond (in preparation)
The Chemistry of the Hydroxyl Group (in press)
The Chemistry of the Carbon–Carbon Triple Bond
The Chemistry of the Azido Group (in preparation)
The Chemistry of Imidoates and Amidines
The Chemistry of the Thiol Group
The Chemistry of the Hydrazo, Azo and Azoxy Groups
The Chemistry of Carbonyl Halides (in preparation)
The Chemistry of the SO, SO_2, —SO_2H and —SO_3H Groups
The Chemistry of the —OCN, —NCO and —SCN Groups
The Chemistry of the —PO_3H_2 and Related Groups

Advice or criticism regarding the plan and execution of this series will be welcomed by the Editor.

The publication of this series would never have started, let alone continued, without the support of many persons. First and foremost among these is Dr. Arnold Weissberger, whose reassurance and trust encouraged me to tackle this task, and who continues to help and advise me. The efficient and patient cooperation of several staff-members of the Publisher also rendered me invaluable aid (but unfortunately their code of ethics does not allow me to thank them by name). Many of my friends and colleagues in Jerusalem helped me in the solution of various major and minor matters, and my thanks are due especially to Prof. Y. Liwschitz, Dr. Z. Rappoport and Dr. J. Zabicky. Carrying out such a long-range project would be quite impossible without the non-professional but none the less essential participation and partnership of my wife.

The Hebrew University,   SAUL PATAI
Jerusalem, ISRAEL

Contributing authors

J. H. Bowie — University of Adelaide, Australia.

S. Forsén — Lund Institute of Technology, Lund, Sweden.

Gordon R. Freeman — University of Alberta, Edmonton, Canada.

Atsushi Kawasaki — Nagoya University, Chikusa-ku, Nagoya, Japan.

M. Nilsson — Royal Institute of Technology, Stockholm, Sweden.

Joseph Niu — Wyandotte Chemical Corporation, Wyandotte, Michigan, U.S.A.

Yoshiro Ogata — Nagoya University, Chikusa-ku, Nagoya, Japan.

Herbert S. Verter — Inter-American University of Puerto Rico, San Germán, Puerto Rico.

Robert West — University of Wisconsin, Madison, Wisconsin, U.S.A.

Contents

Contents

CHAPTER **1**

Equilibrium additions to carbonyl compounds

YOSHIRO OGATA and ATSUSHI KAWASAKI

Nagoya University, Chikusa-ku, Nagoya, Japan

1

I. INTRODUCTION

The reversible additions to carbonyl double bonds (equation 1) at room temperature include reactions such as the formation of *gem*-glycols, hydroxyhydroperoxides, acetals, cyanohydrins, thioacetals, amino alcohols, etc. Some of them have been extensively studied by physical-organic chemists as typical polar reactions to elucidate structural effects or acid–base catalysis. Reviews on acetals[1] and Schiff bases[2] have already appeared in this series, and some physical-organic aspects of the reactions leading to their formation will be emphasized here.

$$\begin{array}{c}\diagdown \\ \diagup\end{array} C{=}O + XY \rightleftharpoons \begin{array}{c}\diagdown \\ \diagup\end{array} C{-}OY \\ \qquad\qquad\qquad\qquad | \\ \qquad\qquad\qquad\qquad X \tag{1}$$

Since the carbonyl group is strongly polarized to $\diagup \overset{\delta+}{C}{\cdots}\overset{\delta-}{O}$, the anionic portion X of the reagent becomes attached to the carbon atom, while the cationic portion Y becomes attached to the oxygen atom (usually Y = H). The polarity of the C=O bond makes this addition much easier than the addition to a non-polar C=C bond. For example, hydrogen cyanide or ammonia does not add to ethylene at room temperature, while they can add to formaldehyde very rapidly, forming methylene cyanohydrin and hexamine, respectively.

Most of these additions and their reverse reactions are subject to general base and/or general acid catalysis. The necessary conditions for the reversibility of addition reaction (1) are that Y should easily be eliminated as a cation and that X should be stable as an anion or a nucleophile, e.g., OH^-, OR^-, CN^-, SO_3H^- or NH_3. If X^- is an unstable anion such as hydride ion or carbanion, the adduct $\diagup CX{-}OY$ is stable, the addition being irreversible. Most of these additions and their reverse reactions are subject to general base and/or general acid catalysis. The basic catalyst is used to abstract cation Y^+ (usually H^+), then the formed anion $\diagup CX{-}O^-$ is stabilized by leaving X^-, if X^- can be eliminated as a stable anion. On the other hand, the acid catalyst acts to protonate (partially) X, pulling to cleave it as XH, if X^- is a stable anion. Hence the adduct which is stable in a neutral state may decompose on addition of acid or base.

It can be expected that the electron-releasing resonance form of OH groups (equation 2) may contribute to the cleavage of the C—X bond (the less electron-releasing halogen group is a poor activator of the cleavage as compared to OH)[3].

$$\begin{matrix} \diagdown \\ \diagup \end{matrix} \underset{X}{\overset{|}{C}} \!\!-\!\! OH \rightleftharpoons \begin{matrix} \diagdown \\ \diagup \end{matrix} \underset{X^-}{C} \!\!=\!\! \overset{+}{O}H \qquad (2)$$

The addition products of XH are α-substituted alcohols, and are sometimes easily dehydrated either intra- or intermolecularly to give a double bond or an ether linkage, for example, in the formation of Schiff bases or acetals.

II. ADDITION OF WATER

When dissolved in water, ketones and aldehydes undergo hydration, yielding *gem*-diols or hydrates[4]:

$$\underset{R^2}{\overset{R^1}{\diagdown}} C\!=\!O + H_2O \rightleftharpoons \underset{R^2}{\overset{R^1}{\diagdown}} C \underset{\diagdown OH}{\overset{\diagup OH}{}} \qquad (3)$$

The extent of hydration and stability of the *gem*-diol depend on the structure of the carbonyl compound. Aldehydes having strong electron-attracting groups (for example chloral) give stable solid *gem*-diols, but even acetaldehyde hydrate may be obtained as crystals at low temperature[5].

The departure from linearity of density and refractive index[5], the molecular weight determined by cryoscopy[6] and ebullioscopy[7,8], and the ultraviolet[9,10] and Raman[11] spectra of the carbonyl group reveal that formaldehyde in aqueous solution is almost completely hydrated. Oxygen isotope exchange is observed between $H_2^{18}O$ and some aldehydes and ketones, suggesting the reversible formation of an hydrate intermediate[12-14].

$$\underset{R^2}{\overset{R^1}{\diagdown}} C\!=\!O + H_2^{18}O \rightleftharpoons \underset{R^2}{\overset{R^1}{\diagdown}} C \underset{\diagdown OH}{\overset{\diagup ^{18}OH}{}} \rightleftharpoons \underset{R^2}{\overset{R^1}{\diagdown}} C\!=\!^{18}O + H_2O \qquad (4)$$

A. Equilibria

The hydration of carbonyl compounds reaches equilibrium very rapidly. The equilibrium constants may be estimated as follows: (a) By measuring the absorbance of carbonyl $(n \to \pi^*)$ at 270–320 mμ which disappears on hydration[15-22]. (b) Since proton magnetic resonance absorptions (aldehyde or α-proton) of free and hydrated carbonyls are different, the ratio of two peak areas gives the ratio of free to hydrated carbonyls[23-28]. E.g., the methyl proton band shifts 0·9 p.p.m. and that of the aldehyde proton 4·9 p.p.m. in the hydrated form as compared to the free form of acetaldehyde[24]. (c) Similarly the ratio of two peak areas in the ^{17}O nuclear magnetic resonance spectrum[28] using carbonyl-^{17}O can be used. (d) The determination of free aldehyde by polarography[29-31]. The hydration equilibrium (3) is very sensitive to temperature; consequently rigorous control is necessary[21,22] when measuring the equilibrium constants. Ultraviolet spectrophotometry is a convenient and extensively used method. Its main shortcomings are (i) that it arbitrarily assumes the identity of the extinction coefficient of carbonyl in aprotic solvents, for example n-hexane, and in water[26], and (ii) that sometimes the overlapping of the $n \to \pi^*$ absorption with others makes measurements difficult.

The dissociation constant of *gem*-diol (the reverse of reaction 3) in an excess of water is given by equation (5). If the carbonyl compounds are insoluble in water, the constant may be determined in a mixed solvent like dioxan–water[18,21].

$$K_d = \frac{[R^1R^2C{=}O]}{[R^1R^2C(OH)_2]} = \frac{1}{K_h} \tag{5}$$

The observed constants K_d (Table 1) tend to increase with increasing electron-releasing power of substituent, for example[28]:

$$Cl_3CCHO < ClCH_2CHO < CH_3CHO$$

$$HCHO < CH_3CHO < CH_3CH_2CHO < (CH_3)_2CHCHO < (CH_3)_3CCHO$$

No simple Taft equation[32] can be applied to the dissociation constants of these aliphatic carbonyls, and steric and hyperconjugative effects have to be taken into account. The former effects are considered by Bell in the following equation[4]:

$$-\log K_d = \rho^* \sum \sigma^* + \delta \sum E_s + C \tag{6}$$

where σ^* is a Taft polar substituent constant, E_s is a steric substituent constant, $\rho^* = 2·12 \pm 0·29$, $\delta = 1·12 \pm 0·26$ and $C = -2·10 \pm$

TABLE 1. ^{17}O magnetic resonance spectra (O.m.r.) and K_d of carbonyl compounds in aqueous solution[28].

| | | O.m.r. chemical shift (p.p.m.)[a] | | |
| | | gem-Diol | Carbonyl | |
No.	Compound	oxygens	oxygen	K_d
1	CH_2O	−51 (140)	[b]	5.0×10^{-4}
2	CH_3CHO	−67 (190)	−550 (45)	0.7
3	CH_3CH_2CHO	−65 (245)	−538 (70)	1.4
4	$CH_3(CH_2)_2CHO$			2.1
5	$(CH_3)_2CHCHO$			2.3
6	$(CH_3)_3CCHO$			4.1
7	CH_2ClCHO			2.7×10^{-2}
8	CCl_3CHO	−56 (360)	[b]	3.6×10^{-5}
9	CH_3COCH_3	[b]	−523 (90)	5×10^2
10	$CH_2ClCOCH_3$	[b]	−528 (235)	9.1
11	$(CH_2Cl)_2CO$	−67 (525)	−521 (440)	0.10
12	$Cl_2CHCOCH_3$			0.35
13	$CH_3COCOCH_3$	−59 (595)	−530	0.50
14	$CH_3COCOOCH_3$			0.32
15	$CH_3COCOOH^c$	−68 (~480)	−555 (780)	0.42
16	$CH_3COCOONa$	[b]	−534 (410)	18.5

[a] Values in brackets are the widths of the bands at half of their maximum intensities.
[b] Not detected.
[c] Chemical shift of carboxylic carbonyl oxygen −245 (400).

0.42. On the other hand, the Greenzaid–Luz–Samuel equation is based on the measurements using ^1H or ^{17}O n.m.r.

$$-\log K_d = \rho^* \sum \sigma^* + B\Delta + C \qquad (7)$$

where Δ is the number of aldehyde hydrogens (ketones, 0; formaldehyde, 2; other aldehydes, 1). $\rho^* = 1.70 \pm 0.07$, $B = 2.03 \pm 0.10$ and $C = -2.81 \pm 0.13$. B represents the stabilization of unhydrated carbonyl group due to substitution by an alkyl group[28,33], i.e. the effect is consistent with the adjacent bond interaction[34a,b] in its direction and magnitude, which is estimated to be 2.8 kcal per alkyl group. A similar effect of alkyl groups has been observed with other carbonyl reactions[35,36]. Table I shows the observed K_d values and Figure 1 the agreement between these values and those calculated according to equation (7), over a range of 10^7.

It is of interest to note that $1/K_d$ or K_h of pyruvic acid is directly proportional to $[H_2O]^3$; this suggests two hydrogen bonds being formed between a molecule of pyruvic acid hydrate and two molecules of water[22,26] (equation 8).

$$CH_3COCO_2H + 3 H_2O \rightleftharpoons CH_3-\overset{\displaystyle OH\cdots OH_2}{\underset{\displaystyle OH\cdots OH_2}{\overset{|}{\underset{|}{C}}}}-CO_2H \qquad (8)$$

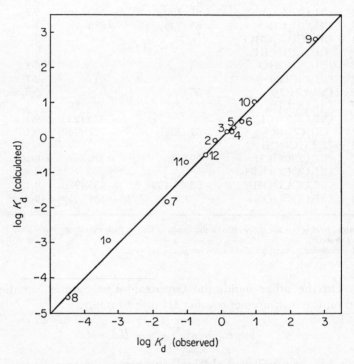

FIGURE 1. Comparison of observed and calculated values of K_d. (The numbers correspond to those in Table 1.)[28]

B. Kinetics and Mechanisms

Both the hydration of aldehydes and ketones and the reverse reaction are very fast but a number of methods have been devised for measuring their rates[37]: (a) measurement of volume change on hydration by dilatometry[38-40]; (b) measurement of temperature change caused by the heat of hydration in an adiabatic system (thermal method)[41-43]; (c) measurement of n.m.r. line-broadening assigned to protons or ^{17}O in free and hydrated carbonyl compounds[44-48]; (d) ultraviolet spectrophotometry of the free carbonyl absorption $(n \rightarrow \pi^*)$[49]; (e) the chemical scavenging method by which a free carbonyl is rapidly converted into a semicarbazone, oxime or phenylhydrazone, its concentration being monitored continuously by ultraviolet spectrophotometry[50]; (f) infrared spectrophotometry of the rate of oxygen exchange via hydration, using the frequency shift between $C=^{16}O$ and $C=^{18}O$ bonds ($C=^{16}O$, 1664 cm^{-1} and $C=^{18}O$, 1635 cm^{-1} in benzophenone[51]); (g) measurement of oxygen exchange rate by mass spectrograph or pressure float apparatus[13]; (h) polarographic determination of free carbonyl[52-60]; (i) refractive index measurements[38]; (j) conductivity measurements of the shift in the equilibrium induced by an abrupt pressure change (pressure jump method)[61]. Thus, for example, the hydroxonium ion-catalysed dehydration of acetaldehyde hydrate has been estimated to be 380 l mole^{-1} s^{-1} by the thermal method, 565 l mole^{-1} s^{-1} by the scavenging method and 490 l mole^{-1} s^{-1} by n.m.r., all of which are virtually identical and show the reliability of these methods[46].

TABLE 2. Acidity constants K_a, acid catalytic constants k_a and base cataltyic constants of conjugate bases k_b (l mole^{-1} s^{-1}) for the hydration of acetaldehyde[42].

Acid	K_a	k_a	k_b	$10^6 K_a k_b / k_a$
1 H$_3$O$^+$	55.5	930	0.00014	8.4
2 Formic	1.77×10^{-4}	1.74	0.065	6.6
3 Phenylacetic	4.88×10^{-5}	0.91	0.054	2.9
4 Acetic	1.75×10^{-5}	0.47	0.157	5.8
5 Trimethylacetic	9.4×10^{-6}	0.33	0.161	4.6
6 Pyridinium	6.0×10^{-6}	0.138	0.177	7.7
7 β-Picolinium	2.9×10^{-6}	0.115	0.26	6.5
8 α-Picolinium	1.10×10^{-6}	0.063	0.20	3.5
9 γ-Picolinium	1.05×10^{-6}	0.087	0.23	2.8
10 2,5-Lutidinium	4.0×10^{-7}	0.046	0.163	1.4
11 2,4-Lutidinium	2.34×10^{-7}	0.040	0.275	1.6
12 2,6-Lutidinium	1.91×10^{-7}	0.029	0.130	0.9
13 2,4-Dichlorophenol	1.8×10^{-8}	—	3.0	—
14 Water	1.8×10^{-16}	(0.00014)	8×10^4	—

The hydration reaction is both general acid- and general base-catalysed, the rate in a single buffer solution being expressed as[13,30-43,49,50,62]:

$$k = k_0 + k_H[H_3O^+] + k_{OH}[OH^-] + \left(k_a + k_b \frac{[B]}{[HA]} \right) [HA] \quad (9)$$

k_0, k_H, k_{OH}, k_a and k_b are catalytic constants of solvent, hydronium ion, hydroxide ion, general acid and base, respectively. Table 2 and Figure 2 show these constants and a Brønsted plot.

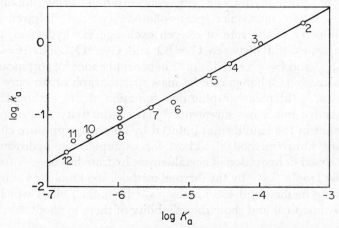

FIGURE 2. Acid catalytic constant k_a versus acidity constant K_a for the hydration of acetaldehyde. (The numbers correspond to those in Table 2.)[42]

The Brønsted catalysis law[63] (equation 10) is applicable to the hydration reaction of acetaldehyde.

$$\log \frac{k_a}{p} = \alpha \log \frac{qK_a}{p} + \log G_a \qquad (10)$$

k_a is the acid catalytic constant and K_a the acidity constant of the acid catalyst, G_a is a constant; p and q are statistical factors, p being the number of protons to be abstracted and q the number of atoms to which the protons add (for example, $p = 1$ and $q = 2$ for monocarboxylic acid). The value of α represents the sensitivity of the catalytic constant to a change of acidity of the acid catalysts, and is a measure of proton transfer from acid to substrate in the transition state; the values of α are 0·54 for acetaldehyde[39,42] and 0·27 for 1,3-dichloroacetone[49]. The corresponding values (β) in the base-catalysed hydration are 0·53 for acetaldehyde[40], 0·55 for ninhydrin[64], while the value for the dehydration of methylene glycol[53–57] is 0·41.

The solvent isotope effect, k_{H_2O}/k_{D_2O}, for the hydration of acetaldehyde has a considerably large value, 3·6–3·9, which suggests the participation of a water molecule as a proton donor or acceptor[65]. These observations lead to the following mechanisms which may operate simultaneously[66].

General acid catalysis

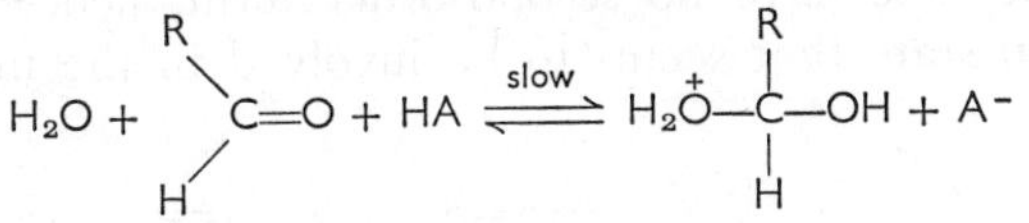

$$H_2O + \quad \overset{R}{\underset{H}{\diagdown}}C{=}O + HA \quad \xrightarrow{\text{slow}} \quad H_2\overset{+}{\underset{H}{\overset{R}{\underset{|}{\overset{|}{C}}}}}{-}OH + A^- \qquad (11)$$

$$H_2\overset{+}{\underset{H}{\overset{R}{\underset{|}{\overset{|}{C}}}}}{-}OH + A^- \quad \xrightarrow{\text{fast}} \quad HO{-}\overset{R}{\underset{H}{\overset{|}{\underset{|}{C}}}}{-}OH + HA \qquad (12)$$

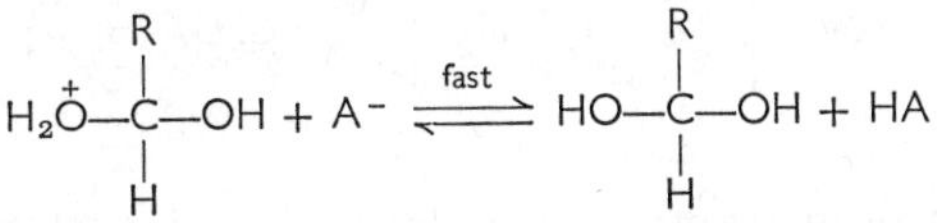

General base catalysis

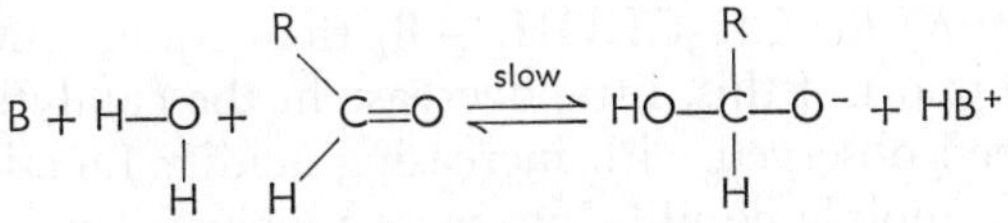

$$B + H{-}\underset{H}{\overset{|}{O}} + \quad \overset{R}{\underset{H}{\diagdown}}C{=}O \quad \xrightarrow{\text{slow}} \quad HO{-}\overset{R}{\underset{H}{\overset{|}{\underset{|}{C}}}}{-}O^- + HB^+ \qquad (13)$$

$$HO{-}\overset{R}{\underset{H}{\overset{|}{\underset{|}{C}}}}{-}O^- + HB^+ \quad \xrightarrow{\text{fast}} \quad HO{-}\overset{R}{\underset{H}{\overset{|}{\underset{|}{C}}}}{-}OH + B \qquad (14)$$

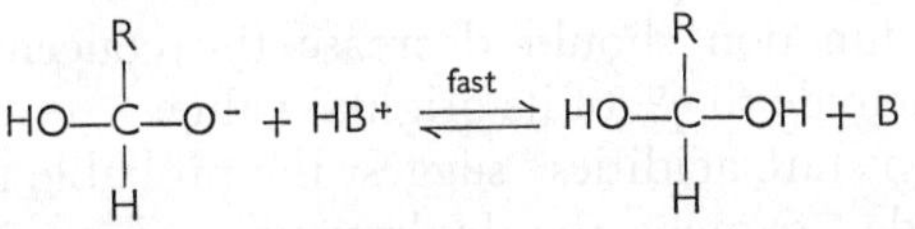

These mechanisms lead to the rate equation consisting of a sum of several terms (e.g., equation 9), but the contribution of a term $k_{ab}[HA][B]$ is little or not observed[41], hence the concerted mechanisms involving a transition state such as **1** cannot be accepted[67].

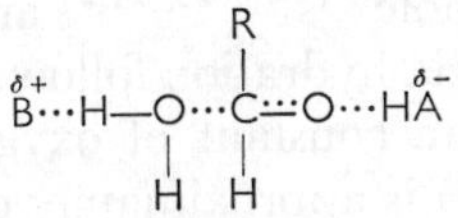

$$\overset{\delta+}{B}\cdots H{-}O\cdots\overset{R}{\underset{H}{\overset{|}{\underset{|}{C}}}}{\cdots}O\cdots H\overset{\delta-}{A}$$

$$\underset{H}{(1)}$$

However, a cyclic mechanism (equation 15) involving a few molecules of water and a molecule of catalytic acid cannot be excluded[68].

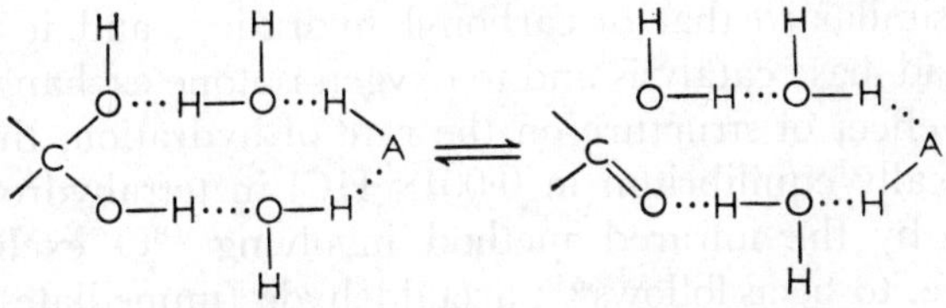

$$(15)$$

The rate constant for the imidazole-catalysed hydration of 1,3-dichloroacetone in 95% dioxan is expressed as follows (Im = imidazole):

$$k = k_1 [Im] + k_2 [Im]^2 \qquad (16)$$

However, other secondary or tertiary amine catalysts including
N-methylimidazole have no second-order term, hence **2** represents
the transition state that seems to be involved in the imidazole catalysis[69].

$$-O\text{---}\overset{|}{\underset{|}{C}}\cdots O\cdots H\cdots N\overset{+}{\diagdown}N\cdots H\cdots N\overset{+}{\diagdown}NH$$

(2)

The slow proton transfer in hydroxonium ion-catalysed hydration
of acetaldehyde is not surprising, because the aldehyde is a much
weaker base (pK_a for $CH_3\overset{+}{C}HOH$, -8) than water (pK_a for H_3O^+,
-1.74). In support of this, little decrease in the catalytic constant of
H_3O^+ has been observed with increasing acidity function from 2.15
to 3.10[47]. If a mobile equilibrium were involved, an increase of 1.0
in the acidity function should decrease the concentration of protonated acetaldehyde to $\frac{1}{10}$ of its original value.

The transition state acidities* suggest the probable transition states
for acetaldehyde, namely the hydroxonium ion-catalysed rate is
determined by proton transfer, while the hydroxide- and water-
catalysed reactions have a concerted mechanism[20b].

The rate constant of oxygen exchange of acetaldehyde (ca. 470
l mole^{-1}s^{-1}) measured by ^{17}O n.m.r. is consistent with the rate constant
of hydration (480–670 l mole^{-1}s^{-1})[42,46,47] and this fact shows that the
oxygen exchange occurs via hydration followed by the dehydration[48].
The *pseudo* first-order rate constant of oxygen exchange in methylene glycol (2×10^{-3} s^{-1}) is approximately one half of that of its dehydration (5×10^{-3} s^{-1}), suggesting that the rate of dehydration
determines the rate of exchange, since the glycol, having two hydroxyl groups, can form an aldehyde by the scission of either of them[48].

The biologically important hydration of carbon dioxide also has a
mechanism similar to that of carbonyl hydration, and it is subject
to general acid–base catalysis and to oxygen isotope exchange[13,70–75].

As to the effect of structure on the rate of hydration, the time to
reach practically equilibrium in 0.001N HCl in tetrahydrofuran has
been found, by the infrared method involving ^{18}O exchange described above, to be as follows[51]: acetaldehyde (immediate) < benz-

---

* If the rate constant is expressed as in equation (9), and K_w is the dissociation
constant of water, the acidities of the hydroxonium ion-catalysed (p$K_H^\ddagger$) and
the uncatalysed (pK_0) transition states are given by:

$$pK_H^\ddagger = \log(k_H/k_0) \quad \text{and} \quad pK_0^\ddagger = \log(k_0/k_H) + pK_w$$

aldehyde (20 min) < 2-naphthaldehyde (25 min) < 1-naphthalde-
hyde (35 min) < 9-anthraldehyde (45 min) < 9-phenanthraldehyde
(55 min). This order agrees with that of the ability to delocalize the
positive charge introduced on protonation of the carbonyl oxygen,
i.e. the longer conjugate systems favour the dispersion of the positive
charge and discourage the nucleophilic attack of water.

The times necessary to virtually attain the pyridine-catalysed
hydration equilibrium of some ketones are[51]: acetone, ninhydrin (3)
(immediate) < cyclohexanone, cyclobutanone (10 min) < 2-cyclo-
pentenone (15 min) < cyclopentanone, fluorenone (1 h) < indanone
(3·5 h).

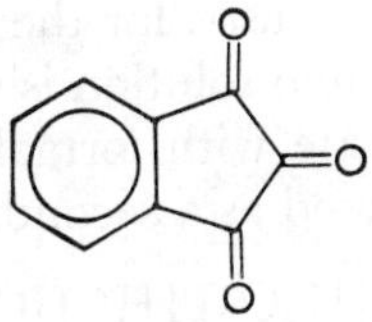

(3)

Acetone is most reactive, due to its low steric hindrance and the
absence of a strongly electron-releasing group; ninhydrin (3) because
it contains three highly electrophilic keto groups. The order of
reactivity of the cyclic ketones suggests that the release of internal
strain by hydration is an important factor in the rate of hydration.
Indeed, cyclopropanone, the most strained cyclic ketone, exists only
as its hydrate[76]. The O—$C_{(1)}$—O bond angle formed in the hydrate
is 110°. Cyclohexanone is more reactive than cyclopentanone probably
because some steric strain is released in the former but acquired in the
latter on formation of the *gem*-diol[51].

III. ADDITION OF HYDROGEN PEROXIDE

Typical reactions of aldehydes with aqueous hydrogen peroxide are
as follows[77]:

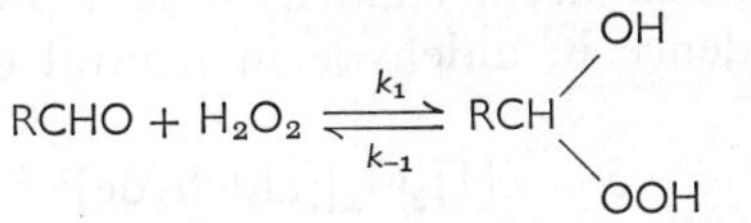

$$RCHO + H_2O_2 \underset{k_{-1}}{\overset{k_1}{\rightleftharpoons}} RCH\begin{smallmatrix}OH\\ \\OOH\end{smallmatrix} \qquad (17)$$

(4)

$$RCH\begin{smallmatrix}OH\\ \\OOH\end{smallmatrix} + RCHO \underset{k_{-2}}{\overset{k_2}{\rightleftharpoons}} RCH-OO-CHR \qquad (18)$$

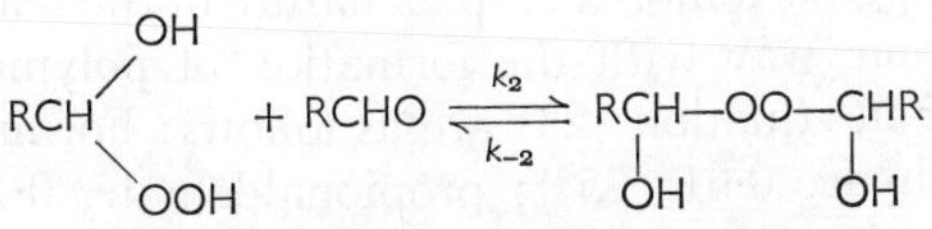

(5)

For example, a mixture of $Me_3CCH_2CHMeCH_2CHO$ (150 g) and 90% H_2O_2 (22 ml) in ethereal solution gives at 5° the hydroperoxide **4** (R = $Me_3CCH_2CHMeCH_2$)[78], while a mixture of 30% H_2O_2 (2 moles) and 6% aqueous HCHO (1 mole) gives at room temperature crystals of **5** (R = H), which on further addition of H_2O_2 (1 mole) followed by standing for several days affords the hydroperoxide **4**[79]. These peroxides generally tend to decompose.

The equilibrium constants of equations (17) and (18) for formaldehyde at 25° are 29 and 23 l mole^{-1}, respectively[79], and those for acetaldehyde at 0° are 48 and 10·2 l mole^{-1}, respectively[80].

The reaction of hydrogen peroxide with formaldehyde in alkaline solution is fast, and it may be used for the analysis of both reactants, while the same reaction in acid solution is slower, the rate being thus easily measurable[81]. The rate with formaldehyde measured by u.v. spectrophotometry is expressed as a second-order equation[79]:

$$v = k_1 [H_2O_2] [HCHO] \tag{19}$$

The reaction is subject to acid catalysis. The rate constant k_1 in an aqueous solution at 25° is expressed as:

$$k_1 = 0.08(1 + 53[H^+]) \text{ l mole}^{-1}\text{min}^{-1} \tag{20}$$

The reverse of reaction (18) with R = H is also subject to acid catalysis, the rate being:

$$k_{-2} = 9.0 \times 10^{-4}(1 + 53[H^+]) \text{ min}^{-1} \tag{20a}$$

The equations for k_1 and k_{-2}, together with the values of K_1 and K_2, give the values of k_{-1} and k_2 as follows[79]:

$$k_{-1} = 0.0024(1 + 53[H^+]) \text{ min}^{-1} \tag{21}$$

$$k_2 = 0.021(1 + 53[H^+]) \text{ l mole}^{-1} \text{ min}^{-1} \tag{22}$$

On the other hand, independent workers[77] postulated for the reactions of formaldehyde, acetaldehyde and propionaldehyde a $\frac{2}{3}$-order rate dependence in aldehyde in neutral or acid solutions at 0–30°:

$$v = k[H_2O_2][\text{aldehyde}]^{2/3} \tag{23}$$

The equation suggests a more complex nature of the reaction, possibly associated in some way with the formation of polymeric materials. The values of k in equation (23) are as follows: Formaldehyde, 0·11 (25°); acetaldehyde, 0·61 (25°); propionaldehyde, 0·75 (20°). This order is not in accordance with the electrophilicity of the carbonyl groups, but if the occurrence of a concurrent hydration reaction is taken into account (formaldehyde > acetaldehyde > propionaldehyde; section II.A), the increasing concentration of free aldehyde reactive towards H_2O_2 could explain the observed results.

The reaction does not have radical nature, i.e. it is insensitive to ferric ion, and it is acid catalysed. Therefore, the acid catalysis may proceed via the transition state **6**.

$$H-O-\overset{\delta+}{\underset{\underset{H}{|}}{O}}\cdots\overset{R}{\underset{\underset{H}{|}}{C}}\cdots O\cdots H\cdots\overset{\delta-}{A} \longrightarrow HOO-\overset{R}{\underset{\underset{H}{|}}{C}}-OH + HA \qquad (24)$$

$$(6)$$

In alkaline solution **5** (R = H) decomposes rapidly evolving hydrogen[82]. The reaction is accelerated at first by the addition of alkali, but it is retarded by an excess of alkali. A mechanism involving a rate-determining attack (26) of a hydrate anion on a molecule of hydrogen peroxide was postulated[82].

$$HCHO + OH^- \rightleftharpoons HCH(OH)O^- \qquad (25)$$

$$HCH(OH)O^- + H_2O_2 \overset{slow}{\rightleftharpoons} HCH(OH)O^{\cdot} + OH^- + {}^{\cdot}OH \qquad (26)$$

$${}^{\cdot}OH + HCH(OH)O^- \longrightarrow HCH(OH)O^{\cdot} + OH^- \qquad (27a)$$

$$2\,HCH(OH)O^{\cdot} \longrightarrow HOCH_2OOCH_2OH \longrightarrow 2\,HCO_2H + H_2 \qquad (27b)$$

The addition of alkali tends to increase the concentration of $HCH(OH)O^-$ but to decrease the concentration of free H_2O_2, hence the rate, as observed, increases with increasing $[OH^-]$ at first and then decreases. However, the following ionic mechanism for the formation of $HOCH_2OOCH_2OH$ seems to be more probable because the nucleophilicity of HOO^- is higher than that of OH^-.

$$HOO^- + HCHO \overset{slow}{\rightleftharpoons} HOOCH_2O^- \rightleftharpoons {}^-OOCH_2OH \qquad (28a)$$

$$HCHO + {}^-OOCH_2OH \rightleftharpoons {}^-OCH_2OOCH_2OH \underset{-H^+}{\overset{H_2O}{\rightleftharpoons}} HOCH_2OOCH_2OH \qquad (28b)$$

The increase of $[OH^-]$ results in an increase of $[HOO^-]$ and at the same time a decrease of $[HCHO]$ by the formation of $HCH(OH)O^-$, hence a maximum in rate with increasing $[OH^-]$ is observed. Both mechanisms agree with the observation that no deuterium is incorporated in the evolved hydrogen gas when the reaction is carried out in deuterium oxide[83].

Some aldehydes and ketones undergo irreversible reactions with H_2O_2, such as the Baeyer–Villiger reaction[84–86], in which esters are produced irreversibly via hydroperoxide adducts, for example.

$$PhCHO + H_2O_2 \rightleftharpoons PhCH(OH)OOH \xrightarrow{\text{rearrangement}} HCOOPh + PhCOOH + H_2O \quad (29)$$

The formed esters are easily hydrolysed to give the corresponding acids and alcohols (or phenols) in aqueous solution. Another reaction is the formation of several peroxides[87–89]; for example, the following is the reaction of acetone at 0° in a 1:1 molar ratio:

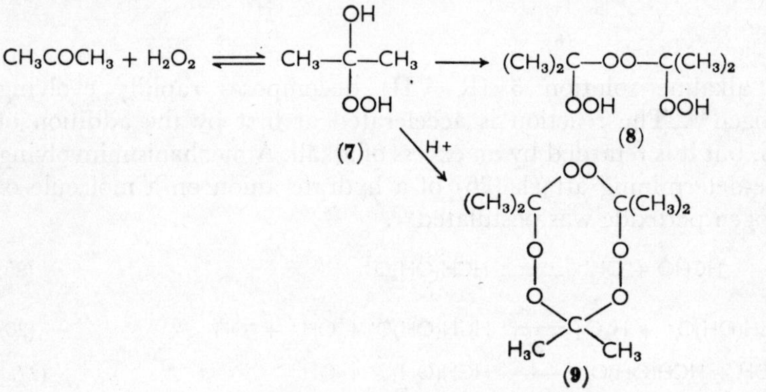

(30a)

(30b)

IV. ADDITION OF ALCOHOLS

For another review in this series see reference 1.

A. Hemiacetal and Hemiketal Formation

When a carbonyl compound is mixed with an alcohol, the alcohol molecule adds to the carbonyl double bond to form a hemiacetal or a hemiketal (equation 31), which, in the presence of acid, readily condenses with another alcohol molecule, dehydrating to an acetal or a ketal respectively (equation 32).

$$R^1R^2C{=}O + ROH \rightleftharpoons R^1R^2C(OH)(OR) \quad (31)$$

$$R^1R^2C(OH)(OR) + ROH \rightleftharpoons R^1R^2C(OR)(OR) + H_2O \quad (32)$$

Hemiacetals, like *gem*-diols, are generally too unstable to isolate. However, their existence is suggested by a number of physical properties of mixtures of carbonyl compounds and alcohols, for example, cryoscopic measurements[90,91], the heat of mixing[92,93], the deviation of refractive index and density from additivity[93–99], the decrease of the carbonyl absorption $(n \to \pi^*)$ in the ultraviolet region[100–107], the decrease in the intensity of carbonyl absorption bands and the increase in alcohol and ether bands in the infrared region[108,109], the shift in n.m.r. spectra[110], and the decrease of carbonyl in the measurement of optical rotatory dispersion[111].

Generally, ketones cannot form hemiketals to an appreciable extent, while aldehydes can form hemiacetals more easily, especially hemiacetals of aldehydes having electron-attracting groups, and hemiacetals with a cyclic-bound ether linkage are stable; for example, hemiacetals of chloral[112], glyoxylate[113], cyclopropanone[114], heterocyclic aldehydes[115] and also five or six-membered cyclic hemiacetals[116–118] are isolable.

The equilibrium constant of reaction (31) is increased by electron-attracting α-substituents in the carbonyl compound[101,119], and decreased by electron-attracting substituents in the alcohol. It also decreases with α-alkyl substituents in the alcohol in the following order, suggesting a steric requirement[94,101,119]:

$$\text{EtOH} > i\text{-PrOH} > t\text{-BuOH}$$

4-Hydroxybutyraldehyde (**10a**, $n = 2$) and 5-hydroxyvaleraldehyde (**10a**, $n = 3$) exist mainly as the cyclic hemiacetal (**10b**), while

$$\text{HOCH}_2(\text{CH}_2)_n\text{CH}{=}\text{O} \rightleftharpoons \text{H}_2\text{C}\underset{\text{O}}{\overset{(\text{CH}_2)_n}{\diagdown}}\text{CH—OH} \tag{33}$$

(**10a**)   (**10b**)

their higher $(n > 3)$ or lower $(n < 2)$ homologues have a lower content of their cyclic forms[104,105,120–122].

Dihydroxyacetone exists mainly as a dimeric hemiacetal (**11**)[103].

(**11**)

Glucose undergoes the mutarotation reaction (34) via an aldehyde intermediate (**12**)[122-128].

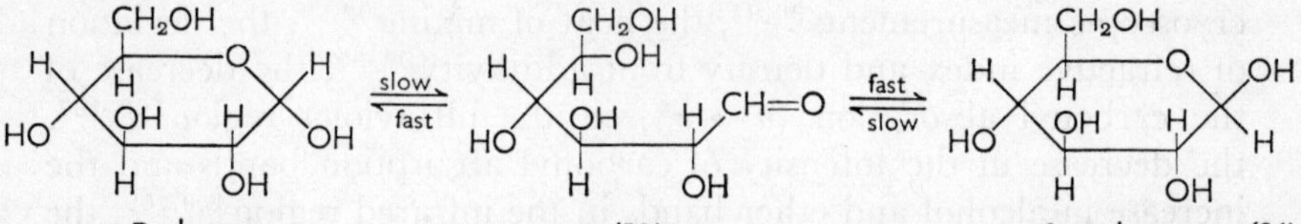

α–D-glucose　　　　　　　　　　(12)　　　　　　　　　β–D-glucose　　(34)

The assumption that cyclohexanone or cyclopentanone can form hemiketals in an acidic methanolic solution[111,129-131] is dubious since virtually no hydration of these ketones occurs and a trace of acid cannot stop the reaction at the stage of hemiketal; the main products should be the ketals.

The formation and decomposition of hemiacetals follow first-order kinetics and are subject to general acid and general base catalysis. They obey the Brønsted catalysis law[93,95,103,132,133]. The mechanisms of hemiacetal formation are analogous to those of hydration and the transition states are illustrated as follows:

General acid catalysis

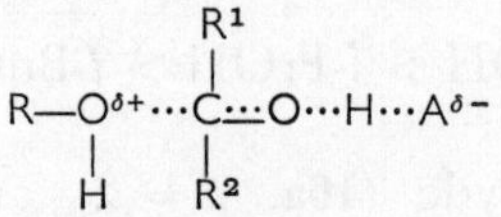

$$R{-}O^{\delta+}{\cdots}\overset{\overset{R^1}{|}}{\underset{\underset{R^2}{|}}{C}}{\cdots}O{\cdots}H{\cdots}A^{\delta-}$$

$$\underset{H}{|}$$

(**13**)

General base catalysis

$$B^{\delta+}{\cdots}H{\cdots}O{\cdots}\overset{\overset{R^1}{|}}{\underset{\underset{R^2}{|}}{C}}{\cdots}O^{\delta-}$$

$$\underset{R}{|}$$

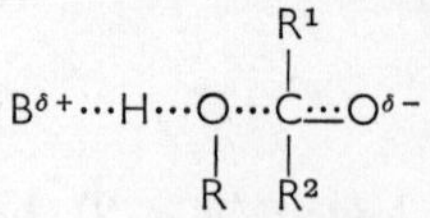

(**14**)

In an aprotic solvent such as benzene, however, the mutarotation of glucose is considered to proceed through a concerted mechanism[122]. For example, 2-hydroxypyridine, which is a ca. 10,000-fold weaker base than pyridine and a ca. 100-fold weaker acid than phenol, has much stronger catalysing ability (ca. 7,000-fold) for the mutarotation of tetramethylglucose than a mixture of equivalent amount of pyridine and phenol, which suggests the cyclic transition state **15**[134].

(15)

B. Acetal and Ketal Formation

Acetals are formed via hemiacetals from mixtures of carbonyl compounds and alcohols in the presence of a trace of acid (equation 35). At room temperature a mixture of aldehyde and alcohol without acid reaches equilibrium only after a long time[110,136].

$$R^1R^2C{=}O + 2\,ROH \underset{}{\overset{H^+}{\rightleftharpoons}} R^1R^2C \underset{OR}{\overset{OR}{<}} + H_2O \qquad (35)$$

Acetals are more stable than hemiacetals, and can be isolated by distillation without decomposition after neutralizing the acid. Ion exchange resins as well as strong acids may be used as catalysts for acetal formation[135-137]. A lower temperature favours the reaction.

Acetals can be identified and determined by infrared[136], mass[137,138], and n.m.r. spectroscopies[110]. The latter has the advantage of enabling the simultaneous determination of aldehyde, alcohol, acetal and hemiacetal, as shown in Figure 3.

When using ultraviolet spectrophotometry, the following procedure may be useful to distinguish whether the decrease in carbonyl absorbance is due to hemiacetal or acetal formation (or both): hemiacetal alone is formed in neutral or basic media, but both hemiacetal and acetal may be formed in acidic media. Consequently the difference in absorbances between neutral and acidic media is assigned to acetal formation[139-141]. Chemical methods such as oximation in neutral or basic media can be used to determine acetals but not hemiacetals[131].

The results from these measurements indicate that most aromatic aldehydes and acyclic ketones form acetals, but hemiacetal formation is very poor[107,136,139-143]. However, strong electron-attracting groups in these carbonyl compounds favour the formation of hemiacetal more than that of acetal[107,142].

2+c.c.g. II

The equilibrium constant, K, for acetal formation tends to increase with increasing electrophilicity of carbonyls and with increasing nucleophilicity of alcohols[144-149]. In the Taft equation, $\log (k/k_0) = \rho^* \sigma^*$, the ρ^* value is $+3.05$ for substituted carbonyls. α,β-Unsaturated and aromatic aldehydes do not undergo facile acetal formation[145,150].

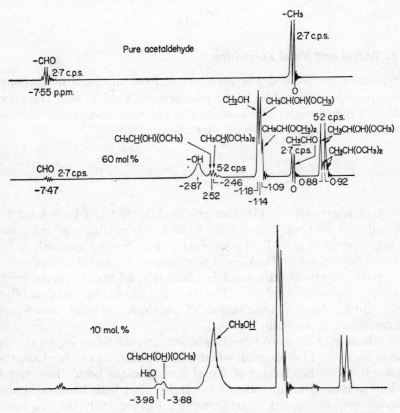

FIGURE 3. N.m.r. spectra of acetaldehyde and its methanol solution[110].

The relationship between the equilibrium constant of ketal formation and the ring size of cyclic ketones is similar to that of cyanohydrin formation (section V.A), i.e. the constant is maximum for cyclobutanone and cyclohexanone and decreases with increasing ring size. Four and six-membered cyclic ketones favour ketal formation by virtue of the release of internal strain, while an increase of internal strain by the ketal formation of five, seven and eight-membered cyclic ketones results in a decrease in the equilibrium constant[130,141]. Thus, 3-methylcyclohexanone (16) yields 93% ketal at equilibrium in methanol, while 3-methylcyclopentanone and heptanone yield only 24 and 21%, respectively[111].

The introduction of a methyl substituent in an α position reduces the yield of ketal, for example, 2,5-dimethylcyclohexanone (17) affords only 25% ketal, and 2,2,5-trimethylcyclohexanone (18) cannot give ketal at all[111]. The conformation of cyclic ketones also affects the equilibrium of ketal formation. Thus, the yield of ketal from 8,9-dimethyl-*trans*-3-decalone (19) is 68%, while the yield from its isomeric 2-decalone (20) is only 9% due to the 1,3-diaxial interaction between 9-methyl and 2-methoxy groups[111].

(16)   (17)   (18)   (19)   (20)

When acetals are hydrolysed in ^{18}O enriched water (reaction 36), the resulting alcohols contain little ^{18}O, suggesting that the fission takes place at the bond between the aldehyde carbon and oxygen[151].

$$R^1 \quad OR \qquad\qquad R^1$$
$$\underset{H}{\overset{}{C}} \underset{OR}{} + H_2{}^{18}O \rightleftharpoons \underset{H}{\overset{}{C}}{=}^{18}O + 2\,ROH \qquad (36)$$

In further support of this, the acetal of optically active $(-)$-α-phenylethyl alcohol is hydrolysed with retention of configuration[152–155].

Kinetic studies on acetal formation are few[90,133,140,151,156–158], but the reverse reaction, i.e. the hydrolysis of acetals, has been extensively studied[155,159–166]. The formation of hemiacetals is subject to general acid and general base catalysis, but acetal formation[157,158] and the hydrolysis of acetals[155,159–166] are subject to specific oxonium ion catalysis, and the following relationship between rate constant and acidity function H_0 is observed[161a,162,166].

$$\log k = -H_0 + \text{constant}$$

The solvent isotope effect, k_{H_2O}/k_{D_2O}, for the hydrolysis of ethylene, diethyl and dimethyl acetals in H_2O and D_2O is observed to be about $\frac{1}{3}$, which is ascribed to the lower concentration of protonated substrate in H_2O than in D_2O[164]. All these results point to a mechanism involving a carbonium ion intermediate (21) as follows:

$$R^1R^2C\diagup^{OR}_{\diagdown OH} + H^+ \rightleftharpoons [R^1R^2C\cdots OR]^+ + H_2O \qquad (37)$$

$$(21)$$

$$[R^1R^2C\cdots OR]^+ + ROH \underset{slow}{\overset{slow}{\rightleftharpoons}} R^1R^2C\diagup^{OR}_{\diagdown \overset{+}{O}R}_{\diagdown H} \qquad (38)$$

$$(21)$$

$$R^1R^2C\diagup^{OR}_{\diagdown \overset{+}{O}R}_{\diagdown H} \rightleftharpoons R^1R^2C\diagup^{OR}_{\diagdown OR} + H^+ \qquad (39)$$

The structural effects on the rate constant for the acid-catalysed hydrolysis of diethyl acetals of aliphatic aldehydes and methyl ketones can be expressed as:

$$\log \frac{k}{k_0} = \rho^*(\sum \sigma^*) + (n - 6)h \qquad (40)$$

n is the number of α-hydrogen atoms, k_0 is the rate constant for acetonal [$Me_2C(OEt)_2$], h is a constant (0.54 ± 0.06), representing the hyperconjugative effect of an α-hydrogen atom on the stabilization of the transition state, and ρ^* value is calculated to be -3.600 for the substituents in carbonyl[161b]. A very large negative Taft ρ^* value (-4.173) has been obtained with substituent R for the hydrolysis of formals, $CH_2(OR)_2$[159,161a]. Hence, electron-releasing groups in both aldehyde and in alcohol promote effectively the hydrolysis of acetals. This may be explained by assuming the electron-releasing groups not only increase the concentration of the conjugate acid of acetal (reverse of equilibrium 39) but also promote the cleavage of the conjugate acid (rate of reverse of 38), by stabilizing the carbonium ion-like transition state. The substituent effects on both the equilibrium and the rate of hydrolysis are in the same direction ($\rho^* < 0$), but for acetal formation the effect of substituents on the equilibrium is positive while that on the rate is negative[161b]. The rates of hydrolysis of ethylene acetals of substituted benzaldehyde are lower by a factor of 30 than those of the corresponding diethyl acetals, because the former have higher entropy of activation[165].

V. ADDITION OF HYDROGEN CYANIDE

The addition of hydrogen cyanide to a carbonyl compound (reaction 41) produces a cyanohydrin.

$$R^1R^2C{=}O + HCN \; \rightleftharpoons \; R^1R^2C \Big\langle {\overset{\displaystyle OH}{\underset{\displaystyle CN}{}}} \qquad (41)$$

The reaction is reversible, but with the lower aldehydes it goes virtually to completion and may be used for the quantitative analysis of formaldehyde (Romijn's procedure)[167a,b]. The addition is both acid and base catalysed. The ordinary preparation of cyanohydrin consists of the addition of aqueous sulphuric acid to a mixture of aqueous alkali metal cyanide and an excess of carbonyl compound at room temperature. Thus acetone cyanohydrin is obtained in 77–78% yield at $20°$ ($R^1 = R^2 = CH_3$ in reaction 41)[168]. Lactonitrile is prepared by the sodium hydroxide-catalysed reaction of hydrogen cyanide with acetaldehyde ($R^1 = CH_3$, $R^2 = H$ in reaction 41)[169].

In general, the rate of reaction satisfies the second-order equation (42a), but alkali metal cyanide alone cannot give an appreciable amount of cyanohydrin (section V.B.1). An abnormal rate law (equation 42b) is often observed for the addition catalysed by quinine in chloroform[170] (section V.C.)

$$v = k[\text{carbonyl}][\text{CN}^-] \qquad (42a)$$

$$v = k[\text{carbonyl}][\text{base}]^2 \qquad (42b)$$

The reaction of benzaldehydes in the presence of catalytic amounts of cyanide in alkaline solution, at rather higher temperatures, leads to the addition of the cyanohydrin to a second molecule of benzaldehyde giving a benzoin (22) reversibly[171–174]:

$$\text{ArCH}{=}\text{O} + \text{CN}^- \; \rightleftharpoons \; \underset{\displaystyle CN}{\text{Ar–CH–O}^-} \; \rightleftharpoons \; \underset{\displaystyle CN}{\text{Ar–}\overset{-}{\text{C}}\text{–OH}} \qquad (43)$$

$$\underset{\displaystyle CN}{\text{Ar–}\overset{-}{\text{C}}\text{–OH}} + \text{ArCH}{=}\text{O} \; \overset{(a)}{\rightleftharpoons} \; \underset{\displaystyle CN \;\; H}{\overset{\displaystyle OH \;\; O^-}{\text{Ar–C——C–Ar}}} \; \overset{(b)}{\rightleftharpoons}$$

$$\underset{\displaystyle H}{\overset{\displaystyle O \;\; OH}{\text{Ar–C–C–Ar}}} + \text{CN}^- \quad ((44)$$

(22)

The rate satisfies the kinetics: $v = k[\text{aldehyde}]^2[\text{CN}^-]$, which suggests that step (a) is probably rate-determining[173]. The reaction in the absence of water is possible and seems to occur on the surface of alkali metal cyanide, the rate equation being the same as that in solution[174]. The reverse reaction, i.e. the decomposition of benzoin to benzaldehyde, occurs smoothly in the presence of alkali metal cyanide, but heating benzoin alone, even to 300°, gives only about 10% benzaldehyde together with other products[175]. Usually, the reaction does not occur in the presence of nucleophiles other than cyanide ion. If a mixture of benzaldehyde and a substituted benzaldehyde is treated with alkali cyanide, a mixed benzoin is formed[171].

A. Equilibria

The formation of cyanohydrin from benzaldehydes (reaction 45) has been studied extensively in view of the substituent effects on rates and equilibria. The rate and equilibrium of the reaction may be followed by determining HCN according to Romijn's method (titration with $AgNO_3$–NH_4SCN)[176,177].

$$XC_6H_4CH{=}O + HCN \underset{k_{-1}}{\overset{k_1}{\rightleftharpoons}} XC_6H_4\underset{\underset{OH}{|}}{C}HCN \qquad (45)$$

TABLE 3. Equilibrium constants ($K = k_{-1}/k_1$), free energy differences (ΔF), and rate constants for substituted benzaldehyde–cyanohydrin formation and dissociation (reaction 45)[178a].

	X = p-Me	F		Cl	
		p-	m-	p-	m-
$10^3 K$ (1 mole^{-1})	8·98	6·30	2·13	3·85	1·97
$10^{-2}\Delta F$ (kcal mole^{-1})[b]	27·5	29·5	35·9	32·4	36·3
$10^8 k_1$ (1 mole^{-1} s^{-1})	0·82	3·59	10·22	6·94	12·45
$10^{10} k_{-1}$ (s^{-1})	0·74	2·26	2·18	2·67	2·45

	Br		I		NO$_2$	
	p-	m-	p-	m-	p-	m-
$10^3 K$ (1 mole^{-1})	3·26	1·88	3·13	2·02	3·4	2·9
$10^{-2}\Delta F$ (kcal mole^{-1})[b]	33·4	36·6	33·6	36·2	33·1	35·9
$10^8 k_1$ (1 mole^{-1} s^{-1})	9·14	14·32	9·38	14·5	37·0	58·4
$10^{10} k_{-1}$ (s^{-1})	2·98	2·69	2·94	2·92	12·6	12·3

[a] Temperature, 20°; equilibrium catalyst, NPr$_3$; rate catalyst, pyridine + pyridinium benzoate; solvent, 96% ethanol.

[b] For dissociation.

The equilibrium constant K for the dissociation of benzaldehyde cyanohydrins (reverse of reaction 45) tends to increase by *meta* and *para* electron-releasing groups, for example CH_3. This tendency, however, is not always present, as shown for example in Table 3, where K decreases in an order[176,178] somewhat different from that of the corresponding Hammett σ values:

p-Me > p-F > H > p-Cl > p-NO$_2$ > p-Br > p-I > m-NO$_2$ > m-F > m-I > m-Cl > m-Br

The order of p-halogen seems to agree with the order of polarizability of carbon halogen bonds (C—X) and electron-releasing resonance effect, but not with that of electron-attracting inductive effect. The abnormally high value for p-F may be explained by its strong resonance effect. The order of m-halogen seems to represent the order of electron-attracting inductive effect, together with a second-order relay of mesomeric effect from a position *ortho* to CHO (**23a** and **23b**), operating in the case of m-F and m-Cl.

(**23a**)   (**23b**)

The ratio of equilibrium constants for *para* and *meta* substituents, K_p/K_m, represents the resonance effect free from the inductive effect[178]. This ratio indicates that the resonance effect for halogens decreases with increasing size of halogen atom, i.e. K_p/K_m : H, 1; F, 2·98; Cl, 1·96: Br, 1·73.

Similarly, the same ratios, K_p/K_m, for methoxy analogues are[179]:

H, 1; MeO, 10·5; MeS, 9·36; MeSe, 8·98

which show that their electron-releasing resonance effects decrease with increasing molecular weight, but the same groups in m-position have increasing electron-attracting inductive effects, i.e. their 10^5K values at 25° are:

H, 4·44; m-MeO, 4·27; m-MeS, 4·07; m-MeSe, 3·95

Alkyl substituents in benzaldehyde tend to increase the 10^3K values in 96% ethanol at 20° as follows[180,181]:

H, 4·47; t-Bu, 7·49; i-Pr, 8·04; Et, 8·18; Me, 8·98

This may be attributed to the stabilization of aldehyde by hyper-conjugation (equation 46)[182] being more effective than the stabilization of aldehyde due to an electron-releasing inductive effect of methyl group.

$$H-\overset{\overset{\displaystyle H}{|}}{\underset{\underset{\displaystyle H}{|}}{C}}-\langle\bigcirc\rangle-CHO \longleftrightarrow H\overset{\overset{\displaystyle H}{|}}{\underset{\underset{\displaystyle H}{|}}{\overset{+}{C}}}=\langle\bigcirc\rangle=CH-\bar{O} \longleftrightarrow etc. \qquad (46)$$

In general, an aryl-substituented cyanohydrin is less stable than the corresponding alkyl-substituted one, due to delocalization of the positive charge of the carbonyl carbon atom bound to the aryl group:

$$\langle\bigcirc\rangle-\overset{\overset{\displaystyle R}{|}}{C}=O \longleftrightarrow {}^{+}\langle\bigcirc\rangle=\overset{\overset{\displaystyle R}{|}}{C}-O^{-} \longleftrightarrow etc. \qquad (47)$$

Thus benzophenone, PhCOPh, cannot form cyanohydrin. This is not due to the steric hindrance, since the substitution of alicyclic groups of similar size (cyclohexyl) for one phenyl group, as in $PhCOC_6H_{11}$, makes the cyanohydrin formation possible again.

The dissociation constant 10^2K for C_6H_5COR, with various alkyl groups at 20°, increases in the order[183]:

H, 0·28; t-Bu, 9; i-Pr, 25; Et, 90; Me, 130.

This behaviour may be explained as before by the hyperconjugative stabilization of the carbonyl form being more effective than the positive inductive stabilization of carbonyl form (equation 46). The relatively large dipole moment[182] and the C—C bond shortening[184] in acetaldehyde support the presence of hyperconjugation contributions such as $H^+CH_2=CH-O^-$.

These substituent effects on the equilibrium have been generalized in a Taft type equation[32,176]:

$$\log (K/K_o) = \sigma^*\rho^* + nh \qquad (48)$$

here, n is the number of C—H bonds available for hyperconjugation, h is a value corresponding to the energy of hyperconjugation, σ^* is a polar substituent constant and ρ^* is a polar reaction constant. When the σ^* value for H is 0, the h values are: Me, 0·07; Et, 0·062; i-Pr, 0·082.

A claim that this effect of alkyl groups is due to a ponderal effect[185] rather than to hyperconjugation[186] was regarded as dubious since, except at quite low temperatures, there is no appreciable ponderal effect calculated by statistical mechanics on the basis of all vibrational frequencies corresponding to the force constants and molecular geometries assumed by the aldehydes and cyanohydrins[187].

The dissociation constants 10^2K of cyclic ketone cyanohydrins in 96% ethanol at 22° vary with the number of ring carbons (n) as follows:

n	5	6	7	8	9	10	11	12	13	14	15	16	17	18	19	20
10^2K	2·1	0·1	13	86	170	∞	112	31	26	6	11	9	12	10	10	7

The higher stability of free ketone for $n = 9$ to 11 has been attributed to the flexibility of the ring bonds[188a].

Carbonyl–cyanohydrin equilibria have been widely used for stereochemical studies, for example in cyclohexanones[188b] and bicyclic ketones[188c].

B. Kinetics and Mechanisms

I. Aldehydes

There have been a number of reports since 1903 on the kinetics of cyanohydrin formation[177,189–192]. The rate constants k_1 in Table 3 for the cyanohydrin formation measured by Romijn's method show that electron-attracting groups apparently favour the reaction. This polar effect suggests a rate-determining addition of nucleophilic cyanide ion, formed by a basic catalyst, on a carbonyl carbon atom[189].

$$\text{ArCH=O} + \text{CN}^- \underset{}{\overset{\text{slow}}{\rightleftharpoons}} \underset{\underset{\text{CN}}{|}}{\text{ArCH—O}^-} \overset{\text{H}^+}{\rightleftharpoons} \underset{\underset{\text{CN}}{|}}{\text{ArCHOH}} \qquad (49)$$

This simple mechanism cannot, however, explain the fact that the addition does not occur with potassium cyanide alone. Furthermore, it does not agree with the relation of rate versus pH observed in acetate buffer, i.e. the change of rate constant with increasing buffer concentration at a constant ratio of [AcOH]/[AcO$^-$] points to general acid catalysis[190a,b]. In addition to this, a specific oxonium ion

2*

catalysis by H_3O^+ has been observed with acetaldehyde and acetone. These facts suggest that one of the mechanisms A or B, involving a general acid or oxonium ion in the transition state, is the correct one.

Mechanism A[190,193]:

$$
\underset{\overset{\delta-}{A}\cdots H\cdots \overset{..}{\underset{..}{O}}}{\overset{\overset{\displaystyle Ar}{|}}{H\!-\!\overset{|}{C}{}^{\delta+}}} + CN^- \underset{}{\overset{slow}{\rightleftharpoons}} \underset{H\!-\!O}{\overset{\overset{\displaystyle Ar}{|}}{H\!-\!\overset{|}{C}\!-\!CN}} + A^- \qquad (50a)
$$

Mechanism B:

$$
A\!-\!H + \underset{O^-}{\overset{\overset{\displaystyle Ar}{|}}{H\!-\!\overset{|}{C}\!-\!CN}} \underset{}{\overset{slow}{\rightleftharpoons}} \underset{H\!-\!O}{\overset{\overset{\displaystyle Ar}{|}}{H\!-\!\overset{|}{C}\!-\!CN}} + A^- \qquad (50b)
$$

Neither the rate data nor consideration of substituent effects for the forward reaction can help to distinguish between the two mechanisms, but, as will be seen later, the substituent effect for the reverse reaction together with the principle of microscopic reversibility seem to favour mechanism B. It is rather unusual that proton abstraction by a weakly basic cyanohydrin anion should be rate-determining, but the cyanohydrin anion seems to be stable enough to attack another benzaldehyde molecule (reaction 44) before becoming protonated. However, it is more probable that the actual activated complex is intermediate between that of mechanisms A and B, i.e. a concerted type (24), in which the $C\cdots CN$ partial bond is more complete than the $H\cdots O$ partial bond

$$
\underset{A^{\delta-}\cdots H^{\delta+}\cdots \overset{..}{O}{}^{\delta-}}{ArHC\cdots CN^{\delta-}}
$$

(24)

As for the effect of substituents in benzaldehyde, it is known that *p*-halogens enhance the rate in the order[174] $H < F < Cl < Br < I$, which is not consistent with the order of electron-attracting inductive effect, thus implying the overlap of an electron-releasing resonance effect. The same order is also observed with *m*-halogens, suggesting the overlap of the inductive effect with the second-order relay of resonance effect, for example, in 23. It is surprising to see that $p\text{-}NO_2 < m\text{-}NO_2$, since $p\text{-}NO_2$ is usually more electrophilic than $m\text{-}NO_2$. This has been explained by assuming that the carbonyl polarization, $\overset{\delta+}{\text{C}}=\overset{\delta-}{\text{O}}$, is suppressed by introducing a strongly electrophilic p-NO$_2$, thus lowering the electrophilicity of the carbonyl carbon, as has been observed in some S_N2 reactions[194]. It seems more probable, however, that the higher solvation (for example, hydration) of a carbonyl group attached to a p-nitrophenyl group results in a lower concentration of free carbonyl to which the cyanide ion can add.

The effect of ring substituents on the reverse reaction, i.e. decomposition of benzaldehyde cyanohydrins, can be calculated from the equation $k_{-1} = Kk_1$; the k_{-1} values have the following order:

$$p\text{-Me} \ll \text{H} < m\text{- and } p\text{-halogen} < m\text{-NO}_2 < p\text{-NO}_2$$

The fact that electron-attracting groups accelerate the reaction suggests that the abstraction of proton from cyanohydrin by a base (or solvent) is rate-determining[174] or, more probably, that in the activated complex (24) the O$\cdots$H bond is stretched more than the C$\cdots$CN bond.

2. Ketones

The rate of formation of ketone cyanohydrin also follows a second-order equation: $v = k_1[\text{ketone}][\text{CN}^-]$. Therefore, the sodium hydroxide-catalysed reaction of acetone with hydrogen cyanide in an aqueous solution follows the rate law[188a,189]: $v = k[\text{acetone}]$ [NaOH]$_{\text{stoich.}}$, but not $v = k_1[\text{acetone}][\text{HCN}]_{\text{stoich.}}$. The k value for weaker bases such as ammonia is less than that for stronger bases like sodium hydroxide.

The reaction is slow in the absence of ionizing solvents like water. As shown in Figure 4, the rate constant increases with increasing dielectric constant[191,193] of solvent, suggesting that the dissociation of cyanide ion is necessary for the reaction to occur.

Table 4[192] lists the second-order rate constants for the formation of ketone cyanohydrin in aqueous ethanol at 10°. Alkyl substituents in acetone tend to lower both the energy and entropy of activation, resulting in small differences in the rate constant, the only exception being t-Bu-CO-Bu-t. α-Substitution favours cyanide addition by lowering hyperconjugation, but discourages the addition by introducing steric hindrance. α-Chloro substitution, except in ClMe$_2$CCO-Pr-i, raises the reactivity, due to the increased electrophilicity of the carbonyl group[192].

TABLE 4. Second-order rate constants (k), entropies of activation (E_a), and Arrhenius factors ($\log A$) for cyanohydrin formation in acetate-buffered 50% aqueous ethanol at $10^{1.92}$.

	Ketone				α-Chloroketone				
RCOR	$10^4 k$ (1 mole^{-1} min^{-1})	$\dfrac{k^a}{k_{ao}}$	E_a (kcal/mole)	$\log A$	$10^2 k$ (1 mole^{-1} min^{-1})	$\dfrac{k}{k_{ao}} \times 10^{-2}$	E_a (kcal/mole)	$\log A$	$\dfrac{k_{Cl}}{k_H}$
MeCOMe	6.0	1	17.7	10.4	7.8	1.3	13.9	9.7	130
EtCOEt	8.5	1.4	14.7	8.3	13	2.1	12.8	9.0	150
i-PrCOPr-i	4.4	0.7	12.7	6.5	<0.15	<0.1	—	—	<2
t-BuCOBu-t	<0.2	<0.1	—	—	—	—	—	—	—
MeCOMe	6.0	1	17.7	10.4	7.8	1.3	13.9	9.7	130
EtCOEt	8.5	1.4	14.7	8.3	13	2.1	12.8	9.0	150
n-PrCOPr-n	4.1	0.7	15.6	8.7	6.0	1.0	9.9	6.4	146
n-BuCOPr-n	6.7	1.1	14.0	7.6	14	2.3	8.9	6.0	205
MeCOMe	6.0	1	17.7	10.4					
EtCOEt	5.9	1	16.3	9.4					
i-PrCOMe	26	4.3	11.9	6.6					
t-BuCOMe	5.1	0.9	13.7	7.3					

^{a} k_{ao} is the rate constant for acetone.

In the reaction with alicyclic ketones in aqueous ethanolic acetate buffer (pH 4) at 27°, the reactivities relative to cyclopentanone are[195]: cyclohexanone, 23·7; *trans*-2-decalone, 23·2; *cis*-2-decalone, 7·5.

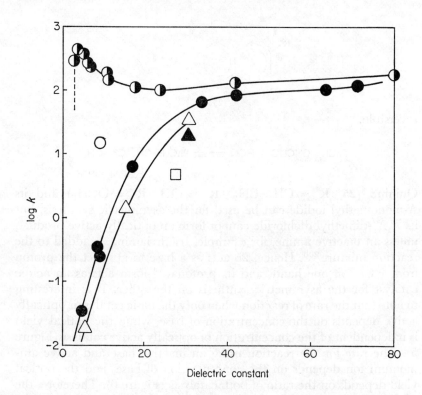

FIGURE 4. Relation between the second-order rate constant (k expressed in terms of $1\,mole^{-1}\,s^{-1}$) and solvent composition for acetone cyanohydrin formation at 20°[193]. ◑ dioxan water; catalyst: NaOH. For the 9:1 mixture ($\epsilon \sim 20$) LiOH, CsOH, $Ba(OH)_2$ and $Sr(OH)_2$ were also used as catalysts. △ dioxan–water; catalyst: $MeBu_3NOH$. ● dioxan–water; catalyst: Et_3N. ▲ methanol; catalyst: cinchonine. □ dioxan–acetonitrile–water (46:46:8); catalyst: Et_3N. ○ Chloroform–water (98:2); catalyst Et_3N.

C. Asymmetric Syntheses of Cyanohydrins

Optically active cyanohydrins may be formed when some optically active bases are used as catalysts for reaction (45)[196–199]. For example, optically active cinchona alkaloids (**25**) lead to the formation of

active benzylidenelactonitrile in chloroform solution. The active
cyanohydrin, however, may be racemized, since the addition is

(25)

reversible.

$$PhCH{=}CHCHO + HCN \underset{CHCl_3}{\overset{}{\rightleftharpoons}} PhCH{=}CH{-}\underset{\underset{CN}{|}}{\overset{\overset{H}{|}}{C}}*{-}OH$$

Quinine (**25**, R′ = CH=CH$_2$, R″ = OH, R‴ = OCH$_3$) and its
N-monomethyl iodide can be used for the asymmetric synthesis, but
its *N,N′*-dimethyl dichloride cannot form an optically active product,
unless an inactive amine, for example, triethylamine, is added to the
reaction mixture[200]. Hence, **25** acts as a base to abstract the proton
from HCN on one hand, and its protonated form acts as an acidic
catalyst for the asymmetric synthesis on the other. It is interesting
to note that the rate of reaction when only the basic catalyst is optically
active depends on the concentration of base, while the optical yield
is independent of the concentration of optically active catalyst (Figure
5). The rate for the reaction using an inactive base and active ammonium ion depends on the concentration of base, and the optical
yield depends on the ratio of both catalysts (Figure 6). Therefore, the
inactive amine abstracts the proton from HCN and accelerates the
reaction, but it cannot give an asymmetric product, while the optically
active ammonium ion can act to form active cyanohydrin[200]. Some
inactive cations, e.g.[200] Et$_4$N$^+$ and Li$^+$, or even an active amine
without a suitable configuration[193], can suppress the asymmetric
synthesis. Prelog[200], on the basis of equation (42b)[170], and of the ion
pair formation (base–H$^+$CN$^-$) in chloroform, postulated a mechanism
involving a rate-determining attack of the ion pair on the complex of
carbonyl–ion pair, where the asymmetric configuration of HB$^+$
enables the stereospecific addition of cyanide ion:

$$(R{-}CH{=}O{\cdots}HB^+CN^-) \overset{slow}{\rightleftharpoons} R{-}\underset{\underset{CN}{|}}{CH}{-}OH + (HB^+CN^-) + B \qquad (51)$$
$$\underset{CN^-BH^+}{}$$

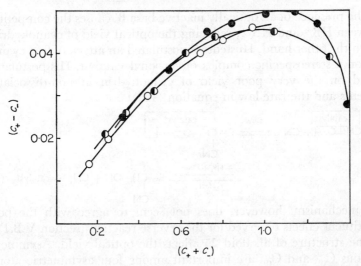

FIGURE 5. Effect of optically active base (X = quinine) and optically inactive acid (Y = cinnamic acid) as catalysts for the asymmetric synthesis of benzylidenelactonitrile[200]. c_+ = [d-cyanohydrin]; c_- = [l-cyanohydrin]; ● [X] = 6·20, [Y] = 5·43; ◑ [X] = 1·20, [Y] = 0·43; ○ [X] = 0·77, [Y] = 0.

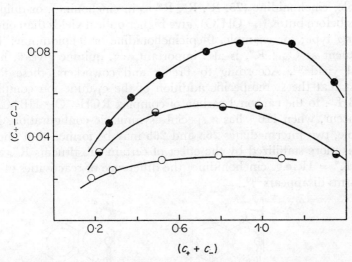

FIGURE 6. Effect of optically inactive base ([Me₃N] = 0·076) and optically active ammonium salt (X = quinine dimethiodide) as catalysts for the asymmetric synthesis of benzylidenelactonitrile[200]. c_+ = [d-cyanohydrin]; c_- = [l-cyanohydrin]. Concentration of quinine dimethiodide: ●, 1·6; ◑, 0·33; ○, 0·16.

The presence of an optically inactive base B′ causes the competition between HB$^+$ and HB′$^+$, reducing the optical yield of cyanohydrin.

On the other hand, Hustedt[193] postulated an attack of free cyanide ion on a stereospecific complex of carbonyl–catalyst. His postulate is based on the very poor yield of cyanohydrin in non-dissociating solvents and the rate law in equation (42a):

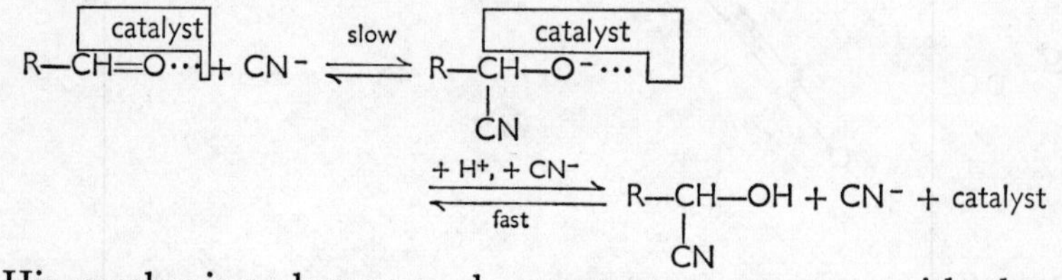

His mechanism, however, does not seem to agree with the polar substituent effects observed for the reverse reaction (section V.B.1).

The structure of alkaloid **25** affects the optical yield. Asymmetric carbons $C_{(8)}$ and $C_{(9)}$ are important among four asymmetric atoms: (a) The loss of asymmetry at $C_{(9)}$ causes failure of the asymmetric synthesis. (b) A catalyst with l configuration at $C_{(9)}$, for example, cinchonine (**25**, R′ = CH=CH$_2$, R″ = OH, R‴ = H), leads to l-cyanohydrin. (c) A catalyst of *erythro* configuration at $C_{(8)}$—$C_{(9)}$, for example, cinchonidine (**25**, R′, R″, R‴ as in cinchonine), or quinine (**25**, as before but R‴ = CH$_3$O), give higher optical yields than one of the *threo* type; for example, 9-epicinchonidine or 9-epiquinine. The substituent at $C_{(6′)}$, R‴, is also important, e.g. quinine gives a high optical yield[200]. According to Prelog and coworkers, these facts suggest that the stereospecific addition of the cyanide ion complex, CN$^-$BH$^+$, to the carbonyl carbon of complex RCH=O···HB$^+$CN$^-$ may occur, when HB$^+$ has a special asymmetric configuration. For example, two intermediates **26a** and **26b** may be formed, but one of them is more stabilized by the effect of certain substituents R‴, and when R‴ = H (e.g. cinchonidine) the difference of reactivities of the two forms disappears[200].

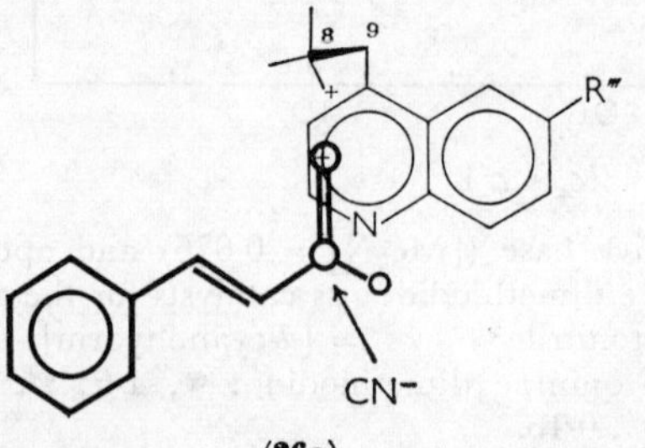

(26a) 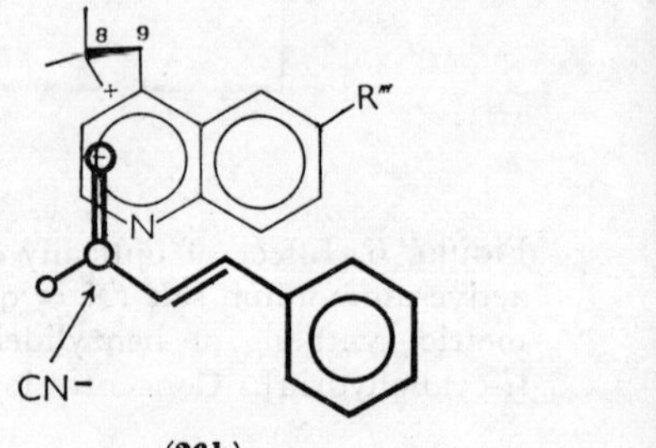   (26b)

VI. ADDITION OF BISULPHITE

The formation of crystalline salts from carbonyl compounds and alkali bisulphite (equation 53) or sulphite (equation 54) is well known. Although the reaction is reversible, the equilibrium often shifts to the right and hence the reaction constitutes one of the most familiar procedures for the isolation, identification and quantitative analysis of carbonyl compounds.

$$R^1R^2C{=}O + NaHSO_3 \rightleftharpoons R^1R^2C\overset{\displaystyle OH}{\underset{\displaystyle SO_3Na}{\Big\langle}} \tag{53}$$

$$R^1R^2C{=}O + Na_2SO_4 + H_2O \rightleftharpoons R^1R^2C\overset{\displaystyle OH}{\underset{\displaystyle SO_3Na}{\Big\langle}} + NaOH \tag{54}$$

For example, the acidimetry of sodium hydroxide produced in reaction (54) is a common method for the determination of formaldehyde[201], and the preparation of ketones from ketoximes can be done in good yields (84–98%) by forming the bisulphite adduct and then destroying it by acidification[202].

There are evidences showing the product to be α-hydroxysulphonic acid and not an α-hydroxy sulphite $[R^1R^2C(OH)OSO_2Na]$:

(a) Preparation of the acetate of the product by the following independent procedures[203,204]:

$$HCHO + KHSO_3 \longrightarrow HOCH_2SO_3K \xrightarrow{(CH_3CO)_2O} CH_3CO_2CH_2SO_3K \tag{55a}$$

$$CHI_3 + K_2SO_3 \longrightarrow ICH_2SO_3K \xrightarrow[\text{fusion}]{CH_3COOK} CH_3CO_2CH_2SO_3K \tag{55b}$$

(b) The presence of the C—S bond by Raman spectroscopy[205].

(c) The equilibrium constant for the sulphur isotope exchange reaction between the carbonyl–bisulphite adduct and labelled bisulphite ion, $H^{34}SO_3^-$, is $1\cdot021$–$1\cdot013$[206,207]. The equilibrium constant may be expressed as $K = (Q_{P34}/Q_{P32})/(Q_{B34}/Q_{B32})$, where Q_P is the partition function of the two isotopic species of the addition compounds, and Q_B is that of the bisulphite ion. If the adduct had a C—O—S bond, the value of K would be unity, while for a C—S bond it should be ca. $1\cdot019$, as observed.

A. Equilibria

Table 5 lists equilibrium constants K, free energy differences ΔF and entropy differences ΔS at various temperatures[208].

TABLE 5. Equilibria data for the addition of sodium bisulphite to carbonyl compounds in water[208].

Carbonyl compound	$10^{-2}K$ (l/mole)[a]			ΔF 30° (kcal/mole)	ΔS 30° (e.u.)
	0°	20°	30°		
CH_3CHO[b]	∞	∞	∞	—	—
C_2H_5CHO	128	120	120	—	—
C_3H_7CHO	720	283	170	6025	7·1
$CH_3(CH_2)_5CHO$	40·6	—	48·0	—	—
CH_3COCH_3[b]	7·90	2·94	1·85	3145	16·0
CH_3COEt[b]	1·58	0·546	0·34	2127	20·7
$CH_3COCH_2CO_2Et$	3·46	0·913	0·500	2370	27·1
$CH_3COC(CH_3)_3$	small	small	small	—	—
PhCHO	17·0	6·57	4·37	3680	12·2
PhCOPh	0	0	0	—	—

[a] $K = [RR'C(OH)SO_3^-]/[RR'CO][HSO_3^-]$.
[b] Concentration of carbonyl compound: 0·05 mole/l. All others: 0·01 mole/l.

The equilibrium constants K' for the reverse reaction determined by iodometry are shown in Table 6. The branching of R in CH_3COR increases the dissociation constant[209], suggesting that steric interaction in the adduct lowers its stability and that the hyperconjugative effect is less important.

TABLE 6. Apparent dissociation constants for the sodium bisulphite adducts of methyl ketones CH_3COR at 0°[a,209].

R	$10^2K'$ (mole/l)	log K'	
		Observed	Calculated
Methyl	0·495	−2·306	−2·306
Ethyl	2·50	−1·602	−1·597
Propyl	3·51	−1·455	−1·360
Amyl	4·34	−1·362	−1·252
Isoamyl	5·96	−1·225	−1·188
Isopropyl	12·6	−0·900	−0·887
t-Butyl	70·0	−0·155	−0·156

[a] Carbonyl concentration: 0·0250 mole/l in 50% aqueous alcohol.

The application of the Branch–Calvin method[210] to the calculation of inductive effect of several types of carbonyl gives reasonable K' values which agree with observation[209,211,212]. The log K' values were calculated for 0·0125–0·0250 mole/l adduct in 50% ethanol at 0°:

$$\text{Aliphatic aldehydes} \qquad \log K' = -0·81 \sum I_c \alpha^n - 2·84 \qquad (56)$$

$$\text{Alkyl methyl ketones} \qquad \log K' = -2·05 \sum I_c \alpha^n - 6·39 \qquad (57)$$

$$\text{Alkyl ethyl ketones} \qquad \log K' = -4·07 \sum I_c \alpha^n - 11·08 \qquad (58)$$

$$\text{Alkyl propyl ketones} \qquad \log K' = -5·02 \sum I_c \alpha^n - 13·72 \qquad (59)$$

I_c is the inductive constant, n is the number of atoms between substituent and reaction site, and $\alpha = \frac{1}{3}$.

Here, the value of $\sum I_c (\frac{1}{3})^n$ is almost proportional to Taft's σ^*, and the variation of the coefficient of $\sum I_c \alpha^n$ with the change of alkyl group is attributed to steric effect.

The dissociation constants K' for p-substituted benzaldehyde–bisulphite adducts obey Hammett's equation, with a negative ρ value. The only exception is p-NO$_2$ which shows a large K' value[213]; this can be explained by the enhanced hydration of carbonyl (or less probably by the suppression of polarization), as was the case with p-nitrobenzaldehyde cyanohydrin (section V.A).

B. Kinetics and Mechanisms

The rates of dissociation of carbonyl–bisulphite adducts in acidic buffers (pH 3–5) show first-order dependence on reactant, in the presence of an oxidizing agent such as iodine which oxidizes the formed sulphurous to sulphuric acid[214,215]. The rate constant increases with pH and it is proportional to $[OH^-]^n$ $(n \leq 1)$[215]. The rate equation for reversible reactions (forward first-order and backward second-order) should, however, be applied in the absence of the oxidizing agent.

$$(60)$$

TABLE 7.　Relative rates[a] for the decomposition of sodium bisulphite adducts of substi
benzaldehydes, in water at $20°$[215].

Substituent	p-NO$_2$	m-NO$_2$	p-Cl	p-Me	p-Me$_2$N	p-MeO	p-
$\Delta \log k$ (pH 3)	$+0.06$	$+0.09$	$+0.09$	$+0.09$	$+0.13$	$+0.22$	$+$

Substituent	o-OH	o-Cl	o-NO$_2$	o-MeO	o-EtO	o-n-PrO	
$\Delta \log k$ (pH 3)	$+0.91$	$+0.01$	-0.04	-0.12	-0.21	-0.24	
$\Delta \log k$ (pH 5)	$+1.02$	—	—	-0.15	-0.22	-0.33	

[a] k is the pseudo first-order rate constant in substrate, Δ represents the increment relative to benzaldeh

Since the adduct XH$^-$ is an acid anion which can undergo further dissociation to X^{2-}, the rate of decomposition is the sum of rates of both decomposition paths in reaction (60):

$$v = k_2[\text{XH}^-] + k_3[\text{X}^{2-}] \tag{61}$$

$$\text{or}\quad v = k_2[\text{XH}^-] + \frac{k_1 k_3[\text{XH}^-]}{k_{-1}[\text{H}^+] + k_3} \tag{62}$$

The above rate law explains the observed dependency on $[\text{OH}^-]^n$ ($n \le 1$). If $k_2 \ll k_3$, the equation is simplified to[215]

$$v = \frac{k_1 k_3[\text{XH}^-]}{k_{-1}[\text{H}^+] + k_3} \tag{63}$$

The relative rate constants for the decomposition of substituted benzaldehyde adducts at pH 3 and 5 are shown in Table 7. The small substituent effects probably reflect their influence on the ionization (k_1) and on the removal of SO$_3^{2-}$ (k_3)[215]. Except for o-OH, o-substituents retard the reaction. The o-OH group accelerates the decomposition, probably by enhancement of the S—C bond fission by hydrogen bonding, as follows:

$$\tag{64}$$

VII. ADDITION OF MISCELLANEOUS ACIDS

Inorganic acids often form rather unstable addition products with carbonyl compounds.

$$\tag{65}$$

Several carbonyl compounds, such as acetophenone, benzophenone, anthraquinone[216] and acetone[217] in sulphuric acid [equation (65), $A = HSO_4^-$] may be protonated to give two ionized particles, as was shown by cryoscopic and ultraviolet absorption studies[216]. Formaldehyde reacts with fuming sulphuric acid to form a ring compound, the so-called methylene sulphate[218] (**27**), probably in stages:

$$HCHO + H_2SO_4 \rightleftharpoons HOCH_2OSO_3H \qquad (66)$$

$$2\ HOCH_2OSO_3 \rightleftharpoons O_2S\ \begin{smallmatrix} O-CH_2-O \\ \\ O-CH_2-O \end{smallmatrix}\ SO_2 \qquad (67)$$

$$(\mathbf{27})$$

The hydrolysis of methylene sulphate can take two routes[219]:

$$O_2S\ \begin{smallmatrix} O-CH_2-O \\ \\ O-CH_2-O \end{smallmatrix} \quad \begin{smallmatrix} \xrightarrow{2\,H_2O}\ 2\ HOCH_2OSO_3H \longrightarrow CH_2O + H_2SO_4 \\ \\ \xrightarrow{H_2O}\ CH_2O + CH_2(OSO_3H)_2 \xrightarrow{H_2O} CH_2O + 2\ H_2SO_4 \end{smallmatrix} \qquad (68)$$

Aromatic and aliphatic sulphinic acids may form adducts with some carbonyl compounds[220–222], for example[220,221]:

$$p\text{-}CH_3C_6H_4SO_2H + HCHO \longrightarrow p\text{-}CH_3C_6H_4\overset{O}{\underset{O}{\overset{\uparrow}{\underset{\downarrow}{S}}}}CH_2OH\ (92\%\ m.p.\ 95°) \qquad (69)$$

The products from formaldehyde and alkanesulphinic acids, α-hydroxymethyl alkyl sulphones are, except for the lower aliphatic homologues, colourless crystals, most of which tend to decompose. The adducts of formaldehyde with aromatic sulphinic acids are relatively stable.

The cryoscopic study of a mixture of acetone and hydrogen bromide in benzene confirmed the formation of a 1:1 complex[223].

The structure of the 1:2 adduct between perchloric acid and mesitaldehyde was studied by X-ray diffraction[224]. The data suggest a structure consisting of layers in which the hydrogen-bonded mesitaldehyde dimers and perchlorate ions occur in a 1:1 ratio.

Carbonyl compounds in 99·5% nitric acid form 1:1 adducts[225] at -40 to $-15°$, their melting points being as follows: benzaldehyde, $-12°$; acetone, $-18°$; acetophenone, $-24°$; benzophenone, $+31°$; di-n-butyl ketone, $-45°$. The infrared spectra of these adducts have broad bands at 2700–2600 cm^{-1} (oxonium ion) and 1400 cm^{-1}

(NO_3^-), which agree with the expected structure: $\diagup C{=}\overset{+}{O}H\ NO_3^-$ [226].

The formaldehyde–HNO_3 complex decomposes at high temperature to give water, carbon dioxide and NO_2 or NO [227].

The complex formation of carbonyl compounds with carboxylic acids has been examined by both thermal and electroconductivity methods [228]. The complex with trifluoroacetic acid is fairly stable. The molar ratio of carbonyl compound vs. CF_3CO_2H in the complex varies with the structure of the carbonyl compounds [229], for example, acetone 2:3 (b.p. 110°); acetaldehyde 1:2 (b.p. 89·5°) and methyl ethyl ketone 2:3 (b.p. 115·5°).

Unstable complex compounds between peroxycarboxylic acids and carbonyls are known, but the complex converts readily to ester by the Baeyer–Villiger reaction (reaction 29). The autoxidation of acetaldehyde generally leads to peracetic acid, which can readily form a complex with unchanged acetaldehyde called acetaldehyde mono-peracetate [230,231a]:

$$CH_3CHO \xrightarrow{O_2} CH_3CO_3H \underset{}{\overset{CH_3CHO}{\rightleftharpoons}} CH_3CH \overset{OH\cdots O}{\underset{OO}{\diagup\diagdown}} C{-}CH_3 \qquad (70)$$

Another review of carbonyl-acid interactions appeared elsewhere in the series [231b].

VIII. ADDITION OF THIOLS AND HYDROGEN SULPHIDE [232–234]

Thiols and hydrogen sulphide may easily add to the carbonyl double bond to form hemithioacetals (28) and gem-thioglycols (29), respectively [235,236]. The product is often too unstable to be isolated; an isolable solid product is often a 2:1 adduct (30) [236–243].

$$R^1R^2C{=}O + RSH \rightleftharpoons R^1R^2C\overset{OH}{\underset{SR}{\diagup\diagdown}} \qquad (71)$$

$$(28)$$

$$R^1R^2C{=}O + H_2S \rightleftharpoons R^1R^2C\overset{OH}{\underset{SH}{\diagup\diagdown}} \qquad (72)$$

$$(29)$$

$$R^1R^2C{\overset{OH}{\underset{SH}{\big<}}} \; + \; R^1R^2C{=}O \; \rightleftharpoons \; R^1R^2C{-}S{-}CR^1R^2 \atop \;\;\;\;\;\;\;\;\;\;\;\; \overset{|}{OH} \;\; \overset{|}{OH} \qquad (73)$$

$$(30)$$

The conversion of furfural to thiofurfural (equation 74) and the formation of its *gem*-dithiol proceed probably via thioglycol as an intermediate[232–234,240,243]:

$$(74)$$

Hemithioacetals or hemithioketals with electron-attracting substituents may be stable enough to be isolated by distillation in vacuum or by crystallization[232–234]. Thus the 1:1 adducts of various perhaloalkyl carbonyls with hydrogen sulphide were isolated[244] and examined as to their stabilities[245]. The reaction of fluoral (CF_3CHO) or chloral (CCl_3CHO) with excess hydrogen sulphide in an autoclave at room temperature gives a mixture of 1:1 (29) and 2:1 adducts (30), but the use of an excess aldehyde gives exclusively the 2:1 adduct. The 1:1 adduct (29) tends to decompose easily at room temperature. *sym*-Dichlorotetrafluoroacetone and hexafluoroacetone give stable adducts (31 and 32), whereas decafluoroethyl ketone gives a very unstable adduct (33), which decomposes completely at room temperature to the component ketone and hydrogen sulphide, possibly because of steric effects.

The reaction products of formaldehyde[246,247] and glyoxal[248,269] with thiols were shown to be hydroxy sulphides 28 and 29 by decomposition reactions[247], infrared and n.m.r. spectroscopy[244,245,250] and X-ray diffraction[251].

A. Equilibria

Table 8 shows the equilibrium constants of hemithioacetals and hemithioketals determined by ultraviolet spectrophotometry[252]. The

TABLE 8. Equilibrium constants for hemithioacetal and hemithioketal formation and hydration at 25° and 1M ionic strength[252].

Carbonyl compound	Thiol	pH	$K_{obs}{}^a$ (1/mole)	$K_{hyd}{}^a$ (1/mole)	$K_{hemi}{}^a$ (1/mole)
CH_3CHO	$CH_3OCH_2CH_2SH$	4·6	17·4	0·85	32
CH_3CHO	$CH_3OCH_2CH_2SD$	4·6 (D_2O)	36·4	0·99	72
CH_3CHO	CH_3OOCCH_2SH	4·6	32·7	0·85	60
CH_3CHO	C_2H_5SH	4·6	19·6	0·85	36
$(CH_3)_2CHCHO$	$CH_3OCH_2CH_2SH$	4·3	11·8	0·24	15
$(CH_3)_3CCHO$	$HOCH_2CH_2SH$	4·6	3·8	< 0·1	3·8
CH_3COCOO^-	CH_3OOCCH_2SH	5·2	5·2	No hydration	5·2
CH_3COCH_3	$CH_3OCH_2CH_2SH$	1·5	< 0·2	No hydration	< 0·2

^{a} See text for explanation of K_{obs}, K_{hyd} and K_{hemi}.

observed equilibrium constant of reaction (71) is obtained as follows, if [H_2O] is very large:

$$
\begin{aligned}
K_{obs} &= \frac{[R^1R^2C(OH)SR]}{([R^1R^2C{=}O] + [R^1R^2C(OH)_2])[RSH]} \\
&= \frac{1}{1 + K_{hyd}} \frac{[R^1R^2C(OH)SR]}{[R^1R^2C{=}O][RSH]} \\
&= \frac{1}{1 + K_{hyd}} K_{hemi}
\end{aligned}
\tag{75}
$$

Here, K_{hyd} is the equilibrium constant of carbonyl hydration and K_{hemi} is the *true* equilibrium constant of hemithioacetal formation from free carbonyls. The K_{hemi} value for acetaldehyde is almost fifty times as large as K_{hyd}. The higher affinity of thiols toward carbonyl compounds as compared with those of the corresponding hydroxy and amino compounds may be due to the stabilization of adducts by double bond–no bond resonance such as:

$$
\overset{|}{R{-}\overset{+}{S}{=}C} \quad \overset{|}{\overline{O}H} \longleftrightarrow R{-}S{-}\overset{|}{\underset{|}{C}}{-}OH \longleftrightarrow R{-}\overline{S} \quad \overset{|}{C}{=}\overset{+}{O}H
$$

Introduction of α-methyl groups in carbonyl compounds reduces the equilibrium constants of formation as was the case with hydration (Table 1); thus we have:

$$
CH_3CHO > (CH_3)_2CHCHO > (CH_3)_3CCHO
$$

On the other hand, the constants decrease with decreasing acidity of the thiols:

$$CH_3OCOCH_2SH > CH_3CH_2SH > CH_3OCH_2CH_2SH$$

B. Kinetics and Mechanisms

Kinetic studies of the addition of thiols to carbonyl compounds are few. The reaction is subject to general acid and specific hydroxide ion catalysis[252-255]. At low buffer concentration the rate is expressed as shown in equation (76).

$$v = k_{H^+}a_{H^+}[\text{carbonyl}][RSH] + k_{RS^-}[\text{carbonyl}][RS^-] \qquad (76)$$

where a_{H^+} is the proton activity. The pH–rate profile gives a minimum which agrees with the following scheme[252]:

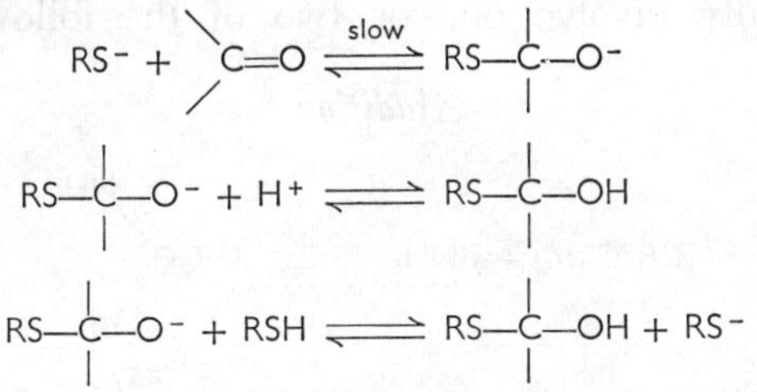

$$RS^- + \ \ \overset{\diagdown}{\underset{\diagup}{C}}=O \ \overset{slow}{\rightleftharpoons} \ RS-\overset{|}{\underset{|}{C}}-O^- \qquad (77)$$

$$RS-\overset{|}{\underset{|}{C}}-O^- + H^+ \rightleftharpoons RS-\overset{|}{\underset{|}{C}}-OH \qquad (78)$$

$$RS-\overset{|}{\underset{|}{C}}-O^- + RSH \rightleftharpoons RS-\overset{|}{\underset{|}{C}}-OH + RS^- \qquad (79)$$

The rate-determining step for the base catalysis is an attack of thiolate ion on the carbonyl group. The rate is almost insensitive to the basicity of the thiolate ion, i.e. a 250-fold increase in basicity results in only a 2-fold increase in rate. This is probably due to the specific stabilization of the transition state by electron release from the carbonyl oxygen to the d orbitals of sulphur (**34**) which needs therefore no electron donation by substituents on the sulphur atom. This is similar to the attacks on acrylonitrile, where the higher nucleophilicity of thiolate ion, as compared to that of amines having the same basicity, is ascribed to the back interaction of π electrons of the double bond with the empty $3d$ orbitals of the sulphur atom[256]. On the other hand, the higher nucleophilicity of thiolate ion in the substitution reaction is ascribed to its polarizability[257], or its tendency to be oxidized[258].

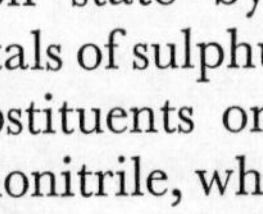

$$\underset{\underset{(\mathbf{34})}{R^1 \diagup \overset{\displaystyle C}{\diagdown} R^2}}{\overset{\displaystyle O \cdots}{\underset{\displaystyle |}{|}} {\scriptstyle -}\ddot{:}SR}$$

The transition state for general acid catalysis is quite similar to that for the addition of other nucleophiles to carbonyl groups, as depicted in **35**. In contrast with the base-catalysed reaction, the rate is much affected by the basicity of thiols[252].

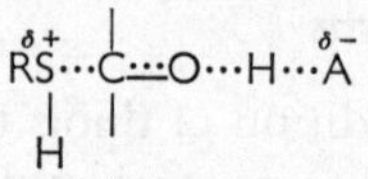

$$\overset{\delta+}{RS}\cdots\underset{\underset{H}{|}}{\overset{|}{C}}\cdots O\cdots H\cdots\overset{\delta-}{A}$$

(**35**)

IX. ADDITION OF AMMONIA AND AMINES[2,66,259-262]

A. General Considerations

The reactions of carbonyl compounds with primary or secondary amines generally involve one or two of the following processes:

Addition

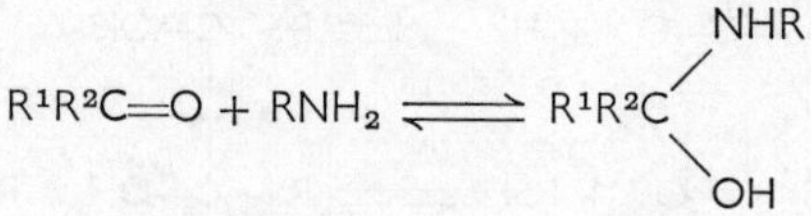

$$R^1R^2C{=}O + RNH_2 \rightleftharpoons R^1R^2C\overset{NHR}{\underset{OH}{\big\langle}} \tag{80}$$

(**36**)

Intramolecular dehydration

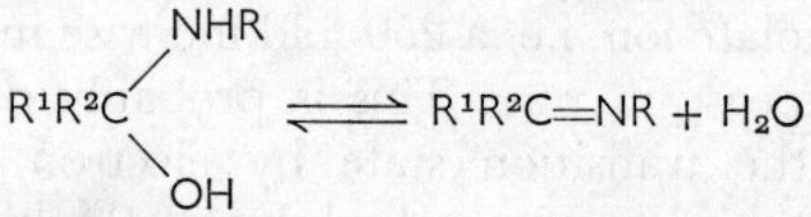

$$R^1R^2C\overset{NHR}{\underset{OH}{\big\langle}} \rightleftharpoons R^1R^2C{=}NR + H_2O \tag{81}$$

(**37**)

Intermolecular dehydration

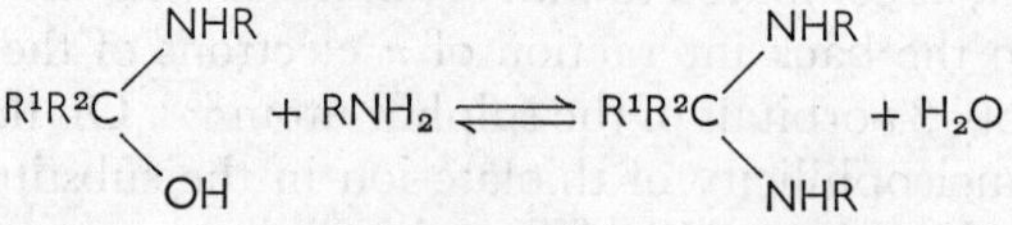

$$R^1R^2C\overset{NHR}{\underset{OH}{\big\langle}} + RNH_2 \rightleftharpoons R^1R^2C\overset{NHR}{\underset{NHR}{\big\langle}} + H_2O \tag{82}$$

(**38**)

Reaction (81) includes the formation of Schiff bases, oximes, semicarbazones and hydrazones and is important for the identification and determination of carbonyl compounds. Reaction (82) represents important industrial processes such as the formation of urea resins, melamine resins and certain fertilizers. The amine–aldehyde reactions are summarized in Scheme 1.

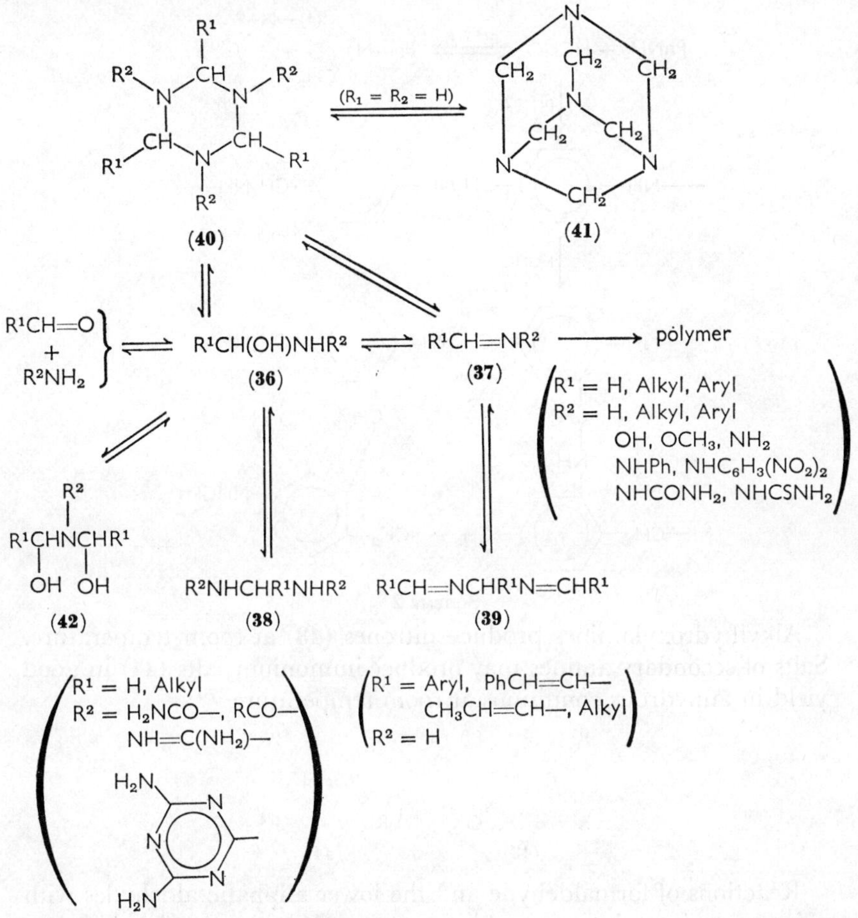

SCHEME 1

Reaction (80) is the fundamental step of all amine–carbonyl reactions. The α-amino alcohol (**36**) is generally too unstable to be isolated, just as a hemiacetal or a *gem*-diol. Primary amines produce Schiff bases (**37**). Schiff bases of aliphatic aldehydes are relatively unstable and readily polymerizable[263,264], while those of aromatic aldehydes, having an effective conjugation system, are more stable[265-267]. The reaction of formaldehyde with anilines (Scheme 2) is rather peculiar: at pH < 2 an irreversible condensation polymerization and cross-linking takes place, while at pH 2–7 a cyclic compound **40** is produced reversibly[268,269].

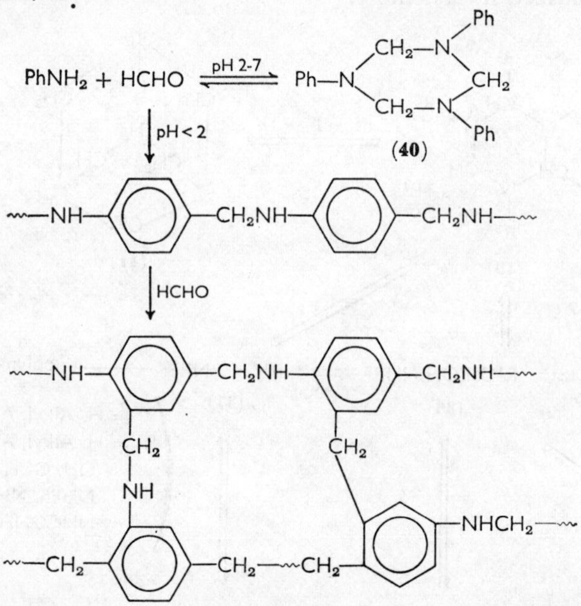

$$PhNH_2 + HCHO \underset{pH\ 2\text{-}7}{\rightleftharpoons}$$

(**40**)

$$\downarrow pH < 2$$

$$\downarrow HCHO$$

SCHEME 2

Alkylhydroxylamines produce nitrones (**43**) at room temperature. Salts of secondary amines may produce immonium salts (**44**) in good yield in anhydrous conditions at room temperature[270]:

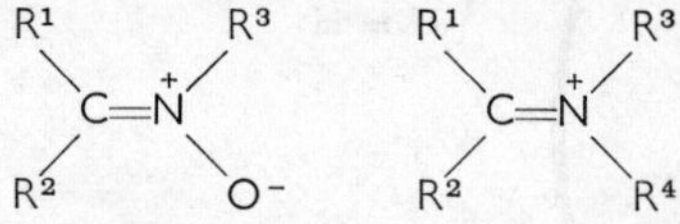

(**43**)   (**44**)

Reactions of formaldehyde and the lower aliphatic aldehydes with primary amines give azomethine compounds (**37**) or their trimerization products, hexahydrotriazines (**40**). E.g., crystalline acetaldehyde–ammonia has structure **40** ($R^1 = CH_3$, $R^2 = H$) with three molecules of water of crystallization[271]. Formaldehyde reacts readily with ammonia yielding hexamethylene-tetramine (hexamine or urotropin) (**41**)[272,273]. The reaction of aromatic and α,β-unsaturated aldehydes and some aliphatic aldehydes with ammonia yields hydroamides (**39**) via Schiff bases (**37**)[274].

Weak bases such as amides, urea and melamine, or secondary amines, produce alkylidenediamines or *gem*-diamines (**38**) by intermolecular dehydration between the α-amino alcohols and the bases[275,276] (reaction 82). For example, ethylidenediurea ($NH_2CONHCHMeNHCONH_2$) is formed from urea and acetaldehyde. Amines having two or more amino groups, for example, urea, can produce polymeric materials by repeating addition (80) and intermolecular condensation (82). α-Amino alcohols (**36**) add further to a carbonyl compound to form α,α'-dihydroxydialkylamines (**42**), this being analogous to reaction (80).

Amine–carbonyl additions are involved in the Mannich[277], Leuckart[278], Skraup[279], Döbner–Miller[280], Pictet–Spingler[281] and Pomeranz–Fritsch[282] reactions which are very important in synthetic chemistry.

Generally, carbonyl containing molecules may be activated by acids, while amines may be activated by bases, i.e. the reactions may be subject to acid and base catalysis. Therefore, the possible reactive species may be: free carbonyl ($R^1R^2C{=}O$), hydrogen bonded carbonyl ($R^1R^2C{=}O\cdots HA$) and/or conjugate acid of carbonyl ($R^1R^2C{\equiv}OH^+$) on the one hand, and free base (RNH_2), hydrogen bonded amine ($RNH–H\cdots B$) and/or conjugate base of amine (RNH^-) on the other. The importance of these species varies, depending on the reagents and on the conditions of reaction.

B. Addition of Melamine, Urea and Amides to Aldehydes

The addition step (equation 80) with these weak bases is often subject to acid and base catalysis, while their condensation step (equation 82) is subject to specific oxonium ion catalysis.

The hydroxymethylation of melamine (an initial stage for the preparation of melamine resins) follows second-order kinetics, while the reverse reaction is of a first-order type[283,284].

$$\text{(83)}$$

The rate of reaction is expressed as:

$$v = k[\text{melamine}][\text{HCHO}] - k'[\text{hydroxymethylmelamine}] \qquad (84)$$

The dependency of k on pH suggests the following mechanisms:

(a) at pH 3–3·8 the reaction occurs between free melamine and protonated formaldehyde ($\overset{+}{C}H_2OH$),

(b) at pH 5·0–6·0 it occurs between free melamine and free formaldehyde, and

(c) at pH 10–10·6 between melamine anion $(C_3H_4N_5NH^-)$ and free formaldehyde[283]. A similar behaviour is observed with the N-hydroxymethylations of urea[276,285–290], substituted phenylureas[291,292], alkylureas[289,293], thioureas[294,295], amides[285,296–302], dicyandiamide[303] and with other N-α-hydroxyalkylations of urea (equation 87)[304–306].

For example, the hydroxymethylation of benzamide in water at pH 2–11 (equation 85) is subject to water, hydronium ion and

$$ArCONH_2 + CH_2O \underset{k_r}{\overset{k_f}{\rightleftharpoons}} ArCONHCH_2OH \qquad (85)$$

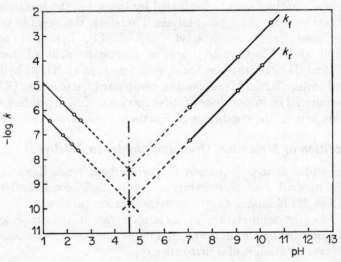

FIGURE 7. Rate constants for the formation (k_f) and decomposition (k_r) of N-methylolbenzamide (at 25° and ionic strength 0·1) as a function of pH[296].

hydroxide ion catalysis and the plot of either log k_f or log k_r versus pH gives a bent line having a minimum at pH 4·6 (Figure 7)[296]. In a buffer solution the reaction is subject to general acid and specific hydroxide ion catalysis, the apparent second-order rate constant being expressed as:

$$k_f = k_0[H_2O] + k_H[H_3O^+] + k_{OH}[OH^-] + k_{HA}[HA] \qquad (86)$$

In contrast, for the reaction of urea with some aliphatic aldehydes (equation 87, R = H, Me, Et, Pr, i-Pr) the second-order rate constant

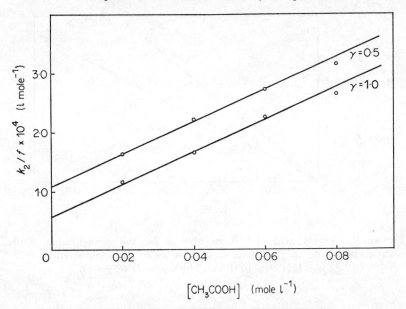

FIGURE 8. Acid catalysis for the condensation of urea with acetaldehyde in acetate buffers at $24 \cdot 2°$ and an ionic strength of $0 \cdot 2$, where γ is $[CH_3COO^-]/[CH_3COOH]$, k_2 the observed second-order rate constant and f the molar fraction of unhydrated aldehyde[305]:

$$f = \frac{[RCHO]}{[RCHO] + [RCH(OH)_2]}$$

increases linearly with the increase in concentration of acid catalyst (Figure 8) and that of base catalyst (Figure 9), the rate constant being expressed as equation (88) (cf. equation 9) which indicates the presence

$$NH_2CONH_2 + RCH{=}O \rightleftharpoons NH_2CONHCHROH \qquad (87)$$

$$k = k_0[H_2O] + k_H[H_3O^+] + k_{OH}[OH^-] + k_{HA}[HA] + k_B[B] \qquad (88)$$

of general acid and general base catalysis[287,304,305]. As shown in Figure 10, the Brønsted α value of general acid catalysis is almost independent of the aldehyde.

As shown in Figure 11, the Taft equation is applicable to the second-order rate constant calculated by assuming that only the unhydrated aldehydes are reactive species $(\rho^* = +3 \cdot 5)$[305].

The reactions of amides with aldehydes are subject to specific hydroxide ion catalysis, while the reaction of urea is subject to general

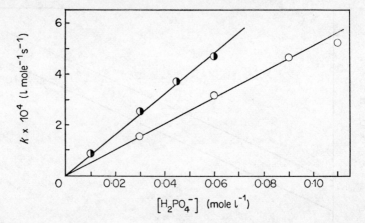

FIGURE 9. General acid and base catalysis for the condensation of urea with acetaldehyde in phosphate buffers at $24.2°$ and ionic strength $0.2\,^{304}$. $\gamma = [HPO_4{}^{2-}]/[H_2PO_4{}^{-}]$. $\bigcirc$, $\gamma = 2$; $\mathbf{\Phi}$, $\gamma = 4$.

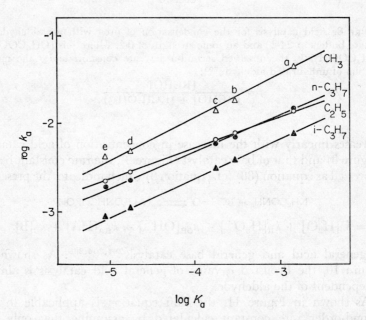

FIGURE 10. Application of the Brønsted catalysis law $\log k_a = \log G_a + \log K_a$ (see equation 10) to the condensation of urea with aliphatic aldehydes RCHO at $24.2°$ and ionic strength $0.2\,^{305}$. a: Chloroacetic acid; b: Methoxyacetic acid; c: Formic acid; d: Acetic acid; e: Pivalic acid.

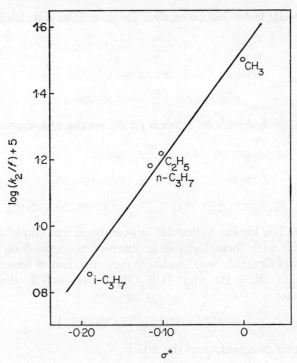

FIGURE 11. Taft plot for the condensation of aliphatic aldehydes with urea in an acetate buffer at 24·2° and ionic strength 0·2[305]. For the meaning of k_2 and f see Figure 8.

base catalysis; the difference may be due to the lower basicity of amides[296]. The most probable mechanisms are[296,304]:

General acid catalysis for the urea–aldehyde reaction

$$RCH{=}O + HA \underset{}{\overset{fast}{\rightleftharpoons}} RCH{=}O{\cdots}HA \tag{89}$$

$$NH_2CONH_2 + RCH{=}O{\cdots}HA \underset{}{\overset{slow}{\rightleftharpoons}} NH_2C\overset{+}{O}H_2CHROH + A^- \tag{90}$$

$$NH_2C\overset{+}{O}H_2CHROH + A^- \underset{}{\overset{fast}{\rightleftharpoons}} NH_2CONH_2CHROH + HA \tag{91}$$

General base catalysis for the urea–aldehyde reaction

$$NH_2CONH_2 + B \underset{}{\overset{fast}{\rightleftharpoons}} NH_2CONH{-}H{\cdots}B \tag{92}$$

$$NH_2CONH{-}H{\cdots}B + RCH{=}O \underset{}{\overset{slow}{\rightleftharpoons}} NH_2CONHCHRO^- + HB^+ \tag{93}$$

$$NH_2CONHCHRO^- + HB^+ \underset{}{\overset{fast}{\rightleftharpoons}} NH_2CONHCHROH + B \tag{94}$$

3+c.c.g. ii

Specific hydronium ion catalysis for the reaction of melamine

$$CH_2{=}O + H^+ \underset{}{\overset{fast}{\rightleftharpoons}} \overset{+}{C}H_2OH \tag{95}$$

$$RNH_2 + \overset{+}{C}H_2OH \underset{}{\overset{slow}{\rightleftharpoons}} R\overset{+}{N}H_2CH_2OH \tag{96}$$

$$R\overset{+}{N}H_2CH_2OH \underset{}{\overset{fast}{\rightleftharpoons}} RNHCH_2OH + H^+ \tag{97}$$

Specific hydroxide ion catalysis for the reaction of benzamide

$$RCONH_2 + OH^- \underset{}{\overset{fast}{\rightleftharpoons}} RCONH^- + H_2O \tag{98}$$

$$RCONH^- + CH_2{=}O \underset{}{\overset{slow}{\rightleftharpoons}} RCONHCH_2O^- \tag{99}$$

$$RCONHCH_2O^- + H_2O \underset{}{\overset{fast}{\rightleftharpoons}} RCONHCH_2OH + OH^- \tag{100}$$

The rate law for the hydroxide ion-catalysed reaction of amides or phenylureas with formaldehyde is observed to depend on the concentration of formaldehyde. At lower concentration of formaldehyde, for example, $R = Ph$ and $[CH_2O]_0 < 0.05$ mole/l, the rate is expressed as:

$$v = k[RCONH_2][CH_2O][OH^-] \tag{101a}$$

but at higher concentrations the rate is:

$$v = k[RCONH_2][OH^-] \tag{101b}$$

The latter expression is explained by a rate-determining formation of amide anion, which reacts rapidly with formaldehyde when its concentration is high. In both cases electron-attracting groups in the amide promote the reaction[292,293,300-302].

The activation energies for the hydroxymethylation of melamine, mono- and bis(hydroxymethyl)melamine are 16, 17 and 15 kcal/mole, respectively[303], and those of urea, monohydroxymethylurea and methylenediurea are 13, 14 and 15 kcal/mole[287], respectively. These analogous values show that the reactivity of the amino end groups is almost independent of the rest of the molecule.

C. Addition of Ammonia to Aldehydes

The formation of crystalline hexamethylenetetramine (**41**) from formaldehyde and ammonia is a fast reaction, which seems to proceed via the following scheme involving the rate-determining formation of bis(hydroxymethyl)amine, as deduced from the kinetic equation[307-309]: $-\mathrm{d}[CH_2O]/\mathrm{d}t = k[NH_3][CH_2O]^2$, and the isolation of intermediates[310,311]:

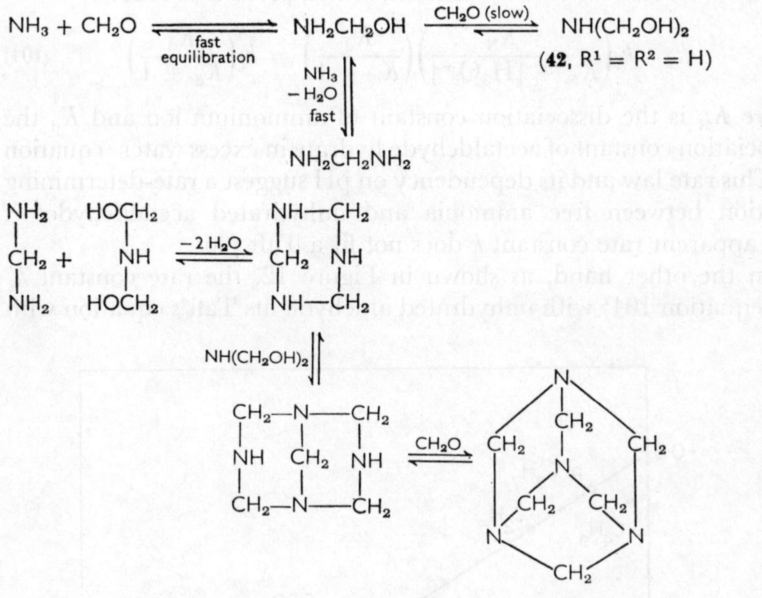

$$NH_3 + CH_2O \underset{\substack{fast \\ equilibration}}{\rightleftharpoons} NH_2CH_2OH \xrightarrow{\underset{}{CH_2O \ (slow)}} NH(CH_2OH)_2$$

(**42**, R¹ = R² = H)

NH₃ / −H₂O fast

$$NH_2CH_2NH_2$$

(102)

$$NH(CH_2OH)_2$$

(**41**)

In acetic acid, the reaction is more complex[312,313].

The reaction of ammonia with aliphatic aldehydes yields crystals of hexahydrotriazines (**47**). In dilute aqueous solutions, on the other hand, they yield mainly α-amino alcohols (**45**) rather than aldimines (**46**) (infrared spectra show only a little C=N band). The determination of the equilibrium constant also accounts for the presence of monomers **45** and **46**[314,315].

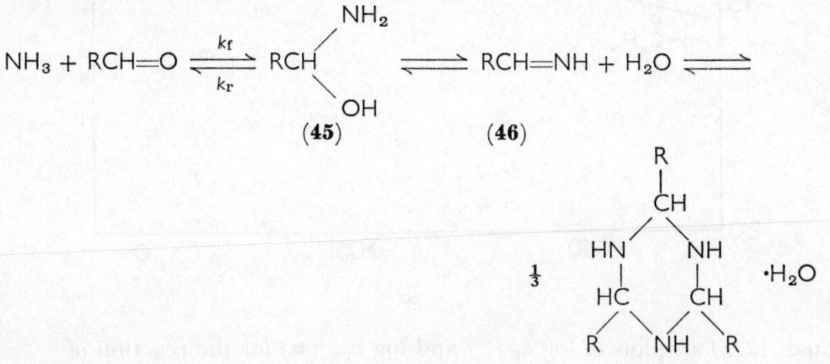

$$NH_3 + RCH{=}O \underset{k_r}{\overset{k_f}{\rightleftharpoons}} RCH\overset{NH_2}{\underset{OH}{\Big\langle}} \rightleftharpoons RCH{=}NH + H_2O \rightleftharpoons$$

(**45**)   (**46**)

·H₂O   (103)

(**47**)

In aqueous solutions at pH 9–11, the formation of **45** and **46** follows second-order kinetics, but the reverse reaction is first-order. The observed second-order rate constant, k, is expressed as follows:

$$k = k_f \left(\frac{K_N}{K_N + [H_3O^+]} \right) \left(\frac{K_d}{K_d + 1} \right) = k_f \left(\frac{K_d}{K_d + 1} \right) \tag{104}$$

where K_N is the dissociation constant of ammonium ion and K_d the dissociation constant of acetaldehyde hydrate in excess water (equation 5). This rate law and its dependency on pH suggest a rate-determining reaction between free ammonia and unhydrated acetaldehyde[314]. The apparent rate constant k does not fit a Taft plot.

On the other hand, as shown in Figure 12, the rate constant k_f' (see equation 104) with unhydrated aldehyde fits Taft's equation with

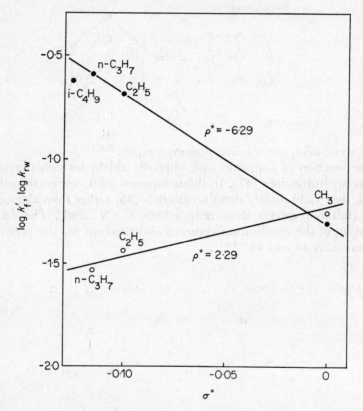

FIGURE 12. Taft plots of $\log k_f$ ($\circ$) and $\log k_{rw}$ ($\bullet$) for the reaction of aliphatic aldehydes RCHO with ammonia[315].

$\rho^* = 2\cdot29$, while the constant, k_{rw}, for the uncatalysed or water-catalysed reverse reaction, has $\rho^* = -6\cdot29$. The acid catalytic constant k_{rH^+} does not fit the Taft equation. The failure of simple Hammett equation with k_{H^+} for the hydrolysis of substituted benzylideneanilines has also been reported[340].

The large negative ρ^* value found above for the reverse reaction is explicable by the contribution of electron releasing substituents both to protonation of the α-amino alcohol (equilibrium 105) and to cleavage of the conjugate acid (reaction 106)[315].

$$\underset{\overset{|}{\text{OH}}}{\overset{\overset{|}{\text{NH}_2}}{\text{RCH}}} + \text{H}_2\text{O} \overset{\text{fast}}{\rightleftharpoons} \underset{\overset{|}{\text{OH}}}{\overset{\overset{|}{\overset{+}{\text{NH}_3}}}{\text{RCH}}} + \text{OH}^- \qquad (105)$$

$$\underset{\overset{|}{\text{OH}}}{\overset{\overset{|}{\overset{+}{\text{NH}_3}}}{\text{RCH}}} + \text{OH}^- \overset{\text{slow}}{\rightleftharpoons} \text{RCH}{=}\text{O} + \text{NH}_3 + \text{H}_2\text{O} \qquad (106)$$

Table 9 shows the equilibrium constant of addition of various amines to carbonyls. The product is mainly α-amino alcohol, but a small amount of aldimine may also exist in equilibrium. As apparent from Table 9, the constants tend to be higher with increasing basicity of amines and electrophilicity of carbonyls.

TABLE 9. Equilibrium constants for amine–carbonyl addition reactions in aqueous solutions.

Amine	Aldehyde	Temperature (°c)	K_{obs} (l/mole)	Reference
Benzamide	Formaldehyde	25	23·2, 22·3	285, 296
Propionamide	Formaldehyde	25	23·5, 21·5	285, 296
Acetamide	Formaldehyde	25	26	297
Chloracetamide	Formaldehyde	25	22, 18·8	297, 296
Urea	Formaldehyde	25	50	287
Urea	Acetaldehyde	24·2	2·05	305
Urea	Propionaldehyde	24·2	1·65	305
Urea	Butyraldehyde	24·2	1·62	305
Urea	Isobutyraldehyde	24·2	1·32	305
Ammonia	Acetaldehyde	20	46·5	315
Ammonia	Propionaldehyde	20	10·5	315
Ammonia	Butyraldehyde	20	7·87	315
Ammonia	Isovaleraldehyde	20	2·06[a]	315

[a] In 60 vol.% dioxan.

The aldimines obtained from some primary aliphatic amines and isobutyraldehyde have mainly *trans* configurations, with one of the C—Me bonds of the isobutylidene group eclipsing the C=N bond in the principal conformations (48)[264].

$$
\begin{array}{c}
\text{H} \\
\text{H} \quad \text{C=N} \quad \text{R} \\
\text{C} \\
\text{Me} \quad \text{Me}
\end{array}
$$

(48)

The reaction of ammonia with aromatic aldehydes yields hydrobenzamides (39) via α-amino alcohols (36) and aldimines (37)[316–318].

$$ ArCH{=}O + NH_3 \underset{}{\overset{a}{\rightleftharpoons}} ArCH\overset{NH_2}{\underset{OH}{\big\langle}} \underset{}{\overset{b}{\rightleftharpoons}} ArCH{=}NH + H_2O \qquad (107) $$

(36, R^1 = Ar, R^2 = H)   (37, R = H)

$$ 2\,ArCH{=}NH + ArCHO \rightleftharpoons ArCH{=}N{-}CHAr{-}N{=}CHAr + H_2O \qquad (108) $$

(39, R^1 = Ar)

The rate of formation of 39 is expressed as:

$$ v = k[ArCHO][NH_3] \qquad (109) $$

Step (107a) is rate-determining for aldehydes having electron-releasing groups, while step (107b) determines the rate for aldehydes having electron-attracting groups. Hence, the Hammett plot does not give a simple straight line as shown in Figure 13. The order of the rates of formation of hydrobenzamides is:

$$ p\text{-}NMe_2 < p\text{-}MeO < H > p\text{-}Cl > p\text{-}CN $$

Also the following facts support the assumption that the rate-determining step changes with substituent[318]:

(a) The rate of disappearance of *p*-cyanobenzaldehyde is higher than that of benzaldehyde ($\rho > 0$ for the addition 107a[321]), while the rate of formation of their hydramides has the reverse order ($\rho < 0$ for the dehydration 107b[321]).

(b) The formation of 39 from *p*-chloro- or *p*-cyanobenzaldehyde has an induction period (equation 107a is fast and equation 107b slow), while no induction period was observed for its formation from *p*-methoxybenzaldehyde (equation 107a is slow and equation 107b fast).

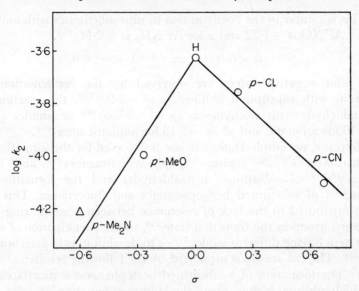

FIGURE 13. Hammett plot for the observed second-order rate constants k_2 of the reaction of aromatic aldehydes with ammonia in methanol at 30° (△ was calculated from data in reference 317.)[318]

D. *Structural Effects on Equilibria and Rates*

Kinetic studies on the formation of Schiff bases, semicarbazones, oximes and hydrazones show that these reactions are subject to specific and general acid catalysis and consist of two steps, i.e. the addition of amines to carbonyls (equation 80) and the dehydration of α-amino alcohols to imines (equation 81)[318-338]. There is evidence that the former is rate-determining in acid media, while the latter is rate-determining in neutral and basic media[319-330]. The reverse reaction, the hydrolysis of Schiff bases, semicarbazones, oximes and hydrazones, proceeds probably via tetrahedral intermediates (**36**)[328,339-349]. These reactions have been discussed elsewhere in this series[2].

Electron-attracting groups in the carbonyl compound or electron-releasing groups in the amine accelerate the amine–carbonyl addition. Thus in reaction (110), the acid-catalysed condensation of substituted benzaldehydes with aniline has a positive ρ value of $+1\cdot54$, while the condensation of benzaldehyde with substituted anilines has a negative ρ value of $-2\cdot00$[350].

$$Ar^1CH{=}O + Ar^2NH_2 \longrightarrow Ar^1CH{=}NAr^2 + H_2O \qquad (110)$$

This is similar to the condensation of nitrosobenzenes with anilines (ρ for Ar^1NO is $+1\cdot22$ and ρ for Ar^2NH_2 is $-2\cdot14$)[351].

$$Ar^1N{=}O + Ar^2NH_2 \rightleftharpoons Ar^1N{=}NAr^2 + H_2O \qquad\qquad (111)$$

Similar negative values are observed for the condensations of benzoin with substituted anilines ($\rho = -2\cdot20$)[352], the reaction of formaldehyde with phenylureas ($\rho = -0\cdot756$)[292] or amides ($\rho = -1\cdot10$ for aromatic and $\rho^* = -2\cdot16$ for aliphatic ones)[299].

However, no simple Hammett law is observed for the formation of semicarbazones[319,321], oximes[353], phenylhydrazones[353] and aldimines[318,354] of substituted benzaldehydes and the formation of oximes[355] of substituted benzophenones and fluorenones. This has been attributed to the lack of resonance between benzene ring and carbonyl group in the transition state[355], to the participation of multiple steps having different values[319] or to the shift in rate-determining step[320]. The last reason is supported by the following results:

(a) The formation of p-substituted acetophenone semicarbazones, in which addition is slow, obeys the Hammett equation[356].

(b) A linear Hammett plot ($\rho^+ = 0\cdot39$) is observed for the reaction of aniline with substituted benzaldehydes at pH $2\cdot55$, where the rate-determining step is the addition, while a non-linear Hammett plot is observed at pH $6\cdot05$, where the rate-determining step varies with substituents[323] (see also Figure 13).

Steric effects also may cause departures from linear free energy relationships. Changes in structure affect both the energy and the entropy of activation of semicarbazone formation[357]. The formation of thiosemicarbazones and oximes of substituted aryl alkyl ketones is retarded by o- or m-isopropyl groups which decrease the entropy of activation[358–361]. The Taft law cannot be applied to aliphatic ketones, but a linear relationship is observed between log k for the formation of their thiosemicarbazones or oximes and log k for the formation of their semicarbazones[357,361,362], suggesting that the same structural effects prevail for these reactions.

The rate of the formation of Schiff bases from piperonal and various aliphatic amines in methanol is related to the free energy of decomposition of the corresponding amine–trimethylboron complexes, but not to the basicity of amines in water[363,364], the basicity depending on the degree of solvation[365].

Table 10 shows that the rate of semicarbazone formation is usually lower for the *para*-substituted benzaldehyde than for the corresponding *ortho* isomer, in spite of the expected steric hindrance[321]. Similar

TABLE 10. Equilibrium and rate constants for the formation of semicarbazones of substituted benzaldehydes in 25% aqueous ethanol at 25°[321].

Substit- uent	$K_1{}^a$ (1 mole^{-1})	$K_2 \times 10^{-5\,b}$ (1 mole^{-1})	$K_{ov} \times 10^{-5\,c}$ (1 mole^{-1})	$k_{ov} \times 10^{-7\,d}$ (l^2 mole^{-2} min^{-1})	k_o/k_p
None	1·32	5·2	6·9	1·08	—
o-MeO	1·67	4·9	8·2	4·75⎫	11·0
p-MeO	0·34	4·5	1·5	0·43⎭	
o-HO	0·33	13·7	4·5	1·03⎫	2·2
p-HO	0·073	15·5	1·1	0·46⎭	
o-Me	1·05	6·7	7·0	1·49⎫	1·6
p-Me	0·62	5·5	3·4	0·92⎭	
o-Cl	19·0	1·2	23·3	1·94⎫	2·3
p-Cl	4·14	2·5	10·5	0·86⎭	
o-NO$_2$	27·1	2·8	75	0·52⎫	0·54
p-NO$_2$	40·1	2·1	83	0·96⎭	
ρ	1·81	−0·17	1·64	0·07	—

^{a} Equilibrium constant for the addition step.
^{b} Equilibrium constant for the dehydration step.
^{c} Overall equilibrium constant.
^{d} Overall rate constant.

results have been observed for the formation of oximes and phenyl-hydrazones[337,353,366]. This may be because the stronger electron-releasing resonance effect of a *para*-substituent is more retarding than the steric effect of the *ortho* substituent[321].

Although it is expected that the α value of Brønsted's catalysis law should vary linearly with Hammett's σ, and/or Taft's σ*, the observed values for the general acid-catalysed condensation of urea with aliphatic aldehydes[305] or that of semicarbazide with aromatic aldehydes[325] vary only slightly with the change of substituent; for example, α values are for acetaldehyde 0·46; propionaldehyde, 0·38; n-butyraldehyde, 0·43; isobutyraldehyde, 0·45. This fact suggests that the structural effect of carbonyls on the electrophilic attack of catalytic acid and the same effect on the nucleophilic attack of amines cancel each other out[325].

On the contrary, the following relationship between the Brønsted α value and the basicity of the nucleophile may be derived[328] from the Brønsted catalysis law and the Swain–Scott equation[367]:

$$\frac{\alpha_N - \alpha_0}{pK_{a0}^N - pK_a^N} = \frac{\beta_A - \beta_0}{pK_a^A - pK_{a0}^A} = \text{constant} \tag{112}$$

where pK_a^N and pK_a^A are the acidity constants of the conjugate acid of nucleophile (N) and catalytic acid (A), α and β the Brønsted and

3*

Swain–Scott constants, respectively, and the subscript 0 denotes a standard. Taking the data for phenylhydrazine as standard, a plot of $(pK_a^N - pK_{a0}^N)$ vs. $(\alpha_N - \alpha_0)$ for the addition reactions of various nucleophiles to carbonyls in aqueous solutions gives a straight line in agreement with equation (112)[328], i.e. higher basicity of nucleophile results in lower α values[326,328,346]. Thus, the α value is nearly zero for strongly basic t-butylamine, the reaction of which requires little acid catalysis, while the α value is 0·54 for weakly basic water, which requires acid catalysis[325]. However, for the attack of thiourea or water on highly reactive carbonyl compounds such as formaldehyde and 1,3-dichloracetone, the α values show negative deviations from the plot.

TABLE 11.   Brønsted α values for the general acid catalysis of the attack of nucleophilic reagents on the carbonyl carbon atom[328].

Reagent	pK_a	Substrate	Solvent	α	Reference
t-Butylamine	10·4	Benzaldehydes	Water	~ 0	344
Hydroxylamine	6·0	Acetone	Water	0·10	326
Phenylhydrazine	5·2	Benzaldehyde	20% Ethanol	0·20	328
Aniline	4·6	p-Chlorobenz-aldehyde	Water	0·25	324
Semicarbazide	3·65	Benzaldehydes	Water	0·25	325
Urea	0·2	Acetaldehyde	Water	0·45	304, 305, 328
Water	$-1·74$	Acetaldehyde	Water	0·54	49

In the general acid or general base catalysis involving proton transfer from or to oxygen, nitrogen and sulphur atoms, it is likely that the acid attacks an atom which becomes more basic in the transition state; similarly the base should abstract a proton which becomes more acidic in the transition state[330].

X. REACTION WITH HALOGENS

Molecular halogen can form a 1:1 complex with carbonyl compounds. Molecular iodine, which is violet in vapour and forms a violet solution when dissolved in aliphatic hydrocarbons or carbon tetrachloride, forms a brown solution when dissolved in carbonyl compounds, ethers or alcohols, attributed to complex formation. For example, the band at λ_{max} 5200 Å for iodine in pure carbon tetrachloride shifts to λ_{max}

4600 Å if acetone is also present[368a]. The complex with acetone is believed to be a charge transfer complex, $Me_2CO^+I_2^-$, which is similar to the complex between aromatic compounds and iodine[368b,c].

In general, the non-bonding oxygen orbital of the complex has its axis perpendicular to the R^1R^2CO plane. Because of its planarity there is a conjugation or hyperconjugation effect between some of the iodine non-bonding π-orbitals and the π-bonding electrons in $C=O$ i.e.[369]

The complex between acetone and molecular bromine[370] has an O-Br distance of 2·82 Å and a Br-O-Br bond angle of about 110°.

The spectrophotometric study[371] of the complex compounds of aldehydes with I_2, ICl and IBr points to the formation of a bond between O and I. The order of electron-accepting ability of these halogens is: ICl < IBr < I_2.

Fluorine, which is quite reactive toward organic compounds, adds irreversibly, aided by catalysts such as AgF_2[372] or CsF[373,374], to the $C=O$ double bond to yield compounds of structure $(R_F)_2CF-OF$ (R_F = fluorocarbon residues), for example[374].

$$F_2C{=}O + F_2 \xrightarrow[\;97\%\;]{\text{CsF}, -196° \text{ to} -78°, \, 4\text{hr}} F_3C{-}OF \quad \text{(b.p. } -95.0°) \tag{113}$$

XI. ACKNOWLEDGEMENTS

The authors are indebted to the American Chemical Society, the Chemical Society (London), the Chemical Society (Japan), The Faraday Society, Pergamon Press Ltd. (*Tetrahedron*), The Royal Dutch Chemical Society, Verlag Chemie (*Liebigs Annalen*) and the editors of *Comptes Rendus de l'Academie des Sciences* and *Helvetica Chimica Acta* for permission to reproduce a number of figures and tables. They also thank Masao Goto, Tetsuo Nakamura, Kazuo Matsuyama, Mototaka Mineno, Fumitoshi Sugiura, Keizo Aoki and Nobuya Okumura for their assistance in the preparation of the manuscript.

XII. REFERENCES

1. E. Schmitz and I. Eichhorn in *The Chemistry of the Ether Linkage*, (Ed. S. Patai), John Wiley and Sons, New York, 1967, pp. 309–351.

2. R. L. Reeves in *The Chemistry of the Carbonyl Group*, Vol. 1 (Ed. S. Patai) John Wiley and Sons, New York, 1966, pp. 567–619.

3. P. Salomaa in *The Chemistry of the Carbonyl Group*, Vol. 1. (Ed. S. Patai), John Wiley and Sons, New York, 1966, pp. 177–232.

4. For the review on hydration, see R. P. Bell in *Advances in Physical Organic Chemistry*, Vol. 4 (Ed. V. Gold), Academic Press, London, 1966, p. 1.

5. I. F. Homfray, *J. Chem. Soc.*, **87**, 1430 (1905).

6. F. Auerbach and H. Barschall, *Arb. Kais. Gesundheitsamt*, **22**, 594 (1905).

7. W. Ledbury and E. W. Blair, *J. Chem. Soc.*, **26**, 2832 (1925).

8. J. F. Walker, *J. Phys. Chem.*, **35**, 1104 (1931); *J. Amer. Chem. Soc.*, **55**, 2821, 2825 (1933).

9. S. A. Schou, *Compt. Rend.*, **182**, 965 (1926); **184**, 1453, 1684 (1927); *J. Chim. Phys.*, **26**, 69 (1929).

10. W. Herold and K. L. Wolf, *Z. Physik. Chem.*, **B5**, 124 (1929); **B12**, 192 (1931).

11. J. H. Hibben, *J. Amer. Chem. Soc.*, **35**, 1104 (1931).

12. J. B. M. Herbert and I. Lauder, *Trans. Faraday Soc.*, **34**, 432, 1219 (1938).

13. M. Cohn and H. C. Urey, *J. Amer. Chem. Soc.*, **60**, 679 (1938).

14. D. Samuel and B. L. Silver in *Advances in Physical Organic Chemistry*, Vol. 3 (Ed. V. Gold), Academic Press, London, 1965, p. 123.

15. W. Herold, *Z. Physik. Chem.*, **B18**, 265 (1932).

16. I. L. Gauditz, *Z. Physik. Chem.*, **B48**, 228 (1941).

17. R. Bieber and G. Trümpler, *Helv. Chim. Acta*, **30**, 1860 (1947).

18. P. Rumpf and C. Bloch, *Compt. Rend.*, **233**, 1364 (1951).

19. R. P. Bell and J. C. Clunie, *Trans. Faraday Soc.*, **48**, 439 (1952).

20a. M. P. Federline, *Compt. Rend.*, **235**, 44 (1952).

20b. J. L. Kurz, *J. Amer. Chem. Soc.*, **89**, 3524, 3528 (1967).

21. R. P. Bell and J. O. McDougall, *Trans. Faraday Soc.*, **56**, 1281 (1960).

22. L. C. Gruen and P. T. McTigue, *J. Chem. Soc.*, **1963**, 5217.

23. E. O. Bishop and R. E. Richards, *Trans. Faraday Soc.*, **55**, 1070 (1959).

24. E. Rombardi and P. B. Sogo, *J. Chem. Phys.*, **32**, 635 (1960).

25. Y. Fujiwara and S. Fujiwara, *Bull. Chem. Soc. Japan*, **36**, 574 (1963).

26. V. Gold, S. Socrates and M. R. Crampton, *J. Chem. Soc.*, **1964**, 5888.

27. J. Hine, J. G. Houston and J. H. Jensen, *J. Org. Chem.*, **30**, 1184 (1965).

28. P. Greenzaid, Z. Luz and D. Samuel, *J. Amer. Chem. Soc.*, **89**, 749 (1967).

29. P. Valenta, *Collection Czech. Chem. Commun.*, **25**, 853 (1960).

30. M. Becker and H. Strehlow, *Z. Elektrochem.*, **64**, 813 (1960).

31. H. Strehlow, *Z. Elektrochem.*, **66**, 392 (1962).

32. R. W. Taft, Jr. in *Steric Effects in Organic Chemistry*, (Ed. M. S. Newman), John Wiley and Sons, New York, 1956, pp. 556–675.

33. R. P. Bell and M. H. Rand, *Bull. Soc. Chim. France*, **115**, 3629 (1955).

34a. M. J. S. Dewar and R. Pettit, *J. Chem. Soc.*, **1954**, 1625.

34b. M. J. S. Dewar and H. N. Schmeising, *Tetrahedron*, **5**, 166 (1959).

35. R. W. Taft, Jr. and I. C. Lewis, *Tetrahedron*, **5**, 210 (1959).

36. M. N. Kreevoy, *Tetrahedron*, **5**, 233 (1959).

37. M. Eigen in *Technique of Organic Chemistry*, Vol. 8, Part Two, (Ed. A. Weissberger), Interscience, New York, 1963, pp. 793–1229.

38. I. Lauder, *Trans. Faraday Soc.*, **44**, 729 (1948).

39. R. P. Bell and W. C. E. Higginson, *Proc. Roy. Soc. (London)*, **A197**, 141 (1949).

40. R. P. Bell and B. deB. Darwent, *Trans. Faraday Soc.*, **46**, 34 (1950).
41. R. P. Bell and J. C. Clunie, *Proc. Roy. Soc. (London)*, **A212**, 33 (1952); *Nature*, **167**, 362 (1951).
42. R. P. Bell, M. H. Rand and K. M. A. Wynne-Jones, *Trans. Faraday Soc.*, **52**, 1093 (1956).
43. L. C. Gruen and P. T. McTigue, *J. Chem. Soc.*, **1963**, 5224.
44. M. Becker, *Z. Elektrochem.*, **68**, 669 (1964).
45. J. Hine and J. G. Houston, *J. Org. Chem.*, **30**, 1328 (1965).
46. P. G. Evans, G. R. Miller and M. M. Kreevoy, *J. Phys. Chem.*, **69**, 4325 (1965).
47. M.-L. Ahrens and H. Strehlow, *Disc. Faraday Soc.*, **39**, 112 (1965).
48. P. Greenzaid, Z. Luz and D. Samuel, *J. Amer. Chem. Soc.*, **89**, 756 (1967).
49. R. P. Bell and M. B. Jensen, *Proc. Roy. Soc. (London)*, **A261**, 38 (1961).
50. R. P. Bell and P. G. Evans, *Proc. Roy. Soc. (London)*, **A291**, 297 (1966).
51. M. Byrn and M. Calvin, *J. Amer. Chem. Soc.*, **88**, 1916 (1966).
52. R. Bieber and G. Trümpler, *Helv. Chim. Acta.*, **30**, 706 (1947).
53. K. Vesely and R. Brdicka, *Collection Czech. Chem. Commun.*, **12**, 313 (1947).
54. R. Brdicka, *Z. Elektrochem.*, **59**, 787 (1955).
55. R. Brdicka, *Collection Czech. Chem. Commun.*, **20**, 387 (1955).
56. J. Koutecky, *Collection Czech. Chem. Commun.*, **21**, 653 (1956).
57. R. Brdicka, *Z. Elektrochem.*, **64**, 16 (1960).
58. K. Vesely and R. Brdicka, *Collection Czech. Chem. Commun.*, **25**, 853 (1960).
59. N. Lindqvist, *Acta Chem. Scand.*, **9**, 867, 1127 (1955).
60. P. L. Henaff, *Compt. Rend.*, **256**, 1752 (1963).
61. S. Ono, M. Takagi and T. Wasa, *Collection Czech. Chem. Commun.*, **26**, 141 (1961).
62. R. Gibert, *J. Chim. Phys.*, **51**, 372 (1954).
63. J. N. Brönsted, *Chem. Rev.*, **5**, 322 (1928).
64. K. Knoche, H. Wendt, M.-L. Ahrens and H. Strehlow, *Collection Czech. Chem. Commun.*, **31**, 388 (1966).
65. Y. Pocker, *Proc. Chem. Soc.*, **1960**, 17.
66. W. P. Jencks, *Progr. Phys. Org. Chem.*, **2**, 63 (1964).
67. C. G. Swain, *J. Amer. Chem. Soc.*, **72**, 4578 (1950).
68. M. Eigen, *Disc. Faraday Soc.*, **39**, 7 (1965).
69. E. H. Cordes and M. Childers, *J. Org. Chem.*, **29**, 968 (1964).
70. V. H. Booth and F. J. W. Roughton, *J. Physiol.*, **92**, 36 (1938).
71. V. H. Booth and F. J. W. Roughton, *Biochem. J.*, **32**, 2049 (1938).
72. M. Kiese and A. B. Hastings, *J. Biol. Chem.*, **132**, 267 (1940).
73. A. Sharma and P. V. Danckwerts, *Trans. Faraday Soc.*, **59**, 386 (1963).
74. C. Ho and J. M. Sturtevant, *J. Biol. Chem.*, **238**, 3499 (1963).
75. B. H. Gibbons and J. T. Edsall, *J. Biol. Chem.*, **238**, 3502 (1963).
76. H. C. Brown, R. S. Fletcher and R. B. Johannesen, *J. Amer. Chem. Soc.*, **73**, 212 (1951).
77. C. N. Satterfield and L. C. Case, *Ind. Eng. Chem.*, **46**, 998 (1954).
78. C. L. Levesque, U.S. Pat. 2,568,682; *Chem. Abstr.*, **46**, 3558 (1952).
79. B. L. Dunicz, D. D. Perrin and D. W. G. Style, *Trans. Faraday Soc.*, **47**, 1210 (1951).
80. P. L. Kooijman and W. L. Ghijsen, *Rec. Trav. Chim.*, **66**, 205 (1947).
81. J. H. Kastle and A. S. Loevenhart, *J. Amer. Chem. Soc.*, **21**, 262 (1899).

82. E. Abel, *Z. Physik. Chem.*, **N7**, 101 (1956).
83. K. Wirtz and K. F. Bonhoeffer, *Z. Physik. Chem.*, **B32**, 108 (1936).
84. C. H. Hassall in *Organic Reactions*, Vol. 9 (Ed. R. Adams), John Wiley and Sons, New York, 1957, p. 73.
85. E. Späth, M. Pailer and G. Gergely, *Ber.*, **73B**, 935 (1940).
86. Y. Ogata, I. Tabushi and H. Akimoto, *J. Org. Chem.*, **26**, 4803 (1961).
87. N. A. Milas and A. Golubovic, *J. Amer. Chem. Soc.*, **81**, 6461 (1959).
88. Y. Furuya and Y. Ogata, *Bull. Chem. Soc. Japan*, **36**, 419 (1963).
89. N. A. Milas and A. Golubovic, *J. Amer. Chem. Soc.*, **81**, 5824 (1959).
90. De Leeuw, *Z. Physik. Chem.*, **77**, 284 (1911).
91. F. E. McKenna, H. V. Tartar and E. C. Lingafelter, *J. Amer. Chem. Soc.*, **71**, 729 (1949) and references therein.
92. M. Mackes, *Compt. Rend.*, **244**, 2726 (1957).
93. K. L. Wolf and K. Merkel, *Z. Physik. Chem.*, **A187**, 61 (1940).
94. F. E. McKenna, H. V. Tartar and E. C. Lingafelter, *J. Amer. Chem. Soc.*, **75**, 604 (1953).
95. I. Lauder, *Trans. Faraday Soc.*, **44**, 729 (1948).
96. V. P. Sumarkov and M. I. Darydova, *J. Appl. Chem.* (*USSR*), **14**, 256 (1941).
97. A. Müller, *Helv. Chim. Acta*, **17**, 1231 (1934); **19**, 225 (1936).
98. H. Adkins and A. E. Broderick, *J. Amer. Chem. Soc.*, **50**, 499 (1936).
99. B. N. Rutovskii and K. S. Zabrodina, *J. Appl. Chem.* (*USSR*), **11**, 302 (1938).
100. W. Herold and K. L. Wolf, *Z. Physik. Chem.*, **B12**, 165 (1931).
101. W. Herold, *Z. Physik. Chem.*, **B18**, 265 (1932).
102. W. Herold, *Z. Elektrochem.*, **38**, 633 (1932); **39**, 566 (1933).
103. R. P. Bell and E. C. Baughan, *J. Chem. Soc.*, **1937**, 1947.
104. C. D. Hurd and J. L. Abernethy, *J. Amer. Chem. Soc.*, **63**, 1966 (1941).
105. C. D. Hurd and W. H. Saunders, Jr., *J. Amer. Chem. Soc.*, **74**, 5324 (1952).
106. A. M. Buswell, E. C. Dunlop, W. H. Rodebush and J. B. Swartz, *J. Amer. Chem. Soc.*, **62**, 325 (1940).
107. N. C. Melchior, *J. Amer. Chem. Soc.*, **71**, 3651 (1949).
108. A. Ashdown and T. A. Kletz, *J. Chem. Soc.*, **1948**, 1454.
109. J. L. Erickson and C. R. Campbell, Jr., *J. Amer. Chem. Soc.*, **76**, 4472 (1954).
110. Y. Fujiwara and S. Fujiwara, *Bull. Chem. Soc. Japan*, **36**, 1106 (1963).
111. C. Djerassi, L. A. Mitscher and B. J. Mitscher, *J. Amer. Chem. Soc.*, **81**, 947 (1959).
112. H. Meerwein, Th. Bersin, and W. Burneleit, *Ber.*, **62**, 1008 (1929).
113. W. Traube, *Ber.*, **40**, 4944 (1907).
114. P. Lipp, J. Buckkremer and H. Seels, *Ann. Chem.*, **499**, 1 (1932).
115. E. Kober and C. Grundmann, *J. Amer. Chem. Soc.*, **80**, 5550 (1958).
116. B. Helferich and H. Köster, *Ber.*, **56**, 2088 (1923).
117. A. Rieche and E. Schmitz, *Chem. Ber.*, **91**, 2693 (1958).
118. W. Treibs and R. Schöllner, *Chem. Ber.*, **94**, 2984 (1961).
119. J. Cantacuzene, *Bull. Soc. Chim. France*, 763 (1962).
120. B. Helferich, *Ber.*, **52**, 1123 (1919).
121. H. Bredereck, *Angew. Chem.*, **69**, 406 (1957).
122. W. Lüttke, *Chem. Ber.*, **83**, 571 (1950).
123. W. Pigmen, *The Carbohydrates*, Academic Press, New York, 1957, p. 35.

124. E. Pacsu and L. A. Hiller, *J. Amer. Chem. Soc.*, **70**, 523 (1948).
125. J. M. Los and K. Wiesner, *J. Amer. Chem. Soc.*, **75**, 6346 (1953).
126. J. N. Brönsted and E. A. Guggenheim, *J. Amer. Chem. Soc.*, **49**, 2554 (1927).
127. T. M. Lowry and I. J. Faulkner, *J. Chem. Soc.*, **127**, 2883 (1925).
128. T. M. Lowry, *J. Chem. Soc.*, **1927**, 2554.
129. O. H. Wheeler and J. L. Mateos, *Anal. Chem.*, **29**, 538 (1957).
130. O. H. Wheeler, *J. Amer. Chem. Soc.*, **79**, 4191 (1957).
131. J. M. Jones and M. L. Bender, *J. Amer. Chem. Soc.*, **82**, 6322 (1960).
132. O. Grobowsky and W. Herold, *Z. Physik. Chem.*, **B28**, 290 (1935).
133. G. W. Meadows and B. de B Darwent, *Trans. Faraday Soc.*, **48**, 1015 (1952).
134. C. G. Swain and J. F. Brown, Jr., *J. Amer. Chem. Soc.*, **74**, 2534, 2538 (1952).
135. E. W. Adams and H. Adkins, *J. Amer. Chem. Soc.*, **47**, 1358 (1925).
136. R. E. McCoy, A. W. Baker and R. S. Gohlke, *J. Org. Chem.*, **22**, 1175 (1957).
137. N. B. Lorette, W. L. Howard and J. H. Brown Jr., *J. Org. Chem.*, **24**, 1731 (1959).
138. H. A. Suter and R. M. Guedin, *Southern Chemist*, **16**, 102 (1956).
139. D. G. Kubler and L. E. Sweeney, *J. Org. Chem.*, **25**, 1437 (1960).
140. J. M. Bell, D. G. Kubler, P. Sartwell and P. G. Zepp, *J. Org. Chem.*, **30**, 4284 (1965).
141. R. Garrett and D. G. Kubler, *J. Org. Chem.*, **31**, 2665 (1966).
142. M. Anteumis, F. Alderwireldt and M. Acke, *Bull. Soc. Chim. Belges*, **72**, 797 (1963).
143. K. G. Shipp and M. E. Hill, *J. Org. Chem.*, **31**, 853 (1966).
144. J. N. Street and H. Adkins, *J. Amer. Chem. Soc.*, **50**, 162 (1928).
145. W. H. Hartung and H. Adkins, *J. Amer. Chem. Soc.*, **49**, 2517 (1927).
146. H. Adkins and E. W. Adams, *J. Amer. Chem. Soc.*, **47**, 1368 (1925).
147. N. Minne and H. Adkins, *J. Amer. Chem. Soc.*, **55**, 299 (1933).
148. H. Adkins, J. Semb and L. B. Bolander, *J. Amer. Chem. Soc.*, **53**, 1853 (1931).
149. R. E. Dunlop and H. Adkins, *J. Amer. Chem. Soc.*, **56**, 442 (1934).
150. G. W. Meadows and B. deB Darwent, *Can. J. Chem.*, **30**, 501 (1952).
151. F. Stasiuk, W. A. Sheppard and A. N. Bourns, *Can. J. Chem.*, **34**, 123 (1956).
152. J. D. Drumheller and L. J. Anderson, *J. Amer. Chem. Soc.*, **77**, 3290 (1955).
153. H. K. Garner and H. J. Lucas, *J. Amer. Chem. Soc.*, **72**, 5497 (1950).
154. E. M. Alexander, H. M. Busch and G. L. Webster, *J. Amer. Chem. Soc.*, **74**, 3173 (1952).
155. J. M. O'Gorman and H. J. Lucas, *J. Amer. Chem. Soc.*, **72**, 5489 (1950).
156. H. Adkins and A. Broderick, *J. Amer. Chem. Soc.*, **50**, 178 (1928).
157. A. J. Deyrup, *J. Amer. Chem. Soc.*, **56**, 60 (1934).
158. R. P. Bell and A. N. Norris, *J. Chem. Soc.*, **1941**, 118.
159. A. Skrabal and H. H. Eger, *Z. Physik. Chem.*, **122**, 349 (1926).
160. J. N. Brönsted and W. F. K. Wynne-Jones, *Trans. Faraday Soc.*, **25**, 59 (1929).
161a. M. M. Kreevoy and R. W. Taft, Jr., *J. Amer. Chem. Soc.*, **77**, 3146 (1955).
161b. M. M. Kreevoy and R. W. Taft, Jr., *J. Amer. Chem. Soc.*, **77**, 5590 (1955).
162. M. M. Kreevoy, *J. Amer. Chem. Soc.*, **78**, 4236 (1956).

163. M. M. Kreevoy, C. R. Morgan and R. W. Taft, Jr., *J. Amer. Chem. Soc.*, **82**, 3064 (1960).
164. M. Kilpatrick, *J. Amer. Chem. Soc.*, **85**, 1037 (1963).
165. T. H. Fife and L. K. Jao, *J. Org. Chem.*, **30**, 1492 (1965).
166. D. McIntyre and F. A. Long, *J. Amer. Chem. Soc.*, **76**, 3240 (1954).
167a. G. Z. Romijn, *Z. Anal. Chem.*, **36**, 19, 21 (1877).
167b. J. F. Walker, *Formaldehyde*, 2nd ed., Reinhold, New York, 1958, p. 388.
168. R. F. B. Cox and R. T. Stormont in *Organic Syntheses*, Coll. Vol. 2 (Ed. A. H. Blatt), John Wiley and Sons, New York, 1943, p. 7.
169. K. Sennewald and K. H. Steil, *Chem. Ing. Tech.*, **30**, 440 (1958); *Chem. Abstr.*, **52**, 18199 (1958).
170. H. Albers and E. Albers, *Z. Naturforsch.*, **9b**, 122 (1954).
171. S. Ide and J. S. Buck in *Organic Reactions*, Vol. 4 (Ed. R. Adams), John Wiley and Sons, New York, 1948, p. 269.
172. R. Adams and C. S. Marvel in *Organic Syntheses*, Coll. Vol. 1, 2nd edn. (Ed. H. Gilman), John Wiley and Sons, New York, 1956, p. 94.
173. E. Stern, *Z. Physik. Chem.*, **50**, 513 (1905).
174. D. R. Nadkarni, S. M. Mehta and T. S. Wheeler, *J. Phys. Chem.*, **39**, 727 (1935).
175. A. Lachman, *J. Amer. Chem. Soc.*, **46**, 708 (1924); **45**, 1529 (1923).
176. A. Lapworth and R. H. F. Manske, *J. Chem. Soc.*, **1928**, 2533.
177. J. W. Baker and M. L. Hemming, *J. Chem. Soc.* **1942**, 191.
178. J. W. Baker and H. B. Hopkins, *J. Chem. Soc.*, **1949**, 1089.
179. J. W. Baker, G. F. C. Barrett and T. W. Tweed, *J. Chem. Soc.*, **1952**, 2831.
180. J. W. Baker, *Tetrahedron*, **5**, 135 (1959).
181. J. W. Baker, *Endeavour*, **1**(1), 8 (1942); *Chem. Abstr.*, **36**, 6514 (1942).
182. E. C. Hurdis and C. P. Smyth, *J. Amer. Chem. Soc.*, **65**, 89 (1943).
183. A. Lapworth, R. H. F. Manske and E. B. Robinson, *J. Chem. Soc.*, **1930**, 1976.
184. D. P. Stevenson, H. D. Burnham and V. Schomaker, *J. Amer. Chem. Soc.*, **61**, 2922 (1939).
185. C. K. Ingold, *Quart Revs.*, **11**, 1 (1957).
186. J. M. W. Scott, *Tetrahedron Letters*, 373 (1962).
187. M. Woltsberg, *Tetrahedron Letters*, 3405 (1964).
188a. V. Prelog and M. Kobelt, *Helv. Chim. Acta*, **32**, 1187 (1949).
188b. O. H. Wheeler and J. Zabicky, *Can. J. Chem.*, **36**, 656 (1958).
188c. O. H. Wheeler, R. Cetina and J. Zabicky, *J. Org. Chem.*, **22**, 1153 (1957).
189. A. Lapworth, *J. Chem. Soc.*, **83**, 995 (1903); **85**, 1206 (1904).
190a. W. J. Svirbeley and J. F. Roth, *J. Amer. Chem. Soc.*, **75**, 3106 (1953).
190b. W. J. Svirbeley and F. H. Broch, *J. Amer. Chem. Soc.*, **77**, 5789 (1955).
191. T. D. Stewart, and B. J. Fontana, *J. Amer. Chem. Soc.*, **62**, 3281 (1940).
192. M. Mousseron, J. Jullien and P. Fauche, *Compt. Rend.*, **241**, 886 (1955).
193. H.-H. Hustedt and E. Pfeil, *Ann. Chem.*, **640**, 15 (1961).
194a. J. W. Baker, *Trans. Faraday Soc.*, **37**, 637 (1941).
194b. E. D. Hughes, C. K. Ingold and U. G. Shapiro, *J. Chem. Soc.*, **1936**, 228.
195. J. Jullien and G. Lamaty, *Tetrahedron Letters*, **1964**, 1023.
196. L. Rosenthaler, *Biochem. Z.*, **14**, 238 (1908); **17**, 257 (1909); **19**, 186 (1909).
197. G. Bredig and P. S. Fiske, *Biochem. Z.*, **46**, 7 (1912).

198. G. Bredig and M. Minaeff, *Biochem. Z.*, **249**, 242 (1932).
199. G. Bredig, F. Gerstner and H. Lang, *Biochem. Z.*, **282**, 88 (1935).
200. V. Prelog and M. Wilhelm., *Helv. Chim. Acta*, **37**, 1634 (1954).
201. J. F. Walker, *Formaldehyde*, 2nd edn., Reinhold, New York, 1958, p. 382.
202. S. H. Pines, J. M. Chemerda and M. A. Kozlowski, *J. Org. Chem.*, **31**, 3446 (1966).
203. W. M. Lauer and C. M. Langkammerer, *J. Amer. Chem. Soc.*, **57**, 2360 (1935).
204. R. L. Shriner and A. H. Land, *J. Org. Chem.*, **6**, 888 (1941).
205. C. N. Caughlan and H. V. Tartar, *J. Amer. Chem. Soc.*, **63**, 1265 (1941).
206. W. A. Sheppard and A. N. Bourns, *Can. J. Chem.*, **32**, 4 (1954).
207. W. A. Sheppard and A. N. Bourns, *Can. J. Chem.*, **32**, 345 (1954).
208. M. A. Gubareva, *Zh. Obshch. Khim.*, **17**, 2259 (1947).
209. K. Shinra, K. Ishikawa and K. Arai, *Nippon Kagaku Zasshi*, **75**, 661 (1954).
210. G. E. K. Branch and M. Calvin, *The Theory of Organic Chemistry*, Prentice-Hall, New York, 1941, p. 201.
211. K. Shinra, K. Ishikawa and K. Arai, *Nippon Kagaku Zasshi*, **75**, 664 (1954).
212. K. Arai, *Nippon Kagaku Zasshi*, **82**, 955 (1961).
213. K. Arai, *Nippon Kagaku Zasshi*, **82**, 1431 (1961).
214. T. D. Stewart and L. H. Donnally, *J. Amer. Chem. Soc.*, **54**, 2333, 3555, 3559 (1932).
215. D. A. Blackadder and C. Hinshelwood, *J. Chem. Soc.*, **1958**, 2720.
216. L. P. Hammett, *Physical Organic Chemistry*, McGraw-Hill, New York, 1940, p. 47.
217. S. Nagakura, A. Minegishi and K. Stanfield, *J. Amer. Chem. Soc.*, **79**, 1033 (1957).
218. D. Delepine, *Compt. Rend.*, **129**, 831 (1899).
219. P. Baumgarten and T. Otto, *Ber.*, **75**, 1687 (1940).
220. E. von Meyer, *J. Prakt. Chem.*, [2] **63**, 167 (1901).
221. L. Field and P. H. Settlage, *J. Amer. Chem. Soc.*, **73**, 5870 (1951).
222. H. Bredereck and E. Bäder, *Chem. Ber.*, **87**, 129 (1954).
223. N. I. Vrzhosek, *Tr. Kievsk. Politekhn. Inst.*, **38**, 106 (1962); *Chem. Abstr.*, **61**, 2534 (1964).
224. C. D. Fisher, L. H. Jensen and W. M. Schubert, *J. Amer. Chem. Soc.*, **87**, 33 (1965).
225. G. Reddelien, *Ber.*, **45**, 2904 (1912); **48**, 1462 (1915); *J. Prakt. Chem.*, **91**, 213 (1915); *Angew. Chem.*, **35**, 580 (1922).
226. W. Hofman, L. Stefaniak and T. Orbanski, *J. Chem. Soc.*, **1962**, 2343.
227. T. V. Healy, *J. Appl. Chem.*, **8**, 553 (1958); *Chem. Abstr.*, **54**, 5315 (1960).
228. B. V. Tronov and N. A. Khitrina, *Izv. Tomsk. Politekhn. Inst.*, **102**, 12 (1959); *Chem. Abstr.*, **59**, 405 (1963).
229. J. Lichtenberger, L. Baumann and J. Breiss, *Bull. Soc. chim. France*, 687 (1954); *Chem. Abstr.*, **48**, 10561 (1954).
230. W. von Doering and E. Dorfman, *J. Amer. Chem. Soc.*, **75**, 5595 (1953).
231a. B. Phillips, F. C. Frostick, Jr. and P. S. Starcher, *J. Amer. Chem. Soc.*, **79**, 5982 (1957).

231b. V. A. Palm, Ü. L. Haldna and A. J. Talvik, in *The Chemistry of the Carbonyl Group*, Part I (Ed. S. Patai), John Wiley and Sons, New York, 1966, pp. 421–460.

232. E. Campaigne, *Chem. Rev.*, **39**, 1 (1946).

233. E. E. Reid in *Organic Chemistry of Bivalent Sulfur*, Vol. 2, Chemical Publishing, New York, 1960, p. 205 and Vol. 3, p. 320.

234. E. Campaigne in *Organic Sulfur Compounds*, Vol. I (Ed. N. Kharasch), Pergamon Press, New York, 1961, p. 134.

235. H. Byasson, *Ber.*, **5**, 482 (1872).

236. E. Baumann, *Ber.*, **18**, 258, 883 (1885); **23**, 60 (1890).

237. E. Patterno and A. Ogialore, *Gazz. Chim. Ital.*, **3**, 533 (1873).

238. G. Wyss, *Ber.*, **7**, 211 (1874).

239. H. Schiff, *Ber.*, **7**, 80 (1874).

240. R. W. Borgeson and J. A. Willkinson, *J. Amer. Chem. Soc.*, **51**, 1435 (1929).

241. E. Campaigne and B. E. Edwards, *J. Org. Chem.*, **27**, 3760 (1962).

242. F. Asinger, M. Thiel and G. Lipfert, *Ann. Chem.*, **627**, 195 (1959).

243. T. L. Cairns, G. L. Evans, A. W. Larchar and B. C. Mckusick, *J. Amer. Chem. Soc.*, **74**, 3982 (1952).

244. J. F. Harris Jr., *J. Org. Chem.*, **25**, 2259 (1960).

245. J. F. Harris Jr., *J. Org. Chem.*, **30**, 2190 (1965).

246. T. G. Levi, *Gazz. Chim. Ital.*, **62**, 775 (1932).

247. H. Böhme and H.-P. Teltz, *Ann. Chem.*, **620**, 1 (1959).

248. M. P. Schubert, *J. Biol. Chem.*, **111**, 671 (1935); **114**, 341 (1936).

249. K. Kipnis and J. Ornfelt, *J. Amer. Chem. Soc.*, **74**, 1068 (1952).

250. K. Griesbaum, A. A. Oswald and B. E. Hudson, Jr., *J. Amer. Chem. Soc.*, **85**, 1969 (1963).

251. R. B. Wooodward, K. Heusler, J. Gosteli, P. Naegeli, W. Oppolzer, P. Ramage, S. Ranganathan and H. Vorbrüggen, *J. Amer. Chem. Soc.*, **88**, 852 (1966).

252. G. E. Lienhard and W. P. Jencks, *J. Amer. Chem. Soc.*, **88**, 3982 (1966).

253. S. Ratner and H. T. Clarke, *J. Amer. Chem. Soc.*, **59**, 200 (1937).

254. G. A. Berchtold, B. E. Edwards, E. Campaigne and M. Carmack, *J. Amer. Chem. Soc.*, **81**, 3148 (1959).

255. R. Mayer, J. Morgenstern and J. Fabian, *Angew. Chem. Int. Ed. Engl.*, **3**, 277 (1964).

256. M. Friedman, J. F. Cavins and J. S. Wall, *J. Amer. Chem. Soc.*, **87**, 3672 (1965).

257. E. S. Gould, *Mechanism and Structure in Organic Chemistry*, Henry Holt, New York, 1959, p. 260.

258. R. E. Davis, H. Nakshbendi and A. Ohno, *J. Org. Chem.*, **31**, 2702 (1966).

259. R. W. Layer, *Chem. Rev.*, **63**, 489 (1963).

260. M. M. Sprung, *Chem. Rev.*, **26**, 297 (1940).

261. P. Zuman, *Chem. Listy*, **46**, 688 (1952); **55**, 261 (1961).

262. P. L. Henaff, *Bull Soc. Chim. France*, 3113 (1965).

263. K. N. Campbell, H. Sommers and B. K. Campbell, *J. Amer. Chem. Soc.*, **66**, 82 (1944).

264. J. Hine and C. Y. Yeh, *J. Amer. Chem. Soc.*, **89**, 2669 (1967).

265. I. A. Savich, *Vestn. Mosk. Univ.*, **11**, 225 (1956); *Chem. Abstr.*, **53**, 1334 (1959).

266. D. N. Robertson, U.S. Pat., 2,920,101 (1960); *Chem. Abstr.*, **54**, 10957 (1960).
267. S. C. Bell, G. L. Conklin and S. J. Childress, *J. Amer. Chem. Soc.*, **85**, 2868 (1963).
268. Y. Ogata, M. Okano and M. Sugawara, *J. Amer. Chem. Soc.*, **73**, 1715 (1951).
269. Y. Ogata and M. Okano, *J. Amer. Chem. Soc.*, **72**, 1459 (1950).
270. N. J. Leonard and J. V. Paukstelis, *J. Org. Chem.*, **28**, 3021 (1963).
271. E. W. Lund, *Acta Chem. Scand.*, **5**, 678 (1951); **12**, 1768 (1958).
272. H. Stetter, *Angew. Chem.*, **66**, 217 (1954); **74**, 361 (1962).
273. J. F. Walker, *Formaldehyde*, 2nd edn., Reinhold, New York, 1953, pp. 404–436.
274. R. H. Hasek, E. U. Elam and J. C. Martin, *J. Org. Chem.*, **26**, 1822 (1961).
275. J. F. Walker, *Formaldehyde*, 2nd edn., Reinhold, New York, 1953, pp. 281–325.
276. M. Imoto, H. Kakiuchi and K. Ko, *Formaldehyde*, (in Japanese), Asakura Co., Tokyo, 1965, pp. 173–232.
277. F. F. Blicke in *Organic Reactions*, Vol. 1 (Ed. R. Adams), John Wiley and Sons, New York, 1942, p. 303.
278. M. L. Moore in *Organic Reactions*, Vol. 5 (Ed. R. Adams), John Wiley and Sons, New York, 1949, p. 301.
279. R. H. F. Manske and M. Kulka in *Organic Reactions*, Vol. 7 (Ed. R. Adams), John Wiley and Sons, New York, 1953, p. 59.
280. F. W. Bergstrom, *Chem. Rev.*, **35**, 153 (1944).
281. W. M. Whaley and T. R. Govindachari in *Organic Reactions*, Vol. 6 (Ed. R. Adams), John Wiley and Sons, New York, 1951, p. 151.
282. W. J. Gensler in *Organic Reactions* Vol. 6, (Ed. R. Adams), John Wiley and Sons, New York, 1951, p. 191.
283. M. Okano and Y. Ogata, *J. Amer. Chem. Soc.*, **74**, 5728 (1952).
284. R. Kveton and F. Hanousek, *Chem. Listy*, **48**, 1205 (1954); **49**, 63 (1955); *Chem. Abstr.*, **49**, 6970 (1955).
285. G. A. Crowe, Jr. and C. C. Linch, *J. Amer. Chem. Soc.*, **70**, 3796 (1948); **71**, 3731 (1949); **72**, 3622 (1950).
286. N. Landquist, *Acta Chem. Scand.*, **9**, 1127, 1459, 1466, 1471, 1477 (1955); **10**, 244 (1956).
287. J. I. de Jong and J. de Jonge, *Rec. Trav. Chim.*, **71**, 643, 661, 890 (1952).
288. A. S. Dunn, *J. Chem. Soc.*, **1957**, 1446.
289. L. E. Smythe, *J. Amer. Chem. Soc.*, **73**, 2735 (1951); **74**, 2713 (1952); **75**, 574 (1953).
290. H. Kadowaki, *Urea Resin*, Kobunshi Kankokai, Tokyo, 1965, pp. 42–99.
291. M. Imoto, T. Tanigaki and T. Maeno, *Kogyo Kagaku Zasshi*, **64**, 389 (1961).
292. M. Imoto, S. Ishida, *Kogyo Kagaku Zasshi*, **67**, 501 (1964).
293. M. Saito and M. Imoto, *Kogyo Kagaku Zasshi*, **67**, 1086 (1964).
294. M. Kawai and M. Inoue, *Nagoyashi Kogyo Kenkyuosho Kenkyu Hokoku*, **17**, 13, 20 (1957); *Chem. Abstr.*, **52**, 8617 (1958).
295. J. Lemaitre, G. Smets and R. Hart, *Bull. Soc. Chim. Belges*, **63**, 182 (1954); *Chem. Abstr.*, **48**, 12530 (1954).
296. J. Ugelstad and J. de Jonge, *Rec. Trav. Chim.*, **76**, 919 (1957).
297. A. Iliceto, *Ann. Chim. Rome*, **43**, 516 (1953); *Chem. Abstr.* **48**, 6208 (1954).

298. J. Koskikallio, *Acta Chem. Scand.*, **10**, 1267 (1956); *Chem. Abstr.*, **52**, 8704 (1958).

299. M. Imoto and M. Kobayashi, *Bull. Chem. Soc. Japan*, **33**, 1651 (1960).

300. M. Imoto and M. Kobayashi, *Bull. Chem. Soc. Japan*, **36**, 1505 (1963).

301. M. Imoto and A. Ninagawa, *Bull. Chem. Soc. Japan*, **36**, 1508 (1963).

302. M. Imoto and M. Saito, *Kogyo Kagaku Zasshi*, **66**, 1296 (1963).

303. R. Kveton, F. Hanousek and M. Kralova, *Chem. Listy*, **46**, 638, 739 (1952); *Chem. Abstr.*, **47**, 8020, 12259 (1953).

304. Y. Ogata, A. Kawasaki and N. Okumura, *J. Org. Chem.*, **30**, 1636 (1965).

305. Y. Ogata, A. Kawasaki and N. Okumura, *Tetrahedron*, **22**, 1731 (1966).

306a. E. Ninagawa, Y. Saiki, Y. Hosono and Y. Okada, *Nippon Kagaku Zasshi*, **88**, 206 (1967).

306b. Y. Ogata, A. Kawasaki and N. Okumura, *Nippon Kagaku Zasshi*, **88**, 1006, (1967).

307. Y. Ogata and A. Kawasaki, *Bull. Chem. Soc. Japan*, **37**, 514 (1964).

308. S. Bose, *J. Indian Chem. Soc.*, **34**, 663 (1957).

309. E. Baur and W. Rüetschi, *Helv. Chim. Acta*, **24**, 754 (1941).

310. H. H. Richmond, G. S. Myers and G. F. Wright, *J. Amer. Chem. Soc.*, **70**, 3659 (1949).

311. P. Duden and M. Scharf, *Ann. Chem.*, **288**, 218 (1895).

312. M. L. Boyd and C. A. Winkler, *Can. J. Res.*, **25B**, 387 (1947).

313. J. R. Polley, C. A. Winkler and R. V. V. Nicolls, *Can. J. Res.*, **25B**, 525 (1947).

314. Y. Ogata and A. Kawasaki, *Tetrahedron*, **20**, 855 (1964).

315. Y. Ogata and A. Kawasaki, *Tetrahedron*, **20**, 1573 (1964).

316. H. H. Strain, *J. Amer. Chem. Soc.*, **49**, 1561 (1927).

317. R. K. McLeod and T. I. Crowell, *J. Org. Chem.*, **26**, 1094 (1961).

318. Y. Ogata, A. Kawasaki and N. Okumura, *J. Org. Chem.*, **29**, 1985 (1964).

319. D. S. Noyce, A. T. Bottini, and S. G. Smith, *J. Org. Chem.*, **23**, 752 (1958).

320. W. P. Jencks, *J. Amer. Chem. Soc.*, **81**, 475 (1959).

321. B. M. Anderson and W. P. Jencks, *J. Amer. Chem. Soc.*, **82**, 1773 (1960).

322. R. Wolfenden and W. P. Jencks, *J. Amer. Chem. Soc.*, **83**, 2763 (1961).

323. E. H. Cordes and W. P. Jencks, *J. Amer. Chem. Soc.*, **84**, 826 (1962).

324. E. H. Cordes and W. P. Jencks, *J. Amer. Chem. Soc.*, **84**, 832 (1962).

325. E. H. Cordes and W. P. Jencks, *J. Amer. Chem. Soc.*, **84**, 4319 (1962).

326. R. B. Martin, *J. Phys. Chem.*, **68**, 1369 (1964).

327. M. Masui and C. Yijima, *J. Chem. Soc.*, *(B)*, **1966**, 56.

328. L. do Amaral, W. A. Sandstrom and E. H. Cordes, *J. Amer. Chem. Soc.*, **88**, 2225 (1966).

329. A. Williams and M. L. Bender, *J. Amer. Chem. Soc.*, **88**, 2508 (1966).

330. J. E. Reiman and W. P. Jencks, *J. Amer. Chem. Soc.*, **88**, 3973 (1966).

331. G. H. Stempel, Jr. and G. S. Schaffel, *J. Amer. Chem. Soc.*, **66**, 1158 (1944).

332. J. B. Conant and P. D. Bartlett, *J. Amer. Chem. Soc.*, **54**, 2881 (1932).

333. F. H. Westheimer, *J. Amer. Chem. Soc.*, **56**, 1964 (1934).

334. E. Barrett and A. Lapworth, *J. Chem. Soc.*, **93**, 85 (1908).

335. S. F. Acree and J. M. Johnson, *Amer. Chem. J.*, **38**, 308 (1932).

336. A. Olander, *Z. Physik. Chem.*, **129**, 1 (1927).

337. D. G. Knore and N. M. Emanuel, *Dokl. Akad. Nauk SSSR*, **91**, 1163 (1953); *Chem. Abstr.*, **49**, 12936 (1955).

338. J. W. Haas, Jr. and R. E. Kadunce, *J. Amer. Chem. Soc.*, **84**, 4910 (1962).
339. A. V. Willi, *Can. J. Chem.*, **31**, 361 (1953).
340. A. V. Willi, *Helv. Chim. Acta*, **39**, 1193 (1956).
341. P. Zuman, *Collection Czech. Chem. Commun.*, **15**, 866 (1950).
342. B. Kastening, L. Holleck and G. A. Melkonian, *Z. Elektrochem.*, **60**, 130 (1956).
343. R. L. Reeves, *J. Amer. Chem. Soc.*, **84**, 3332 (1962).
344. R. L. Reeves and W. F. Smith, *J. Amer. Chem. Soc.*, **85**, 724 (1963).
345. K. Kumar, K. Chatterjee, N. Farrier and B. E. Douglass, *J. Amer. Chem. Soc.*, **85**, 2919 (1963).
346. E. H. Cordes and W. P. Jencks, *J. Amer. Chem. Soc.*, **85**, 2843 (1963).
347. K. Koehler, W. A. Sandstrom and E. H. Cordes, *J. Amer. Chem. Soc.*, **86**, 2413 (1964).
348. R. L. Reeves, *J. Org. Chem.*, **30**, 3129 (1965).
349. W. Bruynell, J. J. Charette and E. De Hoffmann, *J. Amer. Chem. Soc.*, **88**, 3808 (1966).
350. E. F. Pratt and M. J. Kammlet, *J. Org. Chem.*, **26**, 4029 (1961).
351. Y. Ogata and Y. Takagi, *J. Amer. Chem. Soc.*, **80**, 3591 (1958).
352. E. F. Pratt and M. J. Kammlet, *J. Org. Chem.*, **28**, 1366 (1963).
353. G. Vavon and P. Montheard, *Bull. Soc. Chim. France*, **7**, 551 (1940).
354. G. M. Santerre, C. J. Hansrote and T. I. Crowell, *J. Amer. Chem. Soc.*, **80**, 1254 (1958).
355. J. D. Dickinson and C. Eaborn, *J. Chem. Soc.*, **1959**, 3036, 3641.
356. R. P. Cross and P. Fugassi, *J. Amer. Chem. Soc.*, **71**, 22 (1949).
357. F. P. Price, Jr. and L. P. Hammett, *J. Amer. Chem. Soc.*, **63**, 2387 (1951).
358. J. L. Maxwell, M. J. Brownlee and M. P. Holden, *J. Amer. Chem. Soc.*, **83**, 589 (1961).
359. M. J. Craft and C. T. Lester, *J. Amer. Chem. Soc.*, **73**, 1127 (1951).
360. E. C. Suratt, J. R. Proffitt and C. T. Lester, *J. Amer. Chem. Soc.*, **72**, 1561 (1952).
361. I. D. Fairman and J. D. Ge⁺tler, *J. Amer. Chem. Soc.*, **84**, 961 (1962).
362. F. W. Fitzpatrick and J. D. Gettler, *J. Amer. Chem. Soc.*, **78**, 530 (1956).
363. T. I. Crowell and D. W. Peck, *J. Amer. Chem. Soc.*, **75**, 1075 (1953).
364. R. L. Hill and T. I. Crowell, *J. Amer. Chem. Soc.*, **78**, 2284 (1956).
365. R. P. Bell, *J. Phys. Chem.*, **55**, 885 (1951).
366. O. Bloch-Chaude, *Compt. Rend.*, **239**, 804 (1954).
367. C. G. Swain and C. B. Scott, *J. Amer. Chem. Soc.*, **75**, 141 (1953).
368a. F. H. Getman, *J. Amer. Chem. Soc.*, **50**, 2883 (1928).
368b. H. A. Benesi and J. H. Hildebrand, *J. Amer. Chem. Soc.*, **70**, 2832 (1948); **71**, 2703 (1949).
368c. R. L. Strong and J. Parano, *J. Amer. Chem. Soc.*, **89**, 2535 (1967).
369. R. S. Mulliken, *J. Amer. Chem. Soc.*, **72**, 600 (1950).
370. O. Hassel and K. O. Stromme, *Nature*, **182**, 1155 (1958).
371. E. Augdahl and P. Klaboe, *Acta. Chem. Scand.*, **16**, 1637, 1647, 1655 (1962).
372. K. B. Kellogg and G. H. Cady, *J. Amer. Chem. Soc.*, **70**, 3986 (1948).
373. J. H. Prager, *J. Org. Chem.*, **31**, 392 (1966).
374. M. Lustig, A. R. Pitochelli and J. K. Ruff, *J. Amer. Chem. Soc.*, **89**, 2841 (1967).

CHAPTER **2**

Oxidation of aldehydes and ketones

HERBERT S. VERTER

*Inter-American University of Puerto Rico, San Germán,
and the Puerto Rico Nuclear Centre, Mayaguez, Puerto Rico*

I. INTRODUCTION

This section on the oxidation and dehydrogenation of aldehydes and ketones is intended to complement the chapter on the oxidation of aldehydes by transition metals that appeared in Volume 1 of this work[1]. Several types of oxidation, including oxidation reactions equivalent to hydration and intramolecular autoxidation reduction reactions, have been discussed elsewhere in this work. These reactions will not be dealt with in detail in this section. In addition, reactions in which heteroatoms other than oxygen replace hydrogen in the molecule, e.g. brominations, nitrosations, nitrations, and mercurations, will not be discussed, although they are formally oxidations. The sections are arranged by periodic group, by atomic weight within each group, and according to the oxidation number for each element. In general, each section contains a discussion of the scope of the oxidation, followed by a discussion of the mechanism.

II. GROUP III.A

A. Thallium

Thallium(III) is capable only of affecting a two-electron change[2] (equation 1).

$$2e^- + Tl^{3+} \longrightarrow Tl^{1+} \tag{1}$$

The thallium(III)-initiated oxidative rearrangements recently un-covered[3] place this element in a unique category among metal oxidants. In contrast to the report of Littler[4], only a trace of 2-hydroxycyclohexanone was produced in the oxidation of cyclohexa-none by thallium(III). The main product was cyclopentanecarboxylic acid[3] (equation 2). Ring contraction did not occur with cyclopenta-

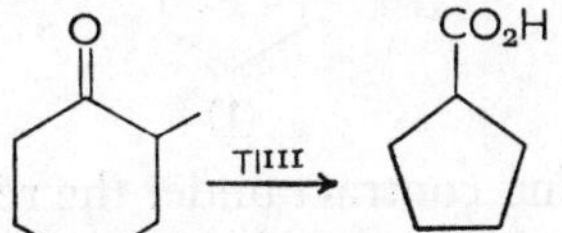

$$
\text{(2)}
$$

none, cycloheptanone, or cyclooctanone; instead thallium containing compounds were produced. Oxidation of 2-methylcyclohexanone yielded a mixture of 2-methylcyclohex-2-en-1-one and 2-methyl-2-hydroxycyclohexanone (equation 3); however, most other cyclo-hexanones studied were oxidized with accompanying ring contraction. Thus 3-methylcyclohexanone yielded mainly 2-methylcyclopen-tanecarboxylic acid, 4-methylcyclohexanone yielded 3-methylcyclo-pentanecarboxylic acid (equation 4), and 2,2-dimethylcyclohexanone yielded 2,2-dimethylcyclopentanecarboxylic acid (equation 5). *Cis* and *trans* β-decalones also underwent ring contraction.

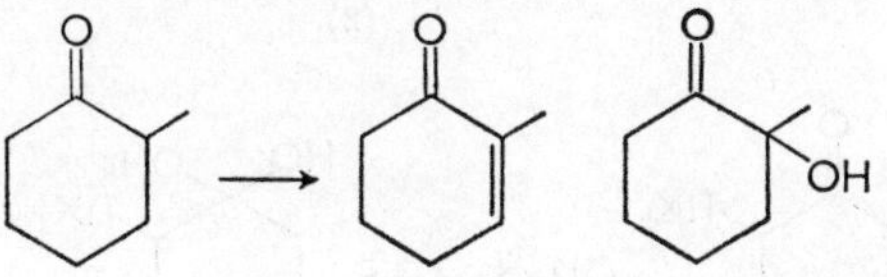

$$
\text{(3)}
$$

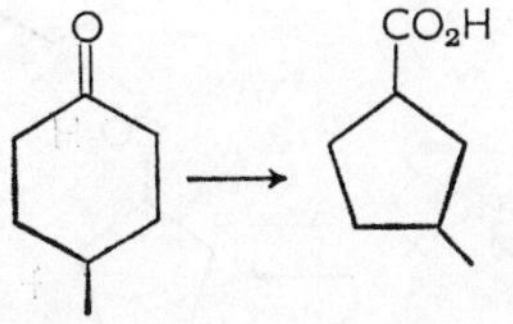

$$
\text{(4)}
$$

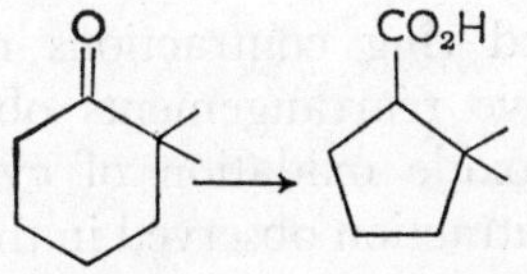

$$
\text{(5)}
$$

Early studies on the kinetics of oxidation of cyclohexanone[4] established that thallium attacks the enol form of the substrate. Using thallium perchlorate, the kinetics were found to follow zero order dependence on oxidant concentration. When tetradeuterated cyclohexanone was oxidized[3], a product retaining most of the deuterium was obtained. This experiment rules out a symmetrical intermediate such as **1** in the ring contraction. Since 2-hydroxycyclo-

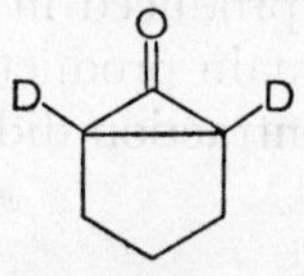

(1)

hexanone does not ring contract under the reaction conditions, it cannot be an intermediate. The reaction is believed to proceed through an intermediate organothallium compound (**2**) (equations 6a–d).

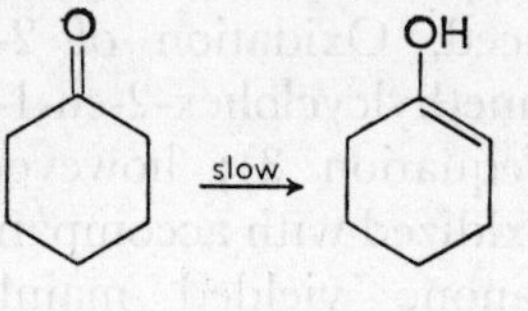

(6a)

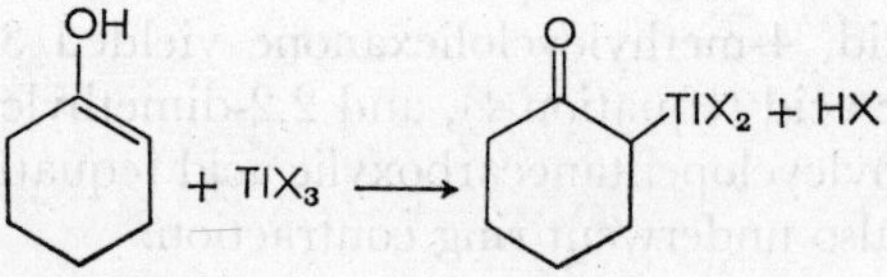

(6b)

(2)

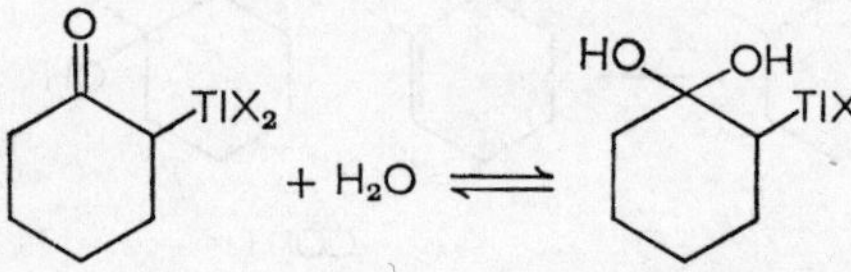

(6c)

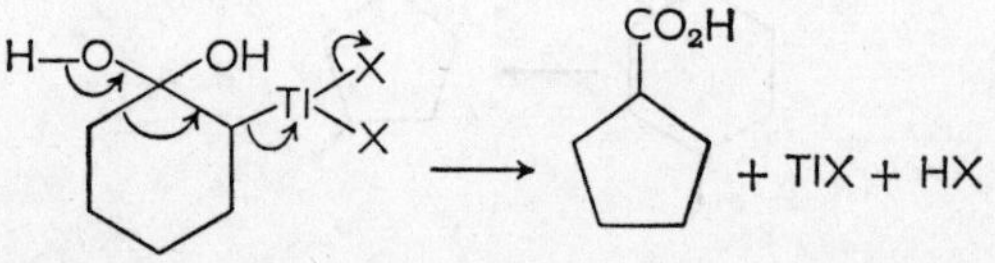

(6d)

Thallium(III)-induced ring contractions may be mechanistically related to the oxidative rearrangements observed in the selenium dioxide–hydrogen peroxide oxidation of cyclic ketones[5] (equation 7). The lack of ring contraction observed in the thallium(III) oxidation

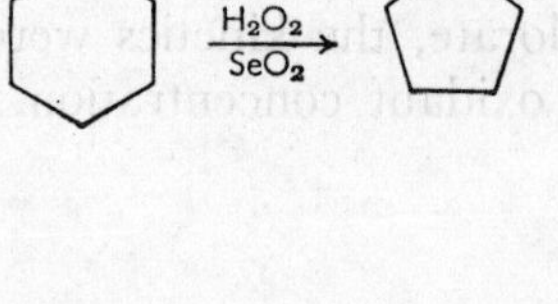

(7)

of 2-methylcyclohexanone can be rationalized on the basis of the proposed mechanism[3]. Enolization would be expected to occur toward the methyl group, and loss of thallium(I) from the derived organothallium intermediate (3) would yield a better stabilized α-keto carbonium ion (4) decreasing the driving force for the rearrangement (equation 8).

$$(8)$$

III. GROUP IV.A

A. Lead

Lead(IV) oxide is primarily a reagent for the oxidative decarboxylation of 1,2-diacids, but it has also been used for the decarboxylation of α-keto acids. Lead(IV) oxide decarboxylation of nepetonic acid (5) was a key step in the determination of the structure of nepetalactone, the physiologically active ingredient in catnip, *Nepeta cataria*[6] (equation 9).

$$(9)$$

Lead(IV) is most commonly employed as its tetraacetate. Lead tetraacetate oxidation of organic compounds[7-9] and, specifically, of sugars[10], has been the subject of several reviews. The oxidant is considered to be more vigorous and less selective than periodic acid. In contrast to periodic acid, which is employed in aqueous solution, lead tetraacetate is usually used in acetic acid solution.

Simple ketones are oxidized to α-acetoxy ketones by lead tetraacetate. Thus dimethyldimedone (6) was converted to a mixture of the stereoisomeric diacetoxy ketones[11] (equation 10).

$$\text{(6)} \qquad \xrightarrow{\text{Pb(OAc)}_4} \qquad \text{(2 stereoisomers)} \qquad (10)$$

Acetoxylation of ketones with lead tetraacetate at room temperature is effectively catalysed by boron trifluoride[12]. Thus cholestan-2- and 3-one gave the 3α-acetoxy and the 2α-acetoxy ketone respectively[12]. Pregnan-20-ones and pregnane-11,20-diones reacted with lead tetraacetate in the presence of boron trifluoride to give the corresponding 21-acetoxy ketones in good yield[13] (equation 11). Acetoxylation of a pregnan-11-one was slower, but at 50° the 9α-acetoxy-11-oxo-compound was obtained (equation 12).

$$(11)$$

$$(12)$$

Occasionally, ring contraction accompanies oxidation. Acetoxylation of 5α-cholestan-3-one in benzene–methanol produced methyl ester **7** as a by-product, in addition to 2α-acetoxycholestanone[14] (equation 13). Rearrangement was favoured if reactions were carried

$$(13)$$

$$(7)$$

out in benzene or methanol. The use of ethanol gave only the α-acetoxylation product.

Enolization of the carbonyl is rate-determining in acetoxylations by lead tetraacetate[12]. It is believed that the intermediate lead ester

(8) initially formed, decomposes with intramolecular donation of an acetoxy group to the adjacent carbon atom (equation 14).

$$Pb(OAc)_4 + \underset{}{\overset{HO}{\underset{}{\diagdown}}}C=C\overset{}{\underset{}{\diagup}} \longrightarrow \underset{}{\overset{(AcO)_3PbO}{\underset{}{\diagdown}}}C=C\overset{}{\underset{}{\diagup}} \longrightarrow$$

(8)

$$Pb(OAc)_2 + \underset{}{\overset{OOAc}{\underset{}{\diagdown}}}C-C- \qquad (14)$$

Dimeric products have been isolated in the oxidation of 1,3-diketones with lead tetraacetate[15]. Thus acetylacetone was oxidatively dimerized to 3,4-diacetylhexa-2,5-dione (equation 15).

$$2\ CH_3COCH_2COCH_3 \xrightarrow[80°]{Pb(OAc)_4} CH_3COCHCHCOCH_3 \qquad (15)$$

with $COCH_3$ groups on the central carbons.

Cleavages of α-ketonic acids and α-hydroxyketones in moist media have been reported[16]. Thus phenylglyoxalic acid was cleaved to benzoic acid and carbon dioxide (equation 16), and benzoin was cleaved to a mixture of benzaldehyde and benzoic acid.

$$PhCOCO_2H + Pb(OAc)_4 + H_2O \longrightarrow PhCO_2H + CO_2 + Pb(OAc)_2 + 2\ AcOH \quad (16)$$

In the presence of alcohols, esters are formed[16]. Thus, when pyruvic acid was oxidized with lead tetraacetate in the presence of ethanol, ethyl acetate and carbon dioxide were produced (equation 17). A mixture of red lead, Pb_3O_4, and acetic acid has been found to be an effective substitute for lead tetraacetate[17,18]. Lead tetra(trifluoroacetate) also shows promise as an oxidant of this type[19].

$$CH_3COCO_2H + EtOH \xrightarrow{Pb(OAc)_4} CH_3CO_2Et + CO_2 \qquad (17)$$

It is believed that ketols and keto acids are cleaved by lead tetraacetate in their hydrated form by a mechanism similar to the cleavage of glycols[16]. A discussion of the mechanism of lead tetraacetate cleavage has appeared in Volume 1 of this work[1].

IV. GROUP V.A

A. Nitrogen

The slow oxidation of formaldehyde by nitrogen peroxide at 128–160° has been studied kinetically[20]. The reaction was found to obey second order kinetics. Above 160° a change in mechanism occurred and at a still higher temperature the reaction became explosive.

More useful from the point of view of the synthetic organic chemist is the oxidation of aldehydes and ketones to carboxylic acids with nitric acid. Acids such as β-chloropropionic acid[21] and diacids such as glutaric acid[22] have been made by nitric acid oxidation of the corresponding aldehyde.

Methyl-p-toluyl ketone was oxidatively cleaved to toluic acid with nitric acid[23] and 2-cyanocyclopentanone was oxidized to glutaric acid[24].

Sometimes vanadium catalysts are used, as in the oxidation of cyclohexanol to adipic acid, which probably proceeds through the intermediate cyclohexanone[25].

The kinetics of the nitric acid oxidation of benzaldehyde to benzoic acid was measured in aqueous dioxan[26]. The rate was found to be independent of the nitric acid concentration, but was first order in aldehyde. Nitrous acid was an effective initiator of the reaction, but it had no influence on the rate function. The rate was proportional to the Hammett acidity function. The mechanism suggested involves a rate-determining hydrogen abstraction from the free and hydrated forms of benzaldehyde by protonated nitrogen dioxide, followed by rapid hydrolysis of the benzoyl nitrites thus formed (equations 18a–d).

$$HNO_3 + HNO_2 \longrightarrow 2\,NO_2 + H_2O \qquad (18a)$$

$$H^+ + NO_2 \longrightarrow HNO_2^+ \qquad (18b)$$

$$H_2O + PhCH{=}O \longrightarrow PhCH(OH)_2 \qquad (18c)$$

then either

$$PhCH(OH)_2 + HNO_2^+ \xrightarrow{\text{slow}} Ph\dot{C}(OH)_2 + H_2NO_2^+ \qquad (18d)$$

$$Ph\dot{C}(OH)_2 + NO_2 \longrightarrow PhC(OH)_2ONO \qquad (18e)$$

$$PhC(OH)_2ONO \longrightarrow PhCO_2H + HNO_2 \qquad (18f)$$

or

$$PhCH{=}O + HNO_2^+ \xrightarrow{\text{slow}} Ph\dot{C}O + H_2NO_2^+ \qquad (18g)$$

$$Ph\dot{C}O + NO_2 \longrightarrow PhCOONO \qquad (18h)$$

$$PhCOONO + H_2O \longrightarrow PhCO_2H + HNO_2 \qquad (18i)$$

B. Bismuth

Aldehydes are not attacked under normal reaction conditions by either bismuth(III) or bismuth(V).

Bismuth(III) oxide (Bi_2O_3) has been shown to be specific for the conversion of acyloins into diketones[27,28] (equation 19). Reaction was effected by warming the reactants in media containing acetic acid sometimes accompanied by a cosolvent. During the reaction bismuth(III) was reduced to metallic bismuth. The reaction was complete in about half an hour at 100°. Thus, by this procedure,

$$Bi^{III} + RCOCHOHR \longrightarrow Bi^0 + RCOCOR \qquad (19)$$

benzoin, anisoin, piperoin, and furoin have been converted to benzil, anisil, piperil, and furil respectively, in yields over 90%. The acyloins acetoin and butyroin have been oxidized to diacetyl and octane-4,5-dione in somewhat poorer yield. The finely divided metallic bismuth produced during the oxidation is very susceptible to air oxidation and may be easily oxidized back to bismuth(III). By the use of catalytic amounts of bismuth carbonate in the presence of air or oxygen, acyloins have been oxidized to 1,2-diketones in good yield[28]. The yields of diketone obtained by this procedure were somewhat lower than if bismuth oxide alone were used, because of further oxidation of the diketones by oxygen.

Production of metallic bismuth has been used as the basis of a qualitative test for acyloins[27].

Presumably the reactive species in bismuth(III) oxidations is bismuth triacetate. A solution of bismuth triacetate in acetic acid can be used instead of the oxide. It is believed that the acyloin and bismuth triacetate react through formation of C—O—Bi linkages, possibly involving the enediol.

Bismuth(V), as sodium bismuthate, $NaBiO_3$, serves as an alternative to periodate and lead tetraacetate as a reagent for the cleavage of acyloins[29,30]. Oxidations are carried out in acidic aqueous solution, or in dioxan–water mixtures. In order to avoid further oxidation of the aldehydes produced, the reaction is often run in the presence of phosphoric acid, which converts the bismuth(III) that is produced to its insoluble phosphate. The reaction progress may also be followed by this method, the initial yellow suspension gradually being replaced by white bismuth(III) phosphate. Under these conditions benzoin was cleaved to benzaldehyde and benzoic acid (equation 20).

$$Bi^V + PhCOCHOHPh \longrightarrow Bi^{III} + PhCHO + PhCO_2H \qquad (20)$$

In the presence of 50% acetic acid, sodium bismuthate selectively cleaves the side chains of corticosteroids[31-33]. Oxidation with bismuth (v) complements periodate oxidation, since oxidation with the latter reagent degrades the dihydroxyacetone side chain only to the α-hydroxyacid stage, whereas oxidation with sodium bismuthate cleaves the dihydroxyacetone side chain to a 17-keto-steroid. In addition, the insolubility of the sodium bismuthate offers an advantage over periodate, since the surplus reagent can be separated easily. The reaction was used to convert desoxycorticosterone to 3-oxo-4-enetioic acid (equation 21) and cortisone to adrenosterone (equation

$$ \tag{21} $$

22). The reaction also has been used as an analytical procedure to

$$ \tag{22} $$

classify corticosteroids as to whether they are formaldehydogenic or ketogenic[33].

It is believed that cleavage of ketols by bismuthate mechanistically parallels similar cleavages by lead tetraacetate and sodium periodate[30]. A cyclic intermediate such as **9** may be involved, in which the bismuth is able to withdraw electrons from the acyloin with concomitant fission of the carbon–carbon bond (equation 23).

$$ \longrightarrow \quad \longrightarrow \quad -CO_2^- + \;>C{=}O + Bi^{III} \text{ species} \tag{23} $$

(9)

V. GROUP VI.A

A. Oxygen

I. Molecular oxygen

The reaction of organic compounds with molecular oxygen has been the subject of several general reviews [34-36]. Reviews of autoxidation in the liquid phase [37-39], in the gaseous phase [40-42] and of inhibition of autoxidation of organic compounds in the liquid phase [43] have appeared. This section discusses both autoxidation, that is, the comparatively slow oxidation effected by free oxygen, and the more rapid processes which set in at higher temperatures. Catalytic oxidation is discussed under the section dealing with the metal used as a catalyst (see especially section XIII.E).

a. *Aldehydes.* Aldehydes readily undergo oxidation to the corresponding acid on standing in air. For certain aldehydes, e.g. (1-phenylcyclopropyl)acetaldehyde, the process is so rapid that exclusion of air is necessary if the aldehyde is to be stored for any long period of time [44]. Autoxidation of aldehydes generally proceeds by a chain mechanism [45] illustrated in equations (24a–c) for benzaldehyde. As

$$PhCHO + \overset{\cdot}{R} \longrightarrow Ph\overset{\cdot}{C}O + RH \qquad (24a)$$

$$Ph\overset{\cdot}{C}O + O_2 \longrightarrow PhCO\overset{\cdot}{O}_2 \qquad (24b)$$

$$PhCO\overset{\cdot}{O}_2 + PhCHO \longrightarrow PhCO_2H + Ph\overset{\cdot}{C}O \qquad (24c)$$

the oxidation proceeds, the benzoyl radical replaces $R^{\cdot}$ as the hydrogen abstractor and propagates the chain. The fate of the initially produced peracid depends on the nature of the reactants and on the reaction conditions. In the absence of water, the peracid usually oxidizes benzaldehyde to benzoic acid. In certain cases the peracid may be isolated (see equation 30).

Autoxidation of aldehydes may be initiated by several catalysts. Metal ions, such as iron or cobalt, convert the aldehyde to a radical by direct electron transfer [46] (equation 25). Radical production in

$$CH_3CHO + Co^{3+} \longrightarrow CH_3\overset{\cdot}{C}O + Co^{2+} + H^+ \qquad (25)$$

autoxidation may also come from heat, light, or ionizing radiation. These initiators usually produce unimolecular decomposition of the substrate [47,48]. For example, the rate of light-induced oxidation of purified n-decanal was independent of the concentration of oxygen

4+c.c.g. II

and was proportional to the concentration of the aldehyde and the intensity of illumination. In contrast, the rate of autoxidation in the absence of light was found to be proportional to the concentration of aldehyde and the concentration of oxygen. The kinetics of the light-induced oxidation of n-decanal suggest a slow homolytic cleavage of the aldehyde as the initiating step[47] (equation 26). Benzoyl peroxide,

$$RCHO \xrightarrow{h\nu} R\overset{.}{C}O + H\cdot \tag{26}$$

benzenediazoacetate, lead tetraacetate, and other materials have been used as catalysts in aldehyde autoxidation[49,50].

The oxidation of several aldehydes has been examined in great detail. The course of the oxidation of formaldehyde in the gas phase depends on the surface of the reaction vessel[51,52]. In pyrex vessels coated with potassium tetraborate, the only detectable products were carbon monoxide, carbon dioxide, and water. The rate of reaction was found to be proportional to the square of the formaldehyde concentration, but was independent of the concentration of oxygen. The reaction was faster on clean silica, and peroxidic products could be isolated. On a borate surface any hydrogen peroxide would be rapidly destroyed[53]. In molybdenum glass, the oxidation occurred with no pressure change in the early stages[54] probably by the overall reaction shown in equation (27). The reaction is slightly autocatalytic,

$$CH_2O + O_2 \longrightarrow CO + H_2O_2 \tag{27}$$

probably due to the formation of H_2O_2 which can react with the formaldehyde (equation 28). Studies on the competitive oxidation

$$H_2O_2 + CH_2O \longrightarrow 2\,H_2O + CO \tag{28}$$

of formaldehyde and glyoxal in the gas phase have shown that the two aldehydes are oxidized by a similar mechanism over the temperature range studied (270–300°)[55]. Both aldehydes were oxidized at a rate proportional to the square of the aldehyde concentration. The rate was almost independent of the oxygen concentration. Carbon monoxide, carbon dioxide, and water were produced along with minor amounts of hydrogen and hydrogen peroxide. A small amount of performic acid was formed, but no perglyoxalic acid. Both oxidations probably involve the formyl radical. Glyoxal is believed to decompose to the formyl radical (equation 29).

$$OCH\overset{.}{C}O \longrightarrow \overset{.}{C}HO + CO \tag{29}$$

Acetaldehyde is oxidized to peracetic acid at low temperatures, to acetic acid at higher temperature, and to a variety of products including formaldehyde and methanol in the gas phase. A synthesis of anhydrous peroxyacetic acid has been described which utilized the autoxidation of acetaldehyde at $0°$ in ethyl acetate in the presence of cobaltous ion[56,57]. An acetaldehyde–peracid adduct (10) was first produced which was then pyrolysed to the peracid (equation 30). Under other

$$CH_3CHO \xrightarrow[0°]{O_2} CH_3\overset{\overset{O}{\|}}{C}\underset{\underset{O—O}{\diagdown\diagup}}{}\overset{\overset{HO}{|}}{C}HCH_3 \xrightarrow{heat} CH_3CO_3H + CH_3CHO \quad (30)$$

$$(10)$$

conditions, the adduct has been decomposed to two molecules of acetic acid by a non-radical rearrangement[58]. Acetaldehyde has been oxidized to acetic anhydride in the presence of cobalt and copper catalysts[59]. Cobalt is believed to function mainly as an initiating catalyst, but copper enhanced the yield of anhydride, possibly by oxidation of the acetyl radical (equations 31a and 31b). The controlled

$$CH_3\dot{C}O + Cu^{2+} \longrightarrow CH_3CO^+ + Cu^+ \quad (31a)$$

$$CH_3CO^+ + CH_3CO_2H \longrightarrow Ac_2O + H^+ \quad (31b)$$

oxidation of acetaldehyde in the gas phase at $315°$ gave methanol as the main product[60]. The authors suggest the scheme illustrated in equation (32). Acetaldehyde has been reacted with air in a flow

$$CH_3\dot{C}O \xrightarrow{-CO} \dot{C}H_3 \xrightarrow{O_2} CH_3OH + \text{other products} \quad (32)$$

reactor of molybdenum glass, which preserves hydrogen peroxide. The main reaction products under these conditions were hydrogen peroxide and formaldehyde[61]. The photo-oxidation of acetaldehyde has been studied[62].

Autoxidations of aliphatic aldehydes branched at $C_{(2)}$ lead to evolution of considerable amounts of carbon monoxide. Initially two moles of carbon monoxide are evolved for every mole of oxygen absorbed[63]. The reaction is catalysed by manganese, cobalt or nickel salts. It has been shown that the loss of carbon monoxide from branched chain aldehydes prodeeds by a homolytic chain reaction mechanism[64,65].

b. *Ketones.* The oxidation of ketones to diketones and ketoaldehydes by oxygen has been discussed in Volume 1 of this work[1]. Ketones are oxidized more slowly, in general, than aldehydes. Oxidation of aldehydes proceeds by abstraction of the aldehydic hydrogen, while ketones generally react by abstraction of the α-hydrogen. Various products can be formed by oxidation of ketones, including aldehydes, diketones and acids; generally, however, these products are formed via an intermediate hydroperoxide. For example, isopropyl mesityl ketone (11) underwent slow autoxidative cleavage on standing in the dark[66]. The products were mesitoic acid (14) and acetone (equations 33). Although authors have represented the α-hydroperoxide as an

$$(33)$$

equilibrium mixture between an open hydroperoxidic form (12) and a closed peroxidic form (13), the equilibrium has been found to greatly favour the 12 form, as shown by infrared and ultraviolet evidence[67]. In the case of diisopropyl ketone in the absence of metal catalysts, the hydroperoxide 15 could be isolated (equation 34). The

$$(34)$$

hydroperoxide underwent non-radical rearrangement to isobutyric acid and acetone[68]. With straight chain ketones, the hydroperoxide decomposed into an acid and an aldehyde. The aldehyde was readily further oxidized to an acid[68] (equations 35a and 35b). The hydro-

$$\overset{\overset{\displaystyle OOH}{|}}{R^1COCHR^2} \longrightarrow R^1CO_2H + R^2CHO \tag{35a}$$

$$R^2CHO \xrightarrow{\ O_2\ } R^2CO_2H \tag{35b}$$

peroxidation proceeds rapidly in basic solution. By shaking 20-ketosteroids with oxygen in the presence of potassium t-butoxide, 17α-hydroperoxidation has been effected[69]. This has been used to introduce a 17α-hydroxyl group into steroidal 20-ketones (equation

$$\tag{36}$$

36). Certain aldehydes have been converted to α-hydroperoxy aldehydes under these conditions. For example, phenylacetaldehyde was oxidized to benzoic acid probably through the intermediacy of hydroperoxides[70]. If the reaction were stopped before completion, benzaldehyde could be isolated. Desoxybenzoin was converted to benzoic acid (74% yield) in only 2·5 min. There is evidence that cyclohexanone was oxidized by oxygen in the presence of t-butoxide to adipaldehydic acid[70]. A transformation of coprostan-3-one (**16**) to A-norcholestane-3-carboxylic acid derivatives was reported. This involved an initial autoxidation, followed by a benzylic acid rearrangement (equation 37). The sequence provides an attractive route to the thermodynamically less stable *trans*-hydrindane series[71].

$$\tag{37}$$

Base-catalysed autoxidation of acetophenone and five- to twelve-membered cyclic aliphatic ketones has been studied in hexamethylphosphoramide[72]. The cyclic ketones were autoxidized to their corresponding dibasic acids in moderate to excellent yields in the presence of either potassium hydroxide or sodium hydroxide. The reaction proceeded more readily in hexamethylphosphoramide than in t-butyl alcohol, and very much more readily than in water. Potassium hydroxide and sodium hydroxide were equally effective catalysts, but lithium hydroxide was extremely poor. The ketone containing the most acidic α-hydrogen was the most readily oxidized. These facts are consistent with a carbanion pathway. The oxidation of conjugated ketones in the γ position by molecular oxygen has been observed[73]. Some α,β-unsaturated ketones with a γ-tertiary hydrogen atom have been oxidized by molecular oxygen to the γ-hydroperoxides, which were reduced to α,β-unsaturated γ-hydroxyketones[74]. Light-catalysed autoxidation of either α,β-unsaturated or β,γ-unsaturated ketones yields α,β-unsaturated γ-hydroperoxyketones[75]. For example 10-hydroperoxy-19-nor-steroids have been prepared from β,γ-unsatu-

$$\text{(38)}$$

rated ketosteroids[76] (equation 38). Similarly Δ^5-cholesten-3-one combined readily with oxygen to give a mixture of equal parts of Δ^4-cholesten-3-one 6β- and 6α-hydroperoxide[77] (equation 39).

$$\text{(39)}$$

The photo-oxidation of acetone has been studied[62]. When acetone was photo-oxidized at 100–250°, the main products were methanol, formaldehyde, carbon monoxide and carbon dioxide, plus traces of methane and acetic acid[78]. At high oxygen pressures, the yield of formaldehyde increased at the expense of methanol. The sequence of reactions (40a–h) is believed to rationalize the evidence.

$$CH_3COCH_3 \xrightarrow{h\nu} CH_3\dot{C}O + \dot{C}H_3 \qquad \text{(40a)}$$

$$CH_3\dot{C}O \longrightarrow \dot{C}H_3 + CO \qquad \text{(40b)}$$

$$CH_3\dot{C}O + O_2 \longrightarrow CH_3\dot{O} + CO_2 \tag{40c}$$

$$\dot{C}H_3 + O_2 \longrightarrow CH_3\dot{O}_2 \tag{40d}$$

$$\dot{C}H_3 + O_2 \longrightarrow CH_2O + H\dot{O} \tag{40e}$$

$$2\,CH_3\dot{O}_2 \longrightarrow 2\,CH_3\dot{O} + O_2 \tag{40f}$$

$$CH_3\dot{O} + CH_3COCH_3 \longrightarrow CH_3OH + CH_3CO\dot{C}H_2 \tag{40g}$$

$$CH_3\dot{O} + O_2 \longrightarrow CH_2O + H\dot{O}_2 \tag{40h}$$

The last reaction is believed to predominate at high oxygen pressures, and to result in a decreased methanol and an increased formaldehyde production.

Benzoin[79], 2,2'-furoin, and 2,2'-thenoin[80] gave purple, blue, and green solutions respectively, when exposed to air in the presence of base. The final products were the corresponding 1,2-diketones. The coloured species is probably a delocalized radical anion or semidone, $RC(\dot{O})=C(O^-)R$[81]. The e.s.r. spectra of semidones produced in this manner have been investigated[80].

Base-catalysed autoxidation of 1,4-diketones to conjugated enediones is also believed to involve radical anion intermediates[82]. The reaction may be illustrated by the oxidation of A-norcoprostane-2,6-dione

(equation 41). Under similar conditions the cyclohexane-1,4-dione **17** was converted to the quinone **18** (equation 42).

(17)   (18)

2. Oxidative ammonolysis

Aldehydes are oxidized in the presence of ammonia in the gas phase in two main directions. One route involves cyclization with the formation of pyridine derivatives at a temperature of about 500°, often without a catalyst, or with catalysts of the aluminium oxide or silica type[83,84] (Chichibabin synthesis). However, ammoni-olysis of aldehydes on oxides of metals of variable valency yields nitriles[85]. In the presence of air, benzonitrile, for example, was formed from benzaldehyde in 93·6% yield[86]; acrolein was converted into acrylonitrile on nearly all types of oxidative catalysts, including vanadium[87] and molybdenum[88]. Two paths have been proposed for the reaction[89]. One postulates oxidation of the aldehyde to an amide, followed by dehydration to the nitrile (equation 43). Alternatively,

$$RCHO \longrightarrow R\dot{C}O$$
$$\longrightarrow RCONH_2 \longrightarrow RCN \quad (43)$$
$$RCHO \longrightarrow RCO_2H \longrightarrow RCO_2^- NH_4^+$$

the reaction may proceed through formation of a Schiff base, followed by dehydrogenation to the nitrile (equation 44). Less is known about

$$RCHO + NH_3 \longrightarrow RCH{=}NH \longrightarrow RCN \quad (44)$$

the vapour phase reaction of ketones with ammonia. Ketones are postulated as intermediates in the oxidative ammonolysis of hydro-carbons[90]. On tin and titanium vanadate catalysts, acetophenone reacted with ammonia by a Chichibabin-type reaction to give 2,4,6-triphenylpyridine in 98% yield, based on ketone consumed[91]. Under other conditions, acetophene was reported to give benzonitrile as the only product, in yields exceeding 70%[86].

3. Photochemical oxidation

The photo-oxidative cleavage of small ring compounds has been investigated[92]. Photo-oxidation of tetramethylcyclobutane-1,3-dione gave a variety of products including acetone and tetramethylethylene oxide[92]. The reaction has been rationalized[93] as illustrated in equa-tion (45). Loss of carbon monoxide from the cyclobutanedione yielded one of two diradical species, which reacted with oxygen to form an intermediate peroxyketone (19) or perlactone (20) respectively. These intermediates decomposed to the observed products.

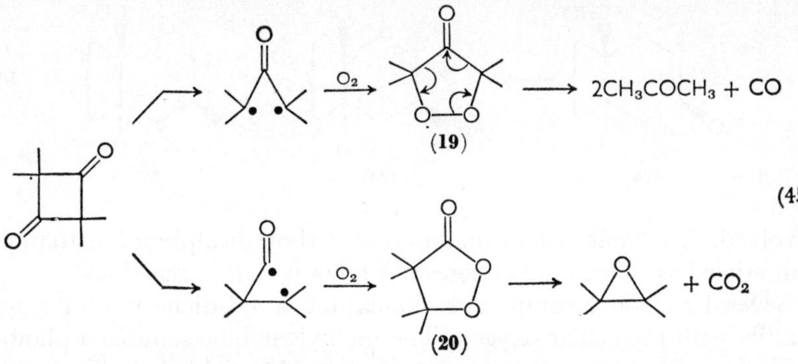

(45)

In a similar manner, dispiro[5.1.5.1]tetradecane-7,14-dione when irradiated in the presence of air yielded cyclohexanone and carbon monoxide as the primary products[92], presumably by a similar route (equation 46). The small-ring ketone 2,2,4,4-tetramethylcyclopro-

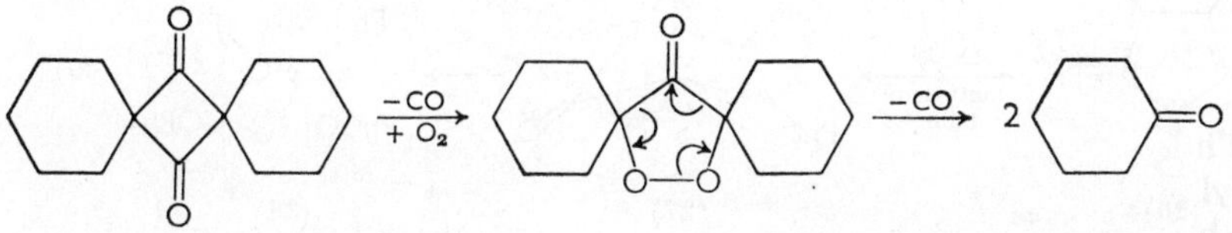

(46)

panone has been prepared and was found to react with oxygen[94]. Photo-oxidative rearrangements have been studied since the turn of the century[95]. For example, solar irradiation of carvone (21) produced two bicyclo[2.1.1]hexanecarboxylic acids as degradation products, probably through the intermediate formation of carvonecamphor (22)[96] (equation 47).

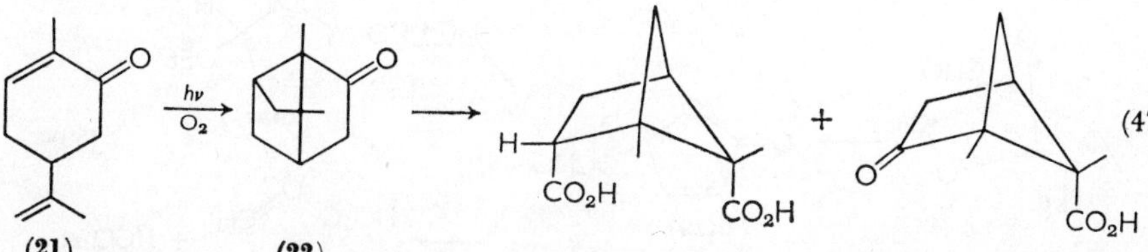

(47)

Photochemical oxidations which formally resemble the Baeyer–Villiger reaction have been reported[97]. The mercury sensitized photo-oxidation of the steroidal 19-aldehyde 23 yielded the nor-alcohol 24 along with the acid 25 (equation 48). The reaction was stereospecific. Apparently the reaction does not proceed by a Baeyer–Villiger mechanism, and a formate intermediate is not believed to be

4*

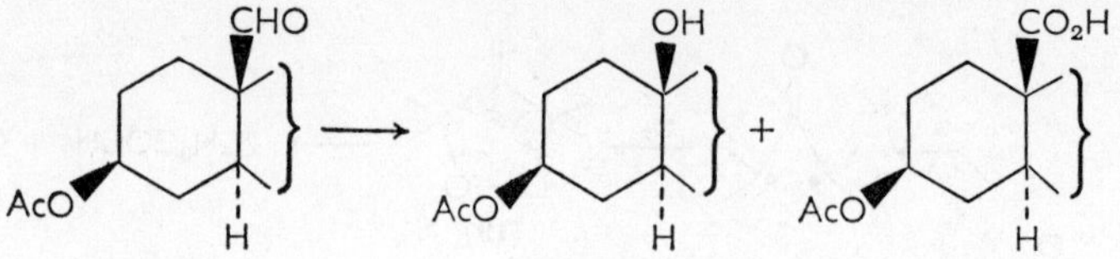

$$(48)$$

(23)   **(24)**   **(25)**

involved. A solvent system made up of carbon disulphide, methanol, and ether has special advantages in photo-oxidation reactions[98].

Several carbonyl compounds containing a 1,3-diene system react readily with molecular oxygen. The methylene blue sensitized photo-oxidation of tetraphenylcyclopentadienone (**26**) yielded *cis*-dibenzoyl-stilbene (**28**), probably through the intermediate peroxyketone **27**[99], (compare with structure **19**) (equation 49).

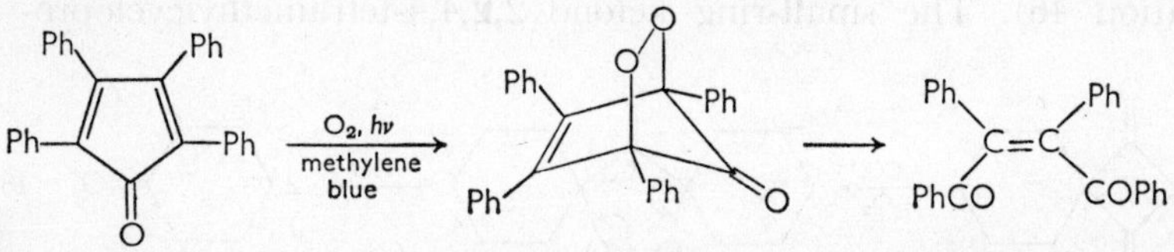

$$(49)$$

(26)   **(27)**   **(28)**

The photo-sensitized oxidation of furfural in ethanol yielded the ethyl ether of the lactol form of 4-oxobut-*cis*-2-enoic acid (**30**)[100]. In the presence of water, a polymer was formed[101,102]. Both reactions probably proceeded through the same intermediate, compound **29** (equation 50). The polymerization is responsible for the familar

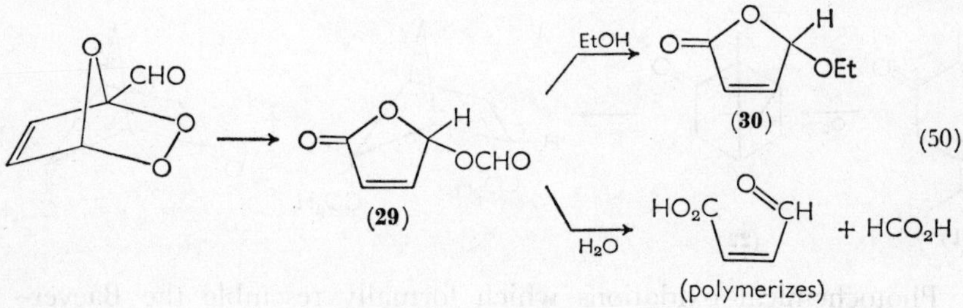

$$(50)$$

(29)

(polymerizes)

darkening of furfuraldehyde on standing. A related cleavage has been observed in the tropolone series[103]. Rose Bengal sensitized photo-oxidation of the benzotropolone (**31**) resulted in cleavage of the tropolone ring and yielded lactone **32** (equation 51). Similarly

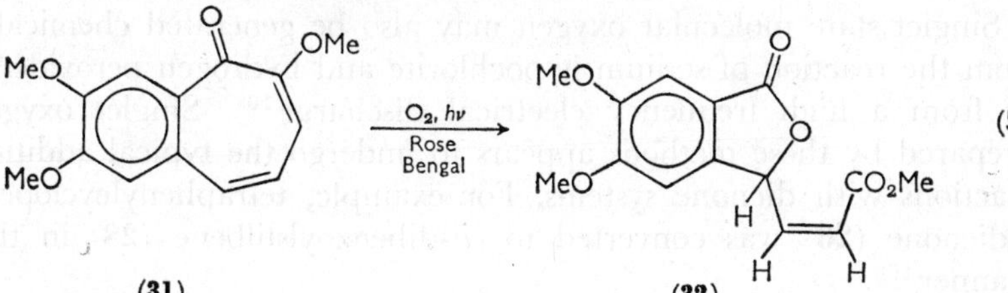

(31) (32) (51)

tropolone methyl ether (**33**) was cleaved in methanol solution to the diester **34** (equation 52).

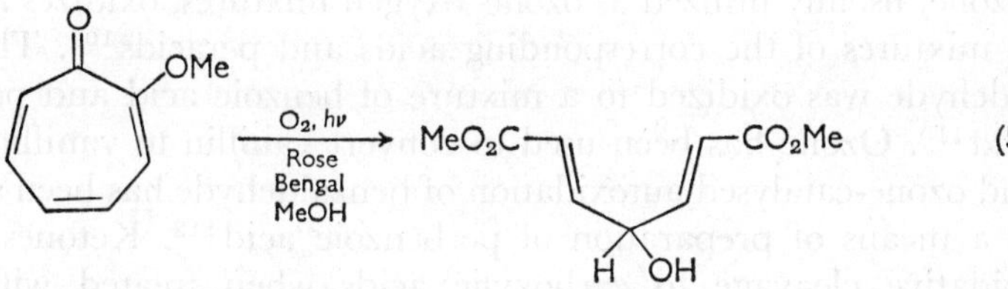

(33) (34) (52)

A number of compounds of this type are believed to undergo oxidation by singlet state oxygen[104]. Diacetylfilicinic acid (**35**), for example, acted both as sensitizer and acceptor of singlet oxygen, and yielded the ring-contracted product **38**[105] (equation 53). The reaction probably proceeded through the intermediacy of compound **36** or **37** which, by conventional benzylic acid rearrangement, yielded the product.

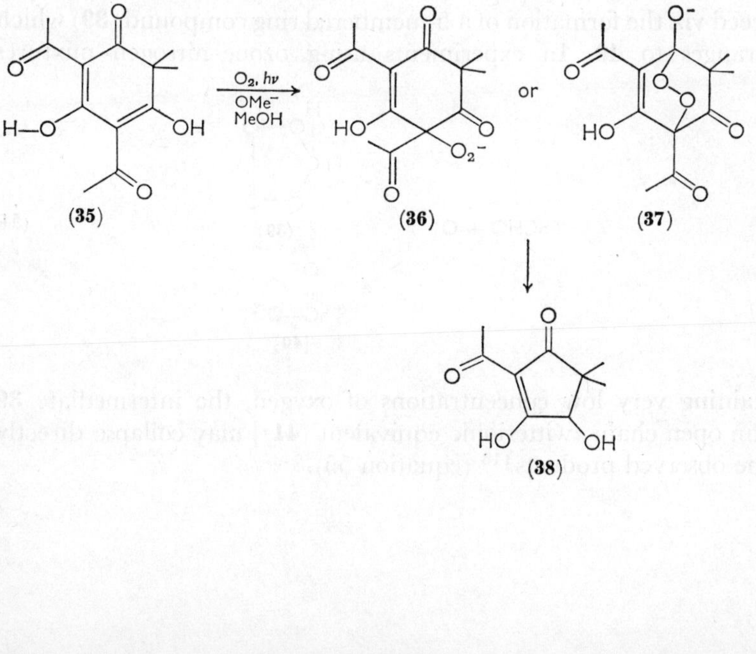

(35) (36) or (37)

(53)

(38)

Singlet state molecular oxygen may also be generated chemically from the reaction of sodium hypochlorite and hydrogen peroxide[106] or from a high frequency electrical discharge[107]. Singlet oxygen prepared by these methods appears to undergo the typical addition reactions with dienone systems. For example, tetraphenylcyclopentadienone (26) was converted to cis-dibenzoylstilbene (28) in this manner[106].

4. Ozone

The reaction of organic compounds with ozone has been reviewed[108]. Ozone, usually utilized as ozone–oxygen mixtures, oxidizes aldehydes to mixtures of the corresponding acids and peracids[109]. Thus benzaldehyde was oxidized to a mixture of benzoic acid and perbenzoic acid[110]. Ozone has been used to convert vanillin to vanillic acid[111], and ozone-catalysed autoxidation of benzaldehyde has been suggested as a means of preparation of perbenzoic acid[112]. Ketones undergo oxidative cleavage to carboxylic acids when treated with ozone. Cyclohexanone was cleaved by ozone to adipic acid[113] and cyclopentadecanone yielded pentadecanedioic acid[114]. Ketone oxidations with ozone are slower than the corresponding aldehyde oxidations.

Aldehyde ozonations showed a deuterium isotope effect when 1-deuteroaldehydes were used[115]. The reactions had a Hammett ρ value of $-1 \cdot 1$ to $-0 \cdot 6$ which indicates electrophilic attack. It has been suggested[115] that the reaction in its initial stages involves the formation of compound 40 by the direct insertion of ozone into the aldehydic C—H bond (equation 54). Alternatively the reaction may proceed via the formation of a 5-membered ring compound (39) which rearranges to 40. In experiments using ozone–nitrogen mixtures

$$PhCHO + O_3 \qquad\qquad (54)$$

containing very low concentrations of oxygen, the intermediate 39 [or an open chain zwitterionic equivalent (41)] may collapse directly to the observed products[116] (equation 55).

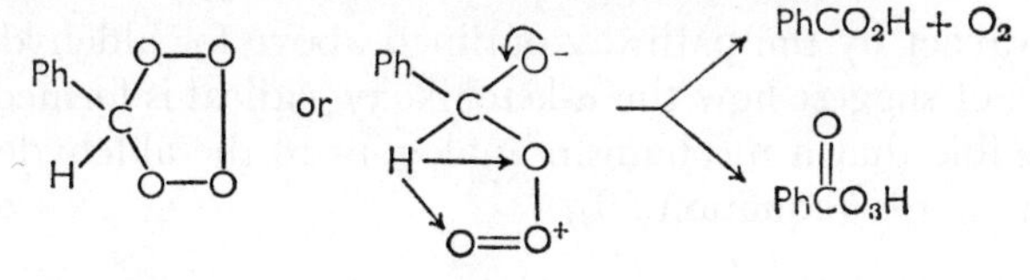
Ph
H
O—O
C
O—O
or
Ph
C
O⁻
H
O
O=O⁺
PhCO₂H + O₂
O
PhCO₃H

(55)

(**39**) (**41**)

However, under the usual reaction conditions, considerable amounts of oxygen are present and it is supposed that the initially formed ozonide decomposes into radical species that catalyse a chain radical mechanism of aldehyde with molecular oxygen[108] (equations 56a–d).

$$\mathrm{R\overset{\overset{O}{\|}}{C}O_3H \longrightarrow R\overset{\overset{O}{\|}}{C}O^{\cdot} + HO_2^{\cdot}} \quad (56a)$$

$$\mathrm{RCO_2^{\cdot} + RCHO \longrightarrow RCO_2H + R\dot{C}O} \quad (56b)$$

$$\mathrm{R\dot{C}O + O_2 \longrightarrow R\overset{\overset{O}{\|}}{C}O_2^{\cdot}} \quad (56c)$$

$$\mathrm{R\overset{\overset{O}{\|}}{C}O_2^{\cdot} + RCHO \longrightarrow R\overset{\overset{O}{\|}}{C}O_2H + R\dot{C}O} \quad (56d)$$

Little conclusive work has been done on the mechanism of oxidation of ketones. Initial suggestions that the reaction involves ozonides of the enol forms of the ketone seem unlikely[114]. A dipolar mechanism analogous to that suggested for aldehydes (equation 57) is inconsistent with the failure to isolate esters among the reaction products[117]. A

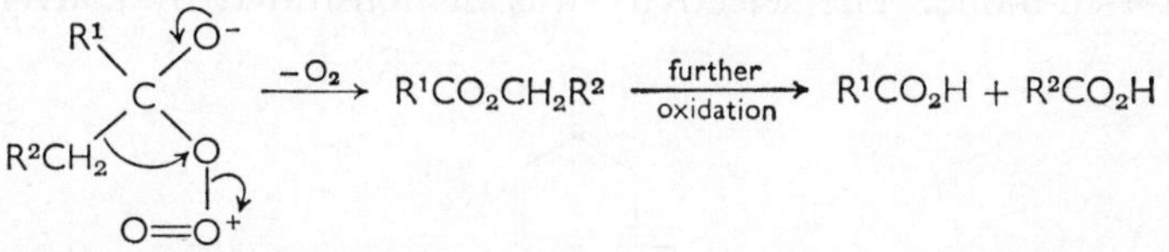
R¹
O⁻
C
R²CH₂
O
O=O⁺
−O₂
R¹CO₂CH₂R²
further
oxidation
R¹CO₂H + R²CO₂H

(57)

radical chain mechanism has also been postulated for the ozonation of ketones in ozone–oxygen mixtures[108] (equations 58a and 58b).

$$\mathrm{R^2CH_2COR^1 + O_3 \longrightarrow R^2\overset{\overset{O^{\cdot}}{|}}{C}HCOR^1 + HO_2^{\cdot}} \quad (58a)$$

$$\mathrm{R^2\overset{\overset{O^{\cdot}}{|}}{C}HCOR^1 \longrightarrow \underset{(\mathbf{42})}{R^2CHO} + \underset{(\mathbf{43})}{R^1\dot{C}O}} \quad (58b)$$

Attack by ozone occurs on the α-carbon to give an α-ketoalkoxy radical which collapses to give aldehyde **42** and acyl radical **43** which continue to react by the pathway outlined above for aldehydes. The authors do not suggest how the α-ketoalkoxy radical is formed, but it may be possible that a mechanism analogous to the aldehyde case is operating here, too (equation 59).

$$R^2CH_2COR^1 + O_3 \longrightarrow \quad \longrightarrow R^2CHCOR^1 \longrightarrow R^2CHCOR^1 + HO_2^{\bullet}$$

$$\tag{59}$$

5. Peroxides

A discussion of the reaction of carbonyl compounds with hydrogen peroxide appears in Chapter 1 and will not be repeated here.

Alkyl hydroperoxides cleave ketones in alkaline solution to the corresponding carboxylic acids[118]. The direction of cleavage sometimes differs from that observed in a Baeyer–Villiger reaction; however, poor yields limit the use of the reaction as a synthetic procedure.

More promising from a synthetic point of view is the use of t-butyl hydroperoxide in the presence of a quaternary ammonium salt, such as Triton B, for the epoxidation of α,β-unsaturated ketones[119]. By this method mesityl oxide, methyl vinyl ketone, methyl isopropenyl ketone, cyclohex-2-en-1-one, and chalcone were converted to epoxides in a nonpolar solvent. The steric requirements of this epoxidizing agent were indicated by its failure to react with isophorone (**44**) and cholest-4-en-3-one. The selectivity was demonstrated by conversion

(**44**)

of 16-dehydroprogesterone to 16α,17α-epoxy-4-pregnene-3,20-dione. Similar reaction conditions have been used for α,β-unsaturated aldehydes, e.g. cinnamaldehyde[120,121] (equation 60).

$$PhCH{=}CHCHO \xrightarrow[\text{pH 8·5}]{t\text{-BuO}_2H} PhCH{-}CHCHO \tag{60}$$

The reaction is believed to proceed by nucleophilic attack of the hydroperoxide anion at the β carbon of the unsaturated ketone, followed by intramolecular displacement of alkoxide [119] ion (equation 61).

$$RO_2^- \; CH_2\!=\!\!=\!\!CH\!-\!CCH_3 \longrightarrow CH_2\!-\!CH\!=\!\!=\!\!CCH_3 \longrightarrow CH_2\!-\!CHCOCH_3 + RO^- \quad (61)$$

Dialkyl peroxides are a source of alkoxy radicals which have been used as homolytic oxidants, e.g. oxidants which react by one-electron transfers to yield radical intermediates. The subject has been reviewed [122,123].

Aliphatic aldehydes are usually decarbonylated in the presence of catalytic amounts of alkoxy radicals by a free-radical chain mechanism [124]. Benzaldehyde, however, was oxidatively dimerized [125] under these conditions (equations 62a–c). In this case, the resonance stabilized radical **45** was sufficiently stable to add to another molecule of benzaldehyde. Dimerization of the addition product **46** yielded dihydrobenzoin dibenzoate (**47**). In the presence of pyridine, benz-

$$RO^{\cdot} + PhCHO \longrightarrow ROH + Ph\overset{\cdot}{C}O \quad (62a)$$
$$(45)$$

$$PhCHO + Ph\overset{\cdot}{C}O \longrightarrow Ph\overset{\cdot}{C}HOCOPh \quad (62b)$$
$$(46)$$

$$2\,Ph\overset{\cdot}{C}HOCOPh \longrightarrow \overset{\displaystyle PhCHOCOPh}{\underset{PhCHOCOPh}{|}} \quad (62c)$$
$$(47)$$

aldehyde reacts with *t*-butyl peroxide to yield **48**, probably formed by the reaction of radical **46** with pyridine [126]. Although all peroxy

$$\overset{\displaystyle Ph}{\underset{\displaystyle}{PhCO_2CH}}\!-\!\!\!\bigcirc\!\!\!N$$

(**48**)

radicals are reactive, there is some evidence that primary peroxy radicals are more reactive than tertiary, probably due to steric factors [127].

Peroxy acids react with ketones to form esters (Baeyer–Villiger reaction) [128,129]. A discussion of this reaction has appeared in Volume

1 of this work[1] and will not be repeated here. Although the Baeyer–
Villiger reaction is probably the most important reaction of peroxy
acids with carbonyl compounds, organic peroxyacids have also been
used in the oxidation of aldehydes to carboxylic acids[130] and have
been suggested as general reagents for this purpose[131]. Oxidation
of testosterone acetate (49) with perbenzoic acid produced several
products corresponding to epoxidation as well as to esterification[132]
(equation 63). The nature of the solvent as well as the nature of the
oxidant affected the course of the reaction. Thus in anhydrous
perchloric acid 49 was oxidized to enol lactone 50, epoxy lactone

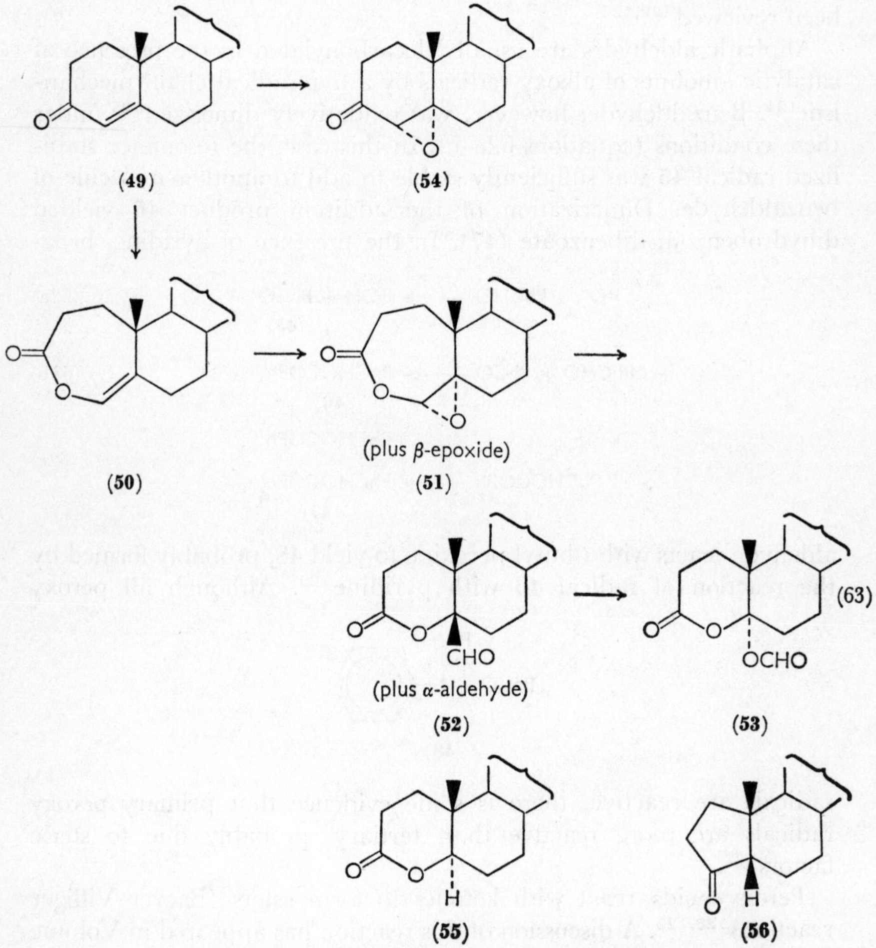

(49)   (54)

(50)   (plus β-epoxide)
                        (51)

                        CHO   OCHO
                        (plus α-aldehyde)
                        (52)   (53)

                        H   O   H
                        (55)   (56)

(63)

51, aldehyde lactone **52**, 5α-formate **53**, and epoxy ketone **54**; while in aqueous perchloric acid, lactone **55** was isolated. When *m*-chloroperbenzoic acid is used as oxidant the A-norketone **56** was also produced.

Peroxyesters, in the absence of catalysts, undergo a reaction with aldehydes similar to that described above for dialkyl peroxides. Thus *t*-butyl perbenzoate reacted with benzaldehyde to produce dihydrobenzoin dibenzoate[133]. In the presence of cuprous bromide, however, anhydrides are produced. Thus benzaldehyde yielded benzoic anhydride, while butyraldehyde reacted to form a mixture of butyric anhydride and benzoic anhydride, probably by disproportionation of the initially formed mixed anhydride. Ketones were little affected under these reaction conditions. Cyclohexanone and 2-methylcyclohexanone yielded small amounts of high-boiling materials, while major portions of these ketones were recovered unchanged.

Diacetyl peroxide has been used to convert ketones of the type **57** into 1,4-diketones[134] (equation 64).

$$2\ R^1R^2CHCOCHR^1R^2 \longrightarrow \begin{array}{c} R^1R^2CCOCHR^1R^2 \\ | \\ R^1R^2CCOCHR^1R^2 \end{array} \qquad (64)$$
$$(\mathbf{57})$$

6. Hydroxide

Several aldehydes have been oxidized to their acids by molten sodium hydroxide. Thus furfural was oxidized to furoic acid[135] and vanillin to vanillic acid[136] in high yields. During the reaction, molecular hydrogen was evolved. The reaction has been applied to ketones, but seems so far to be of little synthetic value[137]. These reactions can formally be regarded as an addition of water, and the term 'oxidative-hydrolytic splitting' has been coined to describe them[138]. A detailed discussion of the reaction of carbonyl compounds with hydroxide ions may be found in Chapter 1. Hydroxide also catalyses a number of autoxidation–reduction reactions such as the benzilic acid rearrangement[1] and the Cannizzaro reaction[1].

B. Sulphur

Dehydrogenation of cyclohexanones to phenols by sulphur or selenium[139] has largely been replaced by catalytic dehydrogenation. The Willgerodt reaction of ammonium polysulphide with ketones to yield amides and the reaction of monoperoxysulphuric acid (Caro's acid, H_2SO_5) with ketones to give esters, a variation of the Baeyer–

Villiger reaction[1], were discussed in Volume 1 of this work[1]. Persulphate in the presence of silver ion has been used to oxidize aldehydes[140,141]. Oxidations by peroxydisulphuric acid, $H_2S_2O_8$, have been reviewed[142]. The peroxydisulphate ion is one of the strongest oxidizing agents known, with a standard redox potential of about 2 v. Reactions involving this ion, however, generally are slow at ordinary temperatures. Peroxydisulphate in the presence of silver ion cleaved acetone to acetic acid and carbon dioxide[143]. The uncatalysed oxidation of formaldehyde and acetaldehyde by this reagent also has been studied[144].

Dimethyl sulphoxide usually does not react with aldehydes, but one case of oxidation of an aldehyde to an ester by dimethyl sulphoxide in the presence of methanol has been reported[145]. Dimethyl sulphoxide in acetic anhydride is an efficient reagent for the oxidation of ketols to α-diketones[146]. Thus benzoin was converted to benzil, anisoin to anisil, and furoin to furil. The procedure also may be applied to aliphatic α-ketols, but the yields are lower. The mechanism of dimethyl sulphoxide oxidation has been the subject of much investigation[147]. A probable oxidation mechanism is illustrated for acyloins by equations (65a–c).

$$\underset{+}{CH_3\overset{O^-}{\underset{|}{S}}CH_3} + Ac_2O \longrightarrow \underset{+}{CH_3\overset{OAc}{\underset{|}{S}}CH_3} + OAc^- \tag{65a}$$

$$\underset{+}{CH_3\overset{OAc}{\underset{|}{S}}CH_3} + Ar\overset{OH}{\underset{|}{C}}HCOAr \longrightarrow \underset{CH_3}{\overset{CH_3}{\underset{+}{}}}\!S\!-\!O\!-\!\overset{H}{\underset{|}{C}}COAr + HOAc \tag{65b}$$

$$\underset{CH_3}{\overset{CH_3}{}}\!\overset{+}{S}\!O\!-\!\overset{\overset{H}{|}}{\underset{|}{C}}COAr \longrightarrow CH_3SCH_3 + H^+ + ArCOCOAr \tag{65c}$$

C. Selenium

Selenium dioxide can convert ketones to diketones or ketoaldehydes[148] and diketones to triketones or tetraketones[11]. These reactions were discussed in Part I of this work[1] and elsewhere[149]. Occasionally rearrangement accompanies oxidation. Thus selenium dioxide oxidation of eucarvone (58) yielded hydroxyketone (60)[150] (equation 66).

The oxidation is believed to proceed through the anion or enol of the bicyclic isomer of eucarvone (59) [151].

$$(66)$$

(58)   (59)   (60)

Dehydrogenation of ketones to α,β-unsaturated systems by selenium dioxide is competitive to the formation of 1,2-diketones. In the case of 1,4-diketones such as 1,2-dibenzoylethane, dehydrogenation is usually the predominant reaction to the exclusion of triketone formation [152]. Dehydrogenation can also be the major reaction in molecules with a single ketone function. Thus selenium dioxide has been used in the conversion of 2,4-cycloheptadienone to tropone [153]. The dehydrogenation reaction has been applied to numerous steroidal molecules. SeO$_2$ has been used for dehydrogenation of saturated 3-ketones to Δ^1-, Δ^4-, and $\Delta^{1,4}$-3-ketones [154,155]; Δ^4-3-ketones to $\Delta^{1,4}$-3-ketones [156], Δ^1-3-ketones to $\Delta^{1,4}$-3-ketones [157] and $\Delta^{4,6}$-3-ketones to $\Delta^{1,4,6}$-3-ketones [158]. Selenium dioxide dehydrogenation of Δ^5-12-ketones to $\Delta^{5,9(11)}$-12-ketones, 12-ketones to $\Delta^{9(11)}$-12-ketones, 7-ketones to Δ^5-7-ketones and D-homo-17-ketones to D-homo-Δ^{15}-17-ketones have been described in the review by Owyang [159].

The dehydrogenations are most often carried out in t-butyl alcohol. Sometimes acetic acid [160] or a little mercury [161] or pyridine [162] is added. Wet pyridine [163], moist benzene [164], acetic acid [165], t-amyl alcohol [166], and dioxane [165] have also been used as solvents.

The use of selenium dioxide for dehydrogenation of ketones often is accompanied by side reactions such as the formation of diketones. Also, selenium tends to enter into the molecule being dehydrogenated [167]. This is unfortunate from the point of view of drug manufacture since selenium compounds are often quite toxic.

Early investigators of selenium dioxide dehydrogenations of 1,4-diketones [168] observed that the reaction of compounds with cis hydrogens was much faster than that of compounds with trans hydrogens. This led to the postulation of a cyclic transition state for dehydrogenation of the form 61 or 62 [169]. However, recent studies on the oxidation of 1α-deutero-5α-androstan-3,17-dione to Δ^1-5α-androsten-3,17-dione proceeded with 93% loss of deuterium and apparently with a small isotope effect[170] (equation 67). This evidence suggests that the net process of introduction of a double bond involves a *trans* diaxial loss of hydrogen ($1\alpha,2\beta$).

$$(67)$$

There still are varying opinions on the detailed mechanism of the dehydrogenation, although current mechanistic thinking favours a rate-determining electrophilic attack on the oxygen of the carbonyl.

Langbein[171], in studies on the dehydrogenation of cortisone, found the reaction to be acid-catalysed and first order in both oxidizer and substrate. He postulated a mechanism which involves abstraction of an α-hydrogen by a protonated selenium dioxide molecule in the slow step (equations 68a and 68b). The intermediate selenium complex **(63)** may either lose an α-hydrogen to yield a diketone or lose a β-hydrogen to yield a dehydrogenated product. Schaefer[152], in studies

$$(68a)$$

$$(68b)$$

on 1,2-dibenzoylethane, found the reaction to be first order in selenium dioxide and diketone, and first order in added acid or acetate ion.

The sole product was *trans*-1,2-dibenzoylethylene, but it was not demonstrated whether the *cis* product was stable under the reaction conditions. Oxidation of the perdeuterated compound indicated that the C—H bond was broken in the rate-determining step. Shaefer postulated a mechanism in which complex formation and enolization occur simultaneously in the rate-determining step (equations 69a–c).

Acid-catalysed:

$$H_3SeO_3^+ + PhCOCH_2CH_2COPh \xrightarrow{\text{slow}} \overset{\displaystyle OSeO_2H}{\overset{|}{PhC}}{=}CHCH_2COPh + H_3O^+ \qquad (69a)$$

Base-catalysed:

$$H_2SeO_3 + PhCOCH_2CH_2COPh \xrightarrow[\text{(AcO}^-)]{\text{slow}} \overset{\displaystyle OSeO_2H}{\overset{|}{PhC}}{=}CHCH_2COPh + H_2O \qquad (69b)$$

$$\overset{\displaystyle OSeO_2H}{\overset{|}{PhC}}{=}CHCH_2COPh \xrightarrow{\hspace{2cm}} \overset{\displaystyle OSeOH}{\overset{|}{PhCOCHCH_2COPh}} \xrightarrow{\hspace{2cm}} PhCOCH{=}CHCOPh + H_2SeO_2 \qquad (69c)$$

Selenium dioxide–hydrogen peroxide mixtures have been found to initiate a series of unusual oxidation reactions. Although selenium dioxide does not normally oxidize aldehydes to acids, selenium dioxide in conjunction with hydrogen peroxide has been used to oxidize several sensitive aldehydes to acids[172]. Thus acrylaldehyde and methylacrylaldehyde were oxidized to acrylic acid and methacrylic acid, respectively. Systems with low water content gave the highest yields. Use of formic acid, acetic acid, or other peroxyacid-forming catalysts in this reaction resulted in polymerization of the reactants. The reagent, which contains selenic acid, H_2SeO_4, has been observed to cause ring contraction in several systems[5]. Thus 5α-cholestan-3-one was converted to a mixture of A-nor (**64**) and A-seco acids (**65**)[173] (equation 70). A mechanism analogous to the Favorskii rearrange-

(**64**)   (**65**)   (70)

ment has been proposed[173] to explain the ring contraction which involves the selenite ester **66**. Addition of hydrogen peroxide across the ketone of **66** yielded intermediate **67**, or its cyclic isomer, which may rearrange to form the ring-contracted product and regenerate selenic acid (equation 71). The secoacid probably was formed by peroxide cleavage of diketone **68**.

$$\text{(71)}$$

Selenious acid, H_2SO_3, also has been proposed as a reagent for the preparation of A-norsteroids by ring contraction[71]. It probably reacts by a similar mechanism.

Selenium dioxide-hydrogen peroxide mixtures have been used to convert steroidal $\varDelta^4$-3-ketones to secolactones[174,175].

VI. GROUP VII.A

A. Halogen

Reviews on the oxidation of organic compounds with halogens[176], on the Lieben haloform reaction[177], and on halogen oxidation of simple carbohydrates[178] have appeared.

Aromatic and aliphatic aldehydes are smoothly oxidized by bromine in aqueous solution. Thus acetaldehyde was converted to acetic acid[179] and benzaldehyde to benzoic acid[180]. Iodine is a weaker oxidant than bromine, and in neutral solution it does not react with aliphatic aldehydes. In base the reaction is rapid, and iodine has been used to oxidize formaldehyde to formic acid[181] under these conditions. In basic solution iodination in the α-position competes with oxidation[182].

In basic solution methyl ketones are converted by halogens to trihalomethyl ketones which are subsequently cleaved by base to carboxylic acids (Lieben haloform reaction). Methylene groups adjacent to a ketone are not easily substituted; thus 4-methyl-4-phenyl-2-pentanone yielded β-phenylisovaleric acid in high yield[183]. Bromine or chlorine in sodium hydroxide, potassium hypochlorite, or commercial bleach all have been used as oxidants in the haloform reaction. Sometimes steam distillation is necessary to complete the cleavage as in the conversion of pinacolone to trimethylacetic acid[184].

Enolizable non-methyl ketones are also oxidatively cleaved with base and halogen. For example 5-n-butyl-2-n-propionylpyridine when treated with hypobromite under mild conditions yielded fusaric acid (69)[185] (equation 72). Cyclic ketones yield dibasic acids[186]. Sometimes

$$
\text{n-Bu} \begin{array}{c} \\ \text{N} \end{array} \text{COEt} \xrightarrow{\text{NaOBr}} \text{n-Bu} \begin{array}{c} \\ \text{N} \end{array} \text{CO}_2\text{H} \qquad (72)
$$

(69)

the oxidation requires molecular oxygen[187]. Sodium hypochlorite has been used as a source of singlet state oxygen[106].

Halogen has also been employed effectively as a reagent for the oxidation of sugars. D-Glucose was converted to D-gluconic acid (equation 73)[188] with chlorine or bromine, and D-fructose was converted to 5-keto-L-gulonic acid[189] (equation 74).

$$
\begin{array}{c}
\text{CH}{=}\text{O} \\
\text{H}{-}\text{C}{-}\text{OH} \\
\text{HO}{-}\text{C}{-}\text{H} \\
\text{H}{-}\text{C}{-}\text{OH} \\
\text{H}{-}\text{C}{-}\text{OH} \\
\text{CH}_2\text{OH}
\end{array}
\longrightarrow
\begin{array}{c}
\text{CO}_2\text{H} \\
\text{H}{-}\text{C}{-}\text{OH} \\
\text{HO}{-}\text{C}{-}\text{H} \\
\text{H}{-}\text{C}{-}\text{OH} \\
\text{H}{-}\text{C}{-}\text{OH} \\
\text{CH}_2\text{OH}
\end{array}
\qquad (73)
$$

$$
\begin{array}{c}
\text{CH}_2\text{OH} \\
\text{C}{=}\text{O} \\
\text{HO}{-}\text{C}{-}\text{H} \\
\text{H}{-}\text{C}{-}\text{OH} \\
\text{H}{-}\text{C}{-}\text{OH} \\
\text{CH}_2\text{OH}
\end{array}
\longrightarrow
\begin{array}{c}
\text{CH}_2\text{OH} \\
\text{C}{=}\text{O} \\
\text{HO}{-}\text{C}{-}\text{H} \\
\text{H}{-}\text{C}{-}\text{OH} \\
\text{H}{-}\text{C}{-}\text{OH} \\
\text{CO}_2\text{H}
\end{array}
\qquad (74)
$$

Aliphatic aldehydes are believed to be oxidized by halogens in the form of their hydrates. Hypohalous acids are not significant oxidants[179,180].

Oxidation of acetaldehyde-1-d by bromine showed a primary deuterium isotope effect which had the same value ($k_H/k_D = 4\cdot3$) as found in the oxidation of ethanol-1,1-d_2[190]. The similar isotope effect suggests that both reactions involve C—H bond cleavage in the rate-determining step. Oxidation was much faster than enolization in acidic solution[180]; the oxidation proceeded more slowly in deuterium oxide than in water[191]. Bromine oxidation of aliphatic aldehydes was found to be subject to general base catalysis[192]. The experimental evidence has been rationalized on the basis of a mechanism which involves simultaneous removal of a proton and a hydride ion from the aldehyde hydrate (equations 75a and 75b). Benzaldehyde is not measurably hydrated in aqueous solution, and may undergo oxidation

$$RCHO + H_2O \rightleftharpoons RCH(OH)_2 \tag{75a}$$

$$Br\!-\!Br \quad H\!-\!\overset{\overset{\displaystyle R}{|}}{\underset{\underset{\displaystyle OH}{|}}{C}}\!-\!OH \quad :B \xrightarrow{\text{slow}} BH^+ + RCO_2H + HBr + Br^- \tag{75b}$$

by a different pathway, possibly involving a benzoyl carbonium ion[180] (equations 76a–c). Previously suggested mechanisms involving

$$Ph\overset{\overset{\displaystyle O}{\|}}{C}\!-\!H \quad Br\!-\!Br \longrightarrow [PhC\!\equiv\!\overset{+}{O} \longleftrightarrow Ph\overset{+}{C}\!=\!O] + HBr + Br^- \tag{76a}$$

$$PhCO^+ + H_2O \longrightarrow PhCO(OH_2)^+ \tag{76b}$$

$$PhCO(OH_2)^+ \longrightarrow PhCO_2H + H^+ \tag{76c}$$

hypobromites[193] such as in equations (77a) and (77b) are ruled out

$$Br_2 + RCH(OH)_2 \longrightarrow RCH\!\!\begin{array}{c}OBr \\ \diagup \\ \diagdown \\ OH\end{array} + H^+ + Br^- \tag{77a}$$

$$\underset{\underset{\displaystyle OH}{|}}{\overset{\overset{\displaystyle B:\;H}{}}{R\!-\!C\!-\!O\!-\!Br}} \longrightarrow RCO_2H + BH^+ + Br^- \tag{77b}$$

by experimental evidence. In this mechanism, if the first step were rate-determining, no isotope effect would be expected with acetaldehyde-1-d; whereas if the second step were rate-determining, no solvent isotope effect would be expected since this step does not involve a proton transfer from the hydroxyl group.

Chloral oxidations showed specific hydroxide ion catalysis, instead of the general base catalysis observed for the other aldehydes studied[192]. It is believed that chloral hydrate reacted as its conjugate base **70** (equations 78a and 78b).

$$CCl_3CH(OH)_2 + OH^- \rightleftharpoons CCl_3CH\underset{\underset{OH}{}}{\overset{\overset{O^-}{}}{<}} + H_2O \qquad (78a)$$
$$(70)$$

$$\underset{}{Br{-}Br} \quad \underset{\underset{OH}{\overset{\overset{CCl_3}{|}}{H{-}C{-}O^-}}{}} \longrightarrow Br^- + HBr + CCl_3CO_2H \qquad (78b)$$

Although bromine oxidation of carbohydrates shows certain similarities to bromine oxidation of simple aldehydes, such as acceleration by added buffer anions[194], neighbouring group participation may be involved in carbohydrate oxidation, and more investigation is needed[195] before definite mechanistic conclusions can be drawn. It has been found that the anions of aldehydes are oxidized about 10^{10} times as fast as the neutral molecules[180,196]. A mechanism involving a hydrated anion similar to the intermediate postulated for the oxidation of chloral is probably involved[179,190,197]. Ketone oxidations, including the haloform reaction, usually proceed by halogenation at the α-carbon, with enolization as the rate controlling step[4]. The halogenated ketone is then either cleaved directly to yield the haloform or hydrolysed to a diketone which is further oxidized[198,199] (equation 79).

$$ArCOCH_2R \xrightarrow{Br_2} ArCOCBr_2R \xrightarrow{NaOH} ArCOCOR \xrightarrow[NaOH]{NaOBr} ArCO_2Na \qquad (79)$$
$$+$$
$$RCO_2Na$$

B. N-Halo imides

The use of N-halo imides for allylic halogenation (Wohl–Ziegler reaction) and for oxidation has been the subject of several reviews[200–202].

Like bromine, N-bromosuccinimide has been used to convert aldehydes to acids. p-Nitrobenzaldehyde was oxidized to p-nitrobenzoic acid[203] by this reagent, and piperonal was oxidized to a mixture of piperonic acid[204] (**71**), 6-bromopiperonal (**72**) and protocatechualdehyde (**73**) (equation 80).

$$(80)$$

(**71**)   (**72**)   (**73**)

Acetals are oxidized to esters by *N*-bromosuccinimide. Thus benzaldehyde diethyl acetal was oxidized to ethyl benzoate[205]. Often radical-producing agents such as benzoyl peroxide[204] and light[205] are employed as catalysts.

Several examples of the use of *N*-bromosuccinimide as a dehydrogenating agent have appeared. The reagent converts certain ketones[206] and diketones[207,208] into α,β-unsaturated ketones without the isolation of an intermediate bromoketone. Thus flavanon-3-ol was converted to flavon-3-ol with spontaneous loss of HBr[206] (equation 81).

$$(81)$$

In one synthesis of colchicine[207,208], the diketone **74** was dehydrogenated by *N*-bromosuccinimide to produce desacetoamidocolchiceine (**75**) (equation 82).

$$(82)$$

(**74**)   (**75**)

Both the ketone dehydrogenations and acetal oxidations probably go through bromine-containing intermediates. In the oxidation of benzaldehyde diethylacetal, the intermediacy of an α-bromoacetal (**76**) has been suggested[205] (equation 83).

$$\text{PhCH(OEt)}_2 \xrightarrow{\text{NBS}} \overset{\overset{\displaystyle Br}{|}}{\text{PhC(OEt)}_2} \longrightarrow \text{PhCO}_2\text{Et} + \text{EtBr(?)} \qquad (83)$$
$$(\textbf{76})$$

The remarkable similarity of the reactions of N-bromosuccinimide and the reactions of bromine have led some authors to postulate that N-bromosuccinimide acts simply as a reservoir capable of sustaining a low steady-state concentration of bromine during the reaction[209]. For example, the change in Hammett ρ in going from 80° to 19° was the same for bromine and certain N-bromo imides, which implies that the reactions have identical enthalpies, entropies (and therefore free energies) of activation[210]. This similarity would be extremely unlikely unless the same reactive species were involved.

There is general agreement that reactions with N-halo imides proceed by a free radical chain mechanism. Assuming molecular bromine to be the reactive species, the propagation step in a typical mechanism involving reaction of an N-halo imide (SNX) with a substrate (RH) may be formulated as in equations (84a–c)[210].

$$X\cdot + RH \longrightarrow HX + R\cdot \qquad (84a)$$

$$HX + SNX \longrightarrow SNH + X_2 \qquad (84b)$$

$$X_2 + R\cdot \longrightarrow RX + X\cdot \qquad (84c)$$

C. Iodoso Compounds

Phenyl iodosodiacetate, PhI(OAc)_2, effects cleavage reactions very much like lead tetraacetate and periodate[9]. Unlike periodate, but like lead tetraacetate, reactions with phenyl iodosodiacetate are usually run in acetic acid. Although it is known that *vicinal* glycols are cleaved through the intermediacy of a cyclic complex[211,212], little is known about the mechanism of ketol and diketone cleavage reactions.

Other iodoso compounds have been used as oxidants. For example o-, m-, and p-iodosobenzoic acid converted ascorbic acid (**77**) to dehydroascorbic acid (**78**) (equation 85) and dialuric acid (**79**) to alloxan[213] (**80**) (equation 86). The m- and p-isomers reacted somewhat more rapidly with the substrates than did the o-isomer. This may be due to the fact that o-iodosobenzoic acid exists mainly as a cyclic isomer[214].

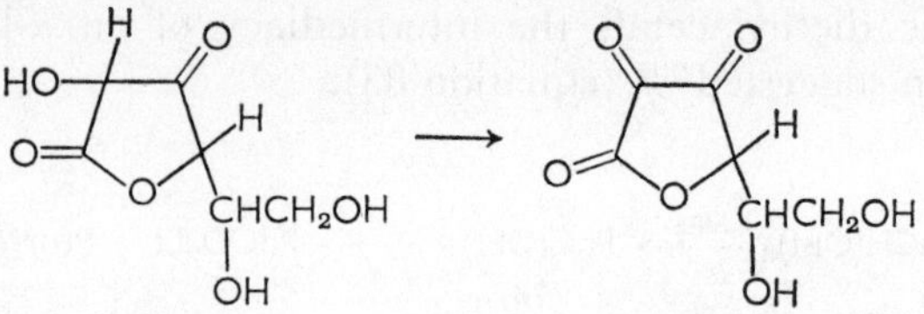

(85)

(77)   (78)

(86)

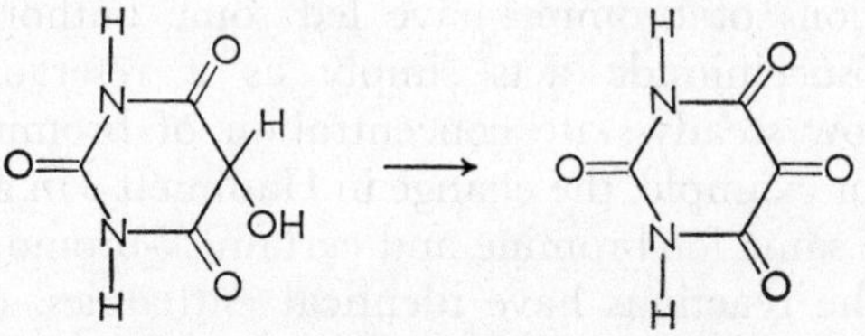

(79)   (80)

Oxidations with o-iodosobenzoic acid showed general acid catalysis, but were also catalysed by copper(II) ion[213]. Ferrous ion was 1/10 as effective as copper as a catalyst. The reaction was first order in oxidant and first order in substrate. It is believed that a protonated form of o-iodosobenzoic acid reacted with ascorbic acid in the rate-determining step.

D. Periodate

Periodate oxidation of organic compounds has been the subject of a review[215]. The use of periodate as a synthetic[216] and analytical[217] reagent in carbohydrate reactions has been discussed.

Periodic acid is ordinarily used in aqueous solution, in contrast to other reagents of this type, such as lead tetraacetate and phenyl iodosodiacetate, which are used in acetic acid.

Under normal reaction conditions aldehydes and ketones are oxidized only slowly by periodate. However, it has been shown that light accelerates the periodate oxidation of simple organic substances such as formaldehyde[218]. Periodate cleaves α-diketones to carboxylic acids although not as rapidly as lead tetraacetate. For example, benzil was only 48% oxidized after 24 hours[219].

Periodate also oxidizes 1,3-diketones[220] and five- or six-membered cyclic 1,3-diketones[221]. Cyclic 1,3-diketones unsubstituted on $C_{(2)}$ reduced four molar equivalents of oxidant yielding one equivalent of carbon dioxide and one equivalent of a dibasic acid (equation 87). Cyclic 1,3-diketones substituted on $C_{(2)}$ reduced three molar equivalents of periodate to yield a mixture of dibasic and monobasic acids (equation 88).

$$(87)$$

$$(88)$$

Cleavage of 1,2-ketols by periodate has been reported[219]. Ketols are postulated to be intermediates in the oxidation of olefins by periodate-permanganate mixtures[222]. Thus oleic acid was oxidized to 9,10-hydroxyketostearic acid, which was then cleaved into carbonyl compounds and acids[223]. If the carbonyl compound is an aldehyde it is usually oxidized further.

Periodate also cleaves α-ketoacids[224]. In the pH range normally used for periodic acid oxidations, several species derived from meta-periodic acid, HIO_4, and paraperiodic acid, H_5IO_6, are in equilibrium[225] (equations 89a and 89b). All species participate in the

$$H_5IO_6 \rightleftharpoons H_4IO_6^- + H^+ \rightleftharpoons H_3IO_6^{2-} + 2H^+ \qquad (89a)$$

$$H_4IO_6^- \rightleftharpoons 2H_2O + IO_4^- \qquad (89b)$$

oxidation, but they do so at different rates. Rate expressions which involve periodic acid therefore tend to be rather complex.

The oxidations of diacetyl, diisobutyryl, benzil, and 1,7,7-trimethyl-bicyclo[2.2.1]hepta-2,3-dione were studied kinetically over a wide pH range[226]. The reactions were first order in diketone and first order in total periodate. The rate constants for the various periodate species increased with increasing negative charge. The reaction was found to be subject to general base catalysis.

The oxidation does not involve the hydrated form of the α-diketone as was previously supposed; rather the reaction is believed to proceed by a nucleophilic attack of each of the various paraperiodate species at the carbonyl carbon atoms. The intermediate cyclic complexes such as 81 thus formed, collapse to the products[227] (equation 90).

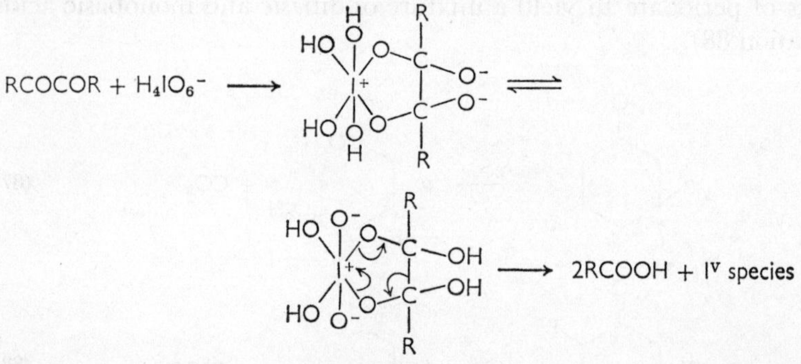

$$RCOCOR + H_4IO_6^- \longrightarrow \text{(structure)} \rightleftharpoons$$

$$\text{(structure, 81)} \longrightarrow 2RCOOH + I^{V} \text{ species} \quad (90)$$

(81)

It is believed that the oxidation of 1,3-diketones proceeds through intermediate 2-ols such as **82** which are further oxidized to the products[221] (equation 91).

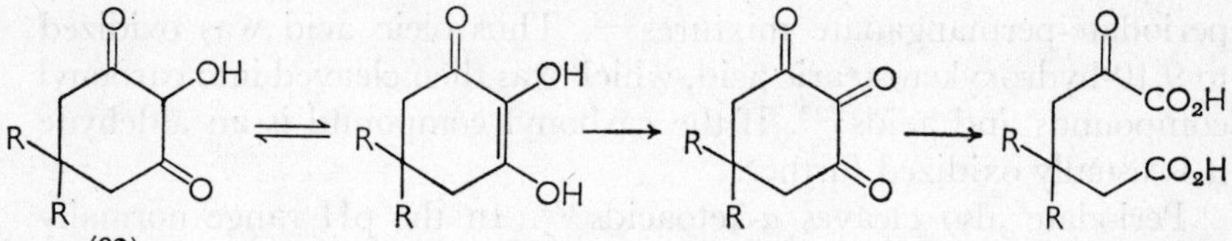

$$\text{(structure 82)} \rightleftharpoons \text{(structure)} \longrightarrow \text{(structure)} \longrightarrow \text{(structure, with } CO_2H) \quad (91)$$

(82)

VII. GROUP I.B

A. Copper

Aliphatic aldehydes are rapidly oxidized to the corresponding acid with Fehling's solution, an alkaline copper(II) solution, containing tartrate to complex with the copper ion and prevent precipitation of the hydroxide[228]. Benedict's solution, a copper citrate complex, has also been used. Aromatic aldehydes are not readily attacked by Fehling's solution, nor does the reagent furnish the corresponding acid in good yield[122]. A copper oxide–silver oxide mixture has been used as a catalyst in the autoxidation of furfural to furoic acid[229].

More useful from a synthetic point of view is the use of copper(II) salts to oxidize 1,2-ketols to 1,2-diketones. Thus 4-hydroxy-3-hexanone was oxidized to dipropionyl by copper acetate in acetic acid[230], and benzoin was oxidized to benzil by copper sulphate in pyridine[231].

Copper(II) acetate has been used as a catalyst in the conversion of α-ketolic side chains of steroids to glyoxal groups (equation 92).

$$(92)$$

Thus 3α,21-dihydroxy-5β-pregnane-11,20-dione was converted to the glyoxal in 94% yield[232]. The oxidation was accelerated by passing air through the mixture, and was retarded by the addition of water. Methanol seemed to be the solvent of choice; with methylene chloride or benzene the glyoxals tended to polymerize. Copper-catalysed autoxidation also has been employed in the oxidation of ascorbic acid to dehydroascorbic acid[233]. Ammonium nitrate in conjunction with catalytic quantities of copper acetate has also been employed in ketol oxidations[234,235]. Copper(II) ion is continuously regenerated, while the ammonium nitrate suffers reductive decomposition to nitrogen and water.

Waters and coworkers have studied the kinetics of oxidation of acetoin to biacetyl[236,237] using a copper citrate complex. They found the reaction to be first order in substrate and hydroxide ion, but

$$(93a)$$

$$(93b)$$

$$(93c)$$

$$(93d)$$

zero order in the oxidant. The reaction showed a brief induction period. They also reported that oxidation occurred much less rapidly than enolization. In order to account for these observations, they proposed a rate-controlling step which involves the displacement of citrate from a copper(I) citrate complex by an enol anion of the acetoin, with the formation of a copper(I)-ketol complex (83) (equation 93a–d). The induction period was assumed to be caused by an initial build up in concentration of copper(I) ion. Once cuprous oxide began to precipitate, the copper(I) ion would be present in solution in constant concentration, and the rate of reaction would be independent of the concentration of copper(II) ion. Wiberg[238] contends that several mechanistically crucial conclusions in the work of Waters were drawn on the basis of experimental artifacts. Wiberg attributes the initial induction period to a failure to thoroughly de-gas the reaction solutions. He also believes that, on the basis of the difficulties in accurately determining hydroxyl ion concentrations in strongly basic dioxan–water mixtures, the rates of enolization and oxidation observed by Waters are the same within experimental error; and finally, that it is unlikely that a ketol could compete effectively with citrate for two of the coordination sites about copper.

Wiberg examined the kinetics of copper(II) ion oxidation of α-hydroxyacetophenone in buffered aqueous pyridine. The reaction followed the rate expression shown in equation (94). The first term

$$\text{rate} = k_1[\text{ketol}] + k_2[\text{ketol}][\text{Cu}^{II}] \tag{94}$$

corresponds to the rate of enolization and the second term is the major one at copper(II) ion concentrations greater than $0\cdot01$ molar. Using deuterated compounds, a kinetic isotope effect was observed. The effect of substituents was correlated with a Hammett ρ of $1\cdot24$. Thus at low copper(II) ion concentrations, the rate corresponds simply to enolization of the ketol. Copper(I) ion has no effect on the rate. This type of mechanism presumably would apply to the Waters copper citrate system. At high copper ion concentrations, the rate-determining step for the second term appears to involve the removal of a proton by pyridine from the copper(II) chelate of the ketol (84) to yield a resonance stabilized radical (85) which is further oxidized to the ketone in subsequent fast steps (equations 95a–c). The proton removal step is formally a Lewis acid-catalysed enolization.

Kinetic data on the copper-catalysed autoxidation of ascorbic acid[233] to dehydroascorbic acid have also been rationalized on the basis of intermediate copper chelates.

$$RCCH_2OH + Cu^{2+} \rightleftharpoons R-C\underset{(84)}{\cdots}CH_2 \quad (95a)$$

$$R-C\cdots C-H \xrightarrow{slow} RCCHOH + Cu^+ \quad (95b)$$
$$(85)$$
$$Py:$$

$$RCOCHOH + Cu^{2+} \longrightarrow RCOCHO + Cu^+ + H^+ \quad (95c)$$

B. Silver

The formation of a silver mirror with Tollen's reagent, i.e. silver(I) ion complexed with ammonia, is a standard test for aldehydes or α-ketols. The reagent is used mainly as an analytical reagent, since the course of the oxidation is often complex. For example, various aldehydic sugars were oxidized by Tollen's reagent to carbon dioxide, formic acid, oxalic acid, and other products, depending on the sugar. Oxalic acid probably was formed from glyoxal or glycol aldehyde[239].

More useful from a preparative point of view is the oxidation of aldehydes to carboxyl groups by silver oxide, usually prepared from silver nitrate and sodium hydroxide. This reagent is especially useful when other oxidizable groups are present. For example, almost quantitative yields of unsaturated carboxylic acids have been obtained from the corresponding aldehydes[240]. This procedure has also been used to prepare thiophene-2-carboxylic acid[241], palmitic acid[242], and anthracene-9-carboxylic acid[243].

The oxidation of vanillin to vanillic acid by silver oxide[244] is interesting in that only ½ mole of silver(I) oxide was required for each molecule of vanillin oxidized. The stoichiometry of the reaction is as shown in equation (96).

$$2\,RCHO + Ag_2O + 2\,NaOH \longrightarrow 2\,RCOONa + 2\,Ag + H_2O + H_2 \quad (96)$$

Argentic or silver(II) ion is a considerably more powerful oxidant than silver(I)[245]. The oxidant is usually employed as the picolinate (86) in water, dimethyl sulphoxide, or dimethoxyethane. Argentic picolinate rapidly converted aldehydes such as butanal, p-tolualdehyde and 2-ethylhexanal to the corresponding acid. In certain cases reactive

5+c.c.g. II

methylene groups were attacked. Thus 5,5-dimethylcyclohexa-1,3-dione gave the 1,2,3-triketone.

(86)

C. Gold

Gold(III) appears to be similar to silver(I) in its ability to oxidize phenolic aldehydes to acids such as vanillic, syringic and o-vanillic acid and has been the subject of a patent for this purpose[246]. In contrast to the silver(I) oxidation of vanillin to vanillic acid, oxidation with gold(III) required equimolar ratios of oxidant and substrate[244]. The mechanism of oxidation by gold may prove to be of interest, for despite apparent similarities in the scope of reactions undergone by the two reagents, gold reacts by a transfer of two electrons (equation 97), while silver reacts by a one-electron transfer (equation 98).

$$2e^- + Au^{3+} \longrightarrow Au^{1+} \tag{97}$$

$$e^- + Ag^{1+} \longrightarrow Ag^0 \tag{98}$$

Differences in the mechanism of the oxidation may therefore be anticipated.

VIII. GROUP II.B

A. Mercury

Aldehydes are converted to the corresponding acid by mercury(II) oxide. For example the phenolic aldehydes vanillin (87), o-vanillin (88) and syringaldehyde (89) have been oxidized to the corresponding

(87)   (88)   (89)

acids by this treatment[247]. Other sensitive aldehydes such as furfural have been oxidized by air[248] in the presence of catalytic amounts of mercuric oxide. Glucose was converted in high yield to calcium gluconate by mercury(II) oxide in the presence of calcium carbonate[249]. Ketones and acyloins have been oxidized to 1,2-diketones by mercury(II) perchlorate in strong acid[4].

Oxidations with mercury often give intermediate organic mercury compounds which can be decomposed by acidifying the solution with sulphur dioxide[247] or by electrolysis[249]. The course of mercury(II) oxidations varies with the physical state of the oxidant. Thus, treatment of vanillin with pure yellow oxide gave vanillic acid in 65% yield while use of technical grade oxide yielded a mixture of 35% vanillic acid and 11% guaiacol[247].

The kinetics of acid-catalysed oxidation of cyclohexanone by mercury(II) perchlorate has been investigated[4]. The reaction was zero order in oxidant and appeared to follow the rate of bromination[250]. The reaction was first order in ketone, first order in acid, and the rate was slightly accelerated by increasing ionic strength. However, addition of sulphate or chloride ion retarded the reaction. In these studies mercury was used as a one equivalent oxidant, that is, mercury (II) was reduced only to mercury(I) (equation 99). No metallic mercury was formed. The main oxidation product appeared to be the diketone which is believed to be formed via an intermediate acyloin.

$$2e^- + 2\,Hg^{2+} \longrightarrow Hg_2^{2+} \tag{99}$$

Acyloins were found to be readily oxidized under the reaction conditions. Enolization appeared to be the rate-determining step which was followed by a fast attack of mercury(II) ion on the enol (equations

(100a)

(100b)

100a–d). The oxidation step is believed to proceed by a two-electron transfer to mercury(II) to give metallic mercury which disproportionates with mercury(II) to yield mercury(I) ion.

The kinetics and mechanism of oxidation of acyloins by mercuric halides has been discussed[251].

$$+ H_2O \longrightarrow \qquad + Hg^0 + H^+ \qquad (100c)$$

$$Hg^0 + Hg^{2+} \rightleftharpoons Hg_2^{2+} \qquad (100d)$$

Mercury(II) acetate has been found to undergo a variety of reactions with ketones. Carvone was dehydrogenated to dehydrocarvacrol by mercuric acetate[252] (equation 101). It is unlikely that the dehydro-

genation proceeds via a free-radical route since carvone is very susceptible to polymerization by light, peroxides, and other radical producing catalysts.

In a similar manner, piperitone (**92**) was formed from menthone (**90**) upon heating the latter with mercury(II) caproate[253] (equation 102). The reaction probably proceeds through an intermediate α-acyloxyketone (**91**).

$$(102)$$

(90)   **(91)**   **(92)**

Mercuric acetate has been used to convert ketones into α-acetoxy-ketones. Thus, cyclohexanone was converted to 2-acetoxycyclo-hexanone[253] and isomenthone (**93**) was acetoxylated with isomeriza-

$$(103)$$

(93)   **(94)**   **(95)**

tion to form 2-acetoxymenthone (**94**), the same product yielded by menthone (**95**)[254] (equation 103). The product which formed was the one that was thermodynamically most stable (all equatorial). Acetoxylation probably proceeded through an intermediate mercury compound **96**[255], formed by attack of mercury(II) acetate on the enol (equations 104a and 104b).

$$
\begin{array}{c}
\diagup \\
C=C \\
\diagup \quad \backslash
\end{array}
\begin{array}{c} OH \\ | \\ \end{array}
+ \ Hg(OAc)_2 \longrightarrow
\begin{array}{c} HgOAc \ O \\ | \quad \diagup\!\!\!/ \\ -C-C \\ | \quad \backslash \end{array}
+ \ HOAc \qquad (104a)
$$

$$(96)$$

$$
\begin{array}{c} HgOAc \ O \\ | \quad \diagup\!\!\!/ \\ -C-C \\ | \quad \backslash \end{array}
\longrightarrow
\begin{array}{c} OAc \ O \\ | \quad \diagup\!\!\!/ \\ -C-C \\ | \quad \backslash \end{array}
+ \ Hg \qquad (104b)
$$

IX. GROUP III.B

A. Cerium

Cerium(IV) is a powerful oxidant, usually used in acid solution. Although it is widely employed as an oxidizing agent in analytical laboratories, only limited application has been made of its oxidizing properties in organic synthesis. For example, although cerium is known to react with many heterocyclic compounds, the products of these oxidations have not been investigated[256].

Formaldehyde was oxidized to formic acid by cerium(IV) ions[257,258], and triphenylacetaldehyde to triphenylcarbinol and carbon monoxide. Isobutyraldehyde (which possesses an α-hydrogen) reacted with cerium(IV) to yield α-hydroxyisobutyraldehyde[259]. However, too few examples are available to be able to make definite generalizations about the scope of the cerium(IV) oxidation of aldehydes. So far, it appears that enolizable carbonyl compounds tend to be attacked in the α-position to yield α-hydroxy carbonyl compounds. Ketones are degraded rapidly in acid solution by cerium(IV). Ketones with an α-hydrogen react with cerium(IV) to give α-hydroxyketones[260]. For example, acetone is converted into α-hydroxyacetone[261].

Like vanadium, but unlike cobalt, cerium(IV) oxidizes ketones more readily than alcohols.

The mechanistic aspects of cerium(IV) oxidation of aldehydes has been discussed in Part I of this work[1]. Like cobalt(III) and manganese(III), cerium(IV) reacts by a one-electron transfer to yield cerium(III) (equation 105).

$$e^- + Ce^{4+} \longrightarrow Ce^{3+} \tag{105}$$

The exact nature of the reactive monomeric cerium(IV) species in acid solution is not known, but coordination with anions in different acids affects the oxidation potential. Probably the species $Ce(H_2O)_n^{4+}$ exists only in perchloric acid. In sulphuric acid–perchloric acid solutions, the most reactive species appears to be $CeSO_4^{2+}$ [262].

As in the case of vanadium(V), cobalt(III), and manganese(III), attack by cerium(IV) apparently proceeds directly on the ketone and not on the enol[263]. The mechanism is still the subject of controversy, however. Shorter[264] has studied the reaction of acetone with cerium(IV) in nitric acid. He prefers to regard the oxidation as proceeding via an attack on the enol. The oxidation was pseudo-first order in cerium(IV) at high acetone concentrations. Some inhibition was observed in the presence of bromine which could be explained on the basis of a competitive reaction of the enol with cerium(IV) and bromine. The mechanism proposed involves the initial rapid formation of a complex between cerium(IV) perchlorate and the enol

$$
\begin{array}{cc}
& \overset{\text{OH}}{\underset{|}{}} \\
CH_3COCH_3 \underset{}{\overset{H^+}{\rightleftharpoons}} CH_3C{=}CH_2 & (106a)
\end{array}
$$

$$
\overset{\text{OH}}{\underset{|}{}} \qquad\qquad \overset{\text{O}}{\underset{\|}{}} \qquad\qquad \overset{\text{O}^{\cdot}}{\underset{|}{}}
$$
$$Ce^{4+} + CH_3C{=}CH_2 \xrightarrow{\text{slow}} [CH_3C{-}\dot{C}H_2 \longleftrightarrow CH_3C{=}CH_2] + Ce^{3+} + H^+ \tag{106b}$$
$$(\mathbf{97})$$

$$Ce^{4+} + CH_3CO\dot{C}H_2 \longrightarrow CH_3CO\overset{+}{C}H_2 + Ce^{3+} \tag{106c}$$

$$CH_3CO\overset{+}{C}H_2 + H_2O \longrightarrow CH_3COCH_2OH + H^+ \tag{106d}$$

form of acetone[265] (equations 106a–d). The rate-controlling step is believed to be decomposition of this complex to form cerium(III) and a ketone radical (**97**). It is hard, however, to see why the enol should be involved in complex formation, since oxidation occurs at a rate considerably faster than the rate of enolization.

The oxidation of cyclohexanone and 2,2,6,6-tetradeuterocyclohexanone in water and deuterium oxide has been studied by Littler[263]. Littler believes that cerium(IV) attacks the ketone. Both the fact that oxidation by cerium(IV) sulphate was found to be faster than enolization and the solvent effect observed support this theory. Oxidation of the deuterated cyclohexanone showed a primary isotope effect and indicated that the α C—H bond was broken in the rate-determining step. Waters favours a mechanism analogous to the style of mechanism proposed for oxidation by vanadium(v). Assuming that the reactive species is $CeOH^{3+}$, the reaction (equations 107a-c) could be formulated as an attack on the ketone by this species (or the

$$Ce^{4+} + H_2O \longrightarrow CeOH^{3+} + H^+ \tag{107a}$$

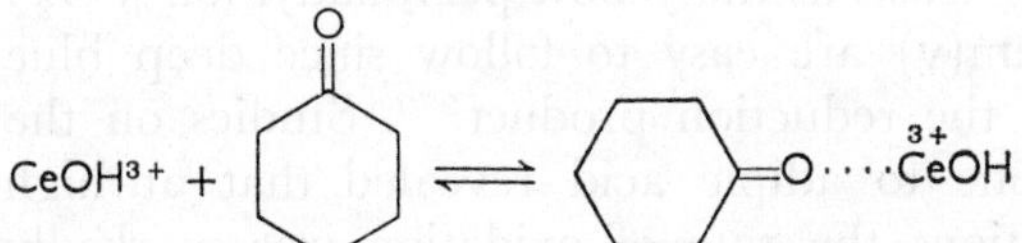

(98)

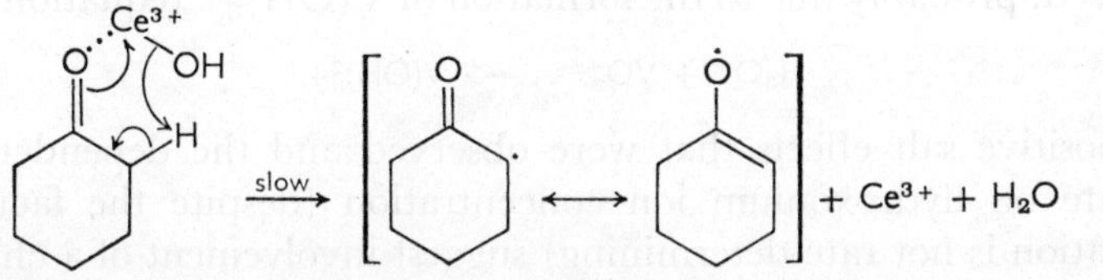

(99)

equivalent sulphate species) to yield a complex (98). The complex decomposes in the rate-determining step to yield the resonance-stabilized radical (99) which is further oxidized (or disproportionates) to the ketol.

X. GROUP V.B

A. Vanadium

Pentavalent vanadium has seen limited use in synthetic organic chemistry, although it has been employed as an analytical tool[266]. The oxidizing power of vanadium(v) is less than that of chromium(vi) or manganese(iii)[267]. A general qualitative survey has been made of vanadium(v) oxidations[268]. Oxidations are usually carried out in acid solution. Aldehydes and ketones which can enolize are readily attacked by vanadium(v). Thus chloral was not attacked and form-aldehyde was attacked only slowly, while acetaldehyde, propional-dehyde and butyraldehyde were rapidly attacked. Acetone, ethyl methyl ketone, diethyl ketone, diisopropyl ketone, cyclopentanone, cyclohexanone, acetonylacetone, and diacetyl were all rapidly oxidized.

Vanadium has been used as a catalyst in nitric acid oxidations (see section IV.A). It has been shown that benzil and benzaldehyde on oxides of vanadium at 350–420° reacted with water in the absence of air, to form benzoic acid and other oxygenated compounds[269].

In general, vanadium(v) oxidizes carbonyl compounds more readily than alcohols[270].

The mechanism of vanadium(v) oxidation of aldehydes has been discussed in Volume 1 of this work[1] and will not be repeated here. In acid solution vanadium(v) exists as the yellow pervanadyl ion VO_2^+. Oxidations with vanadium(v) are easy to follow since deep blue VO^{2+} [vanadium(IV)] is the reduction product[270]. Studies on the oxidation of cyclohexanone to adipic acid revealed that at high sulphuric acid concentrations the rate of oxidation was markedly increased, probably due to the formation of $V(OH)_3^{2+}$ (equation 108).

$$H_3O^+ + VO_2^+ \longrightarrow V(OH)_3^{2+} \tag{108}$$

The positive salt effects that were observed and the dependence of the rate on hydroxonium ion concentration (despite the fact that enolization is not rate-determining) suggest involvement of a charged species such as $V(OH)_3^{2+}$. In sulphuric acid the species may be $V(OH)_3HSO_4^+$.

Reactions are first order each in vanadium(v) and substrate; the reaction mixture induced the polymerization of acrylonitrile and the reduction of mercurous chloride[268]. This induced polymerization and reduction points to a free-radical intermediate which can initiate the polymerization of acrylonitrile, and can reduce mercury by a one-electron exchange. The rate of oxidation of cyclohexanone and 2,2,6,6-tetradeuterocyclohexanone by vanadium(v) in water and in deuterium oxide has been studied[263]. The solvent effect observed suggests that attack of vanadium occurred directly on the ketone rather than on the enol. Observation of a sizeable isotope effect ($k_H/k_D = 4.2$ at 50°) suggests that an α C—H bond was broken in the rate-determining step. The mechanism illustrated in equations (109a-d) has been proposed[270]. The initially formed vanadium–cyclohexanone complex (100) decomposes in the rate-determining step by a proton or a hydride transfer to yield the resonance stabilized radical 101. It is believed that 101 is the radical that can initiate the polymerization of acrylonitrile and reduce mercury. The radical is further oxidized (or disproportionates) to the ketol.

Oxidation of 3-hydroxy-3-methylbutan-2-one[271] required two equivalents of vanadium(v) per mole, and yielded a mixture of acetone and acetic acid. The reaction was shown to be independent of H_0 at low acid concentrations, and to be first order in vanadium(v)

$$VO_2^+ + H_3O^+ \rightleftharpoons V(OH)_3^{2+} \tag{109a}$$

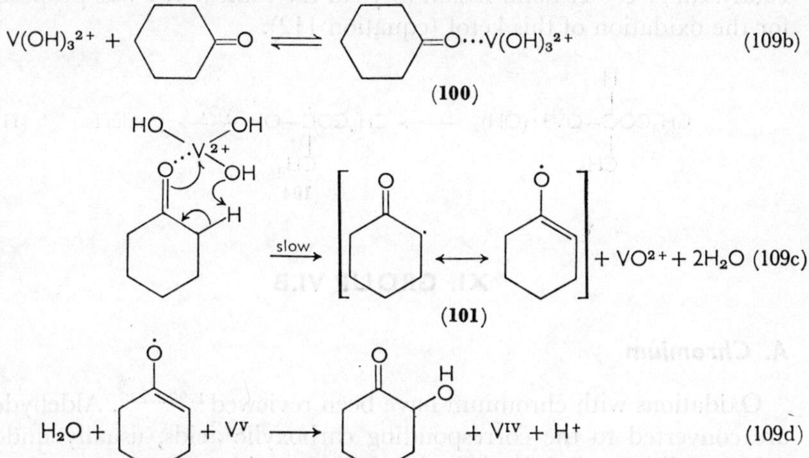

$$V(OH)_3^{2+} + \text{(cyclohexanone)} =O \rightleftharpoons \text{(cyclohexanone)}=O \cdots V(OH)_3^{2+} \tag{109b}$$

(100)

$$\xrightarrow{\text{slow}} \left[\text{(radical)} \longleftrightarrow \text{(radical)} \right] + VO^{2+} + 2H_2O \tag{109c}$$

(101)

$$H_2O + \text{(enol radical)} + V^V \longrightarrow \text{(ketol)} + V^{IV} + H^+ \tag{109d}$$

and first order in ketol. Alternate acid-dependent and acid-independent pathways were proposed (equation 110). The acid-independent

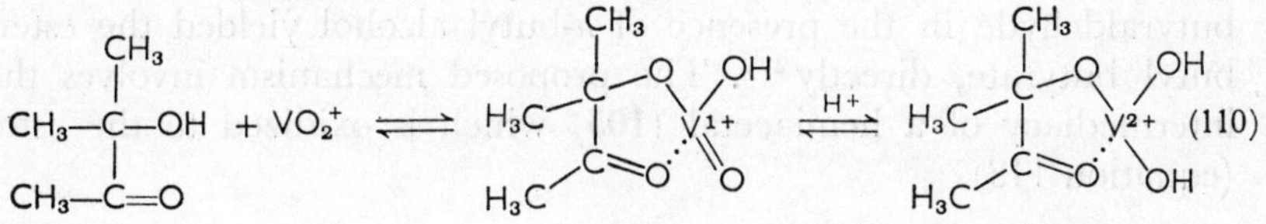

$$\begin{array}{c} CH_3 \\ | \\ CH_3-C-OH \\ | \\ CH_3-C=O \end{array} + VO_2^+ \rightleftharpoons \text{(102)} \xrightarrow{H^+} \text{(103)} \tag{110}$$

(102)   (103)

decomposition involves intermediate **102**, while the acid-dependent route involves intermediate **103**. The intermediates decompose to an

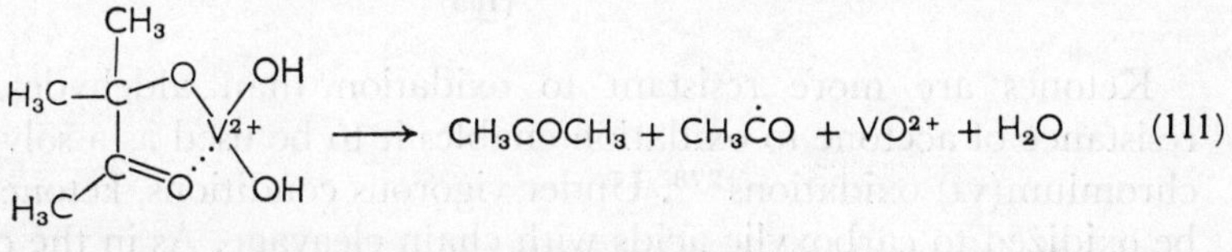

$$\text{(intermediate)} \longrightarrow CH_3COCH_3 + CH_3\dot{C}O + VO^{2+} + H_2O \tag{111}$$

acetyl radical which would be oxidized further (or disporportionate) to acetic acid (equation 111).

Acetoin consumed four equivalents of vanadium(v) per mole and thus must have undergone oxidation by a different route[271]. If carbon–carbon bond cleavage occurred in a manner analogous to 3-hydroxy-3-methylbutan-2-one, acetaldehyde would be produced initially. However, since acetaldehyde was found to be oxidized at a slower rate than acetoin, it cannot be an intermediate. The oxidation was acid-catalysed. A C—H bond fission to yield the radical **104** was proposed for the oxidation of this ketol (equation 112).

$$CH_3CO\overset{\overset{\displaystyle H}{|}}{\underset{\underset{\displaystyle CH_3}{|}}{C}}-OV^{2+}(OH)_2 \longrightarrow CH_3CO\overset{\cdot}{\underset{\underset{\displaystyle CH_3}{|}}{C}}-OH \xrightarrow{VO_2^+} products \qquad (112)$$

(104)

XI. GROUP VI.B

A. Chromium

Oxidations with chromium have been reviewed[1,272,273]. Aldehydes are converted to the corresponding carboxylic acids, usually under acid conditions. Sulphuric acid–chromic acid mixtures have been used to prepare heptanoic acid from the aldehyde[274] and furoic acid from furfural[275]. Yields are seldom quantitative; carbon dioxide and lower acids are the principal by-products[276]. Oxidation of *n*-butyraldehyde in the presence of *n*-butyl alcohol yielded the ester, butyl butyrate, directly[277]. The proposed mechanism involves the intermediacy of a hemiacetal (**105**) which is oxidized to the ester (equation 113).

$$PrCHO + BuOH \longrightarrow \left[Pr\overset{\overset{\displaystyle H}{|}}{\underset{\underset{\displaystyle OH}{|}}{C}}-OBu \right] \xrightarrow{CrO_3} PrCO_2Bu \qquad (113)$$

(105)

Ketones are more resistant to oxidation than aldehydes. The resistance of acetone to oxidation enables it to be used as a solvent in chromium(VI) oxidations[278]. Under vigorous conditions, ketones may be oxidized to carboxylic acids with chain cleavage. As in the case of permanganate oxidations, oxidative attack occurs preferentially at a methinyl carbon[279]. Trimethylacetic acid was prepared from pina-colone, $(CH_3)_3CCOCH_3$, by oxidation with chromic anhydride in acetic acid[280] and diphenic acid has been prepared by dichromate oxidation of phenanthraquinone[281]. When applied to more complex systems such as steroids[282], a variety of products can be obtained. Fieser[283] investigated the chromic acid oxidation of cholesterol, which yields, via the ketone **106**, diketones, ketols, dicarboxylic acids and acetals (equation 114).

Certain alicyclic 1,2-diketones, for example 3,3,6,6-tetramethyl-1,2-cyclohexanedione[284], have been prepared by oxidation of the ketol with chromic anhydride although milder reagents are usually used.

Chromic acid is usually employed in aqueous or acetic acid solution, but pyridine, t-butyl alcohol, acetic anhydride, acetone, and dimethylformamide have also been used as solvents[1]. Oxidations are usually carried out at room temperature or with cooling. Generally, oxidation of aldehydes to acids proceeds more slowly than oxidation of the corresponding alcohols to the aldehydes[285]. Changing the solvent will also affect the relative rates of oxidation of aldehyde and alcohol. Alcohol oxidations proceed more rapidly in acetic acid than in water, whereas the rate of aldehyde oxidation is little affected[286]. The presence of electron-withdrawing groups on the molecule speeds aldehyde oxidations but slows alcohol oxidations[286].

The mechanistic aspects of oxidation of aldehydes with chromium (VI) has been discussed in Volume 1[1].

Studies on the comparative rate of reaction of cyclohexanone and 2,2,6,6-tetradeuterocyclohexanone in water and in deuterium oxide have led to the conclusion that the oxidative cleavage of ketones proceeds by attack on the enol[287]. The resistance to oxidation of certain terpenoid ketones such as **107** has been attributed to steric hindrance to attack of the enol[288]. Recent kinetic studies on isobutyro-

(107)

phenone and 2-chlorocyclohexanone provide further support to the theory that chromic acid oxidation proceeds via an enol interme- diate[289]. Although the eventual reduction product in oxidations with chromium(VI) oxide and dichromate is chromium(III), valence states other than chromium(VI) are apparently involved in the oxidation. In the fission of t-butyl phenyl ketone by chromic acid, there is evi- dence that a two-electron change is involved from chromium(V) to chromium(III)[290]. Wiberg[258] contends that oxidation with chromic acid is in fact an oxidation by chromium(V) to 67% extent. He offers the scheme illustrated in equations (115) to explain the formation of chromium(V) by reaction of chromium(VI) with a substrate S, yielding products P. Chromium(V) appears to be a more powerful and less

$$\mathrm{Cr^{VI}} + S \longrightarrow \mathrm{Cr^{IV}} + P \tag{115a}$$

$$\mathrm{Cr^{IV}} + \mathrm{Cr^{VI}} \longrightarrow 2\,\mathrm{Cr^{V}} \tag{115b}$$

$$2\,\mathrm{Cr^{V}} + 2S \longrightarrow 2\,\mathrm{Cr^{III}} + 2P \tag{115c}$$

selective oxidant than chromium(VI)[258]. It has been estimated that cyclohexanone is 15,000 times as reactive to chromium(VI) than adipic acid, but both compounds are equally reactive toward chro- mium(V). Roček[276] has shown that the production of glutaric and succinic acid as by-products in the chromic acid oxidation of cyclo- hexanone is due to attack by chromium(V). He found that although adipic acid itself was unreactive to chromic acid, when the reaction was carried out in the presence of 3-methyl-2-butanone, adipic acid was oxidized to glutaric and succinic acid (equation 116). Evidently reaction of 3-methyl-2-butanone with chromic acid produces chro- mium(V) which then attacks the adipic acid.

$$HO_2C(CH_2)_4CO_2H \xrightarrow[\text{i-PrCOMe}]{CrO_3} HO_2C(CH_2)_2CO_2H + HO_2C(CH_2)_3CO_2H \quad (116)$$

XII. GROUP VII.B

A. Manganese

I. Permanganic acid and manganese(III)

Manganese oxidations have been reviewed[272,291,292]. Aromatic and aliphatic aldehydes are smoothly converted to the corresponding carboxylic acids by permanganate in acid or alkali. Alkaline permanganate has been employed in the preparation of 2-ethylcaproic acid from the aldehyde[293]. Acid permanganate has been used for the oxidation of heptaldehyde to heptanoic acid[294]. The aromatic aldehyde piperonal has been oxidized to piperonylic acid by permanganate[295]. In general, manganese(III) pyrophosphate attacks only aldehydes that can enolize; however, certain α,β-unsaturated aldehydes such as acrolein are known to react[296]. On the other hand manganese(III) sulphate is a more powerful oxidant which will react with aldehydes which cannot form an enol, like formaldehyde[297].

Ketones are much more resistant to oxidation than aldehydes. Thus acetone can be used as a solvent for permanganate oxidations[298]. Under vigorous conditions ketones are oxidized with chain cleavage to carboxylic acids. Oxidative attack occurs preferentially at a methinyl carbon, rather than at a methylene one[299]. Manganese(III) pyrophosphate has been used to oxidize enolizable ketones[300] and α-keto acids[301]. In the case of pyruvic acid, oxidative decarboxylation to acetic acid occurred[300].

Electron-withdrawing substituents have little effect on the rate of permanganate oxidation in neutral solution, whereas they greatly accelerate the rate[302] in basic solution. Aldehydes may sometimes be oxidized more rapidly than alcohols by running the oxidation in neutral solution. This is because alcohols are oxidized rapidly only via their alkoxide ions, while aldehydes are oxidized at reasonable rates even in neutral solution[286].

Manganese dioxide is the usual reduction product in neutral and in all but strongly basic solution; manganese dioxide is usually not reduced further since it is a weak oxidant (see section XII.A.2). Acid permanganate and manganese(III) salts are usually reduced to manganese(II).

The mechanistic aspects of manganese oxidations of aldehydes have been discussed in Volume 1 of this work[1]. The most common mode of oxidation of aliphatic ketones involves reaction at the α-position to yield an acyloin[303] or, in some cases, possible direct oxidation to a 1,2-dicarbonyl compound[304].

The control of rate by enolization in the acid-catalysed oxidation by permanganate has been suggested in studies on the oxidation of acetophenone[305]. The oxidation of cyclohexanone by acid permanganate is zero order in permanganate and first order in ketone, which also points to a rate-controlling enolization[4].

In acid solution, oxidation by manganese(III) presents a more complex mechanistic picture. Manganese(III) is stable only in very acidic solution, where it is employed either as the pyrophosphate, $Mn(H_2P_2O_7)_3^{3-}$, or as the sulphate. As has been pointed out above, the nature of the ion associated with the manganese affects its oxidizing power, and very possibly alters the mechanism by which it reacts.

Manganic pyrophosphate oxidations of enolizable ketones were found to be zero order in oxidant, but first order in carbonyl compound and in acid[300]. Presumably manganese(III) reacts with the enol to form the resonance stabilized radical 108, which can react with another manganese(III) ion and water to form a ketol[303]. Alternatively the radical may disproportionate to form a ketol and starting material (equations 117a–e).

$$RCH_2\overset{\overset{\text{O}}{\|}}{C}R \rightleftharpoons RCH=\overset{\overset{\text{OH}}{|}}{C}R \tag{117a}$$

$$RCH=\overset{\overset{\text{OH}}{|}}{C}R + Mn^{3+} \longrightarrow [RCH=\overset{\overset{\text{O}\cdot}{|}}{C}R \longleftrightarrow R\dot{C}H\overset{\overset{\text{O}}{\|}}{C}R] + Mn^{2+} + H^+ \tag{117b}$$
$$(108)$$

$$R\dot{C}HCRO + Mn^{3+} \longrightarrow R\overset{+}{C}HCRO + Mn^{2+} \tag{117c}$$

$$R\overset{+}{C}HCRO + H_2O \longrightarrow R\overset{\overset{\text{OH}}{|}}{C}HCRO + H^+ \tag{117d}$$

or

$$2\,R\dot{C}HCRO + H_2O \longrightarrow R\overset{\overset{\text{OH}}{|}}{C}HCRO + RCH_2CRO \tag{117e}$$

The oxidation of pyruvic acid by manganic pyrophosphate apparently does not proceed by enolization, but by the cyclic intermediate 109[301] (equation 118). The acyl radical produced by decomposition of the intermediate reacts further to form acetic acid.

In contrast with oxidation with manganic pyrophosphate, oxidation of cyclohexanone with manganese(III) sulphate proceeded at a rate greater than enolization[263]. When 2,2,6,6-tetradeuterocyclohexanone was oxidized, a primary isotope effect was observed. It is believed that oxidation with manganese(III) sulphate involves a direct attack on the ketone, with breakage of an α C—H bond in the rate-determining step.

$$\text{(109)} \quad \longrightarrow \quad R\dot{C}O + CO_2 + Mn(H_2P_2O_7)_2^{2-} \qquad (118)$$

The kinetics of permanganate oxidation of acetone has been examined in basic aqueous solution[304]. The authors interpret the data in terms of a mechanism in which the enolate is formed in an equilibrium step, followed by its rate-determining oxidation (equations 119a–c). It may be possible that methylglyoxal can also be

$$CH_3COCH_3 + OH^- \rightleftharpoons CH_3\overset{O^-}{\underset{|}{C}}=CH_2 \qquad (119a)$$

$$CH_3\overset{O^-}{\underset{|}{C}}=CH_2 + MnO_4^- \xrightarrow{\text{slow}} CH_3CO\dot{C}H_2 + MnO_4^{2-} \qquad (119b)$$

$$CH_3CO\dot{C}H_2 + MnO_4^- + OH^- \longrightarrow CH_3COCH_2OH + MnO_4^{2-} \qquad (119c)$$

formed directly from acetone without going through the ketol, as illustrated in equation (120). All the postulated intermediates (e.g.

$$CH_3CO\dot{C}H_2 + MnO_4^- \longrightarrow CH_3COCH-O-MnO_3^- \longrightarrow CH_3COCHO \qquad (120)$$
$$\underset{HO^-}{\overset{|}{H}} \qquad + \\ Mn^{IV} \text{ species}$$

pyruvate, methylglyoxal, etc.) are oxidized faster than acetone itself and thus satisfy the kinetic evidence.

2. Manganese dioxide

The oxidation of organic compounds by manganese dioxide has been the subject of a review[291]. Normally, specially prepared manganese dioxide of high activity is employed[306]. Activity may be graded by measuring the yield of cinnamaldehyde (as the 2,4-dinitrophenylhydrazone) from cinnamyl alcohol under standard conditions[307]. Although manganese dioxide has a fairly high oxidation potential, about 1·3 v, it is only a very mild oxidizing agent. Manganese dioxide is mainly used for the oxidation of alcohols to aldehydes, although, at higher temperatures, aldehydes are further oxidized to acids. Thus, benzaldehyde yielded benzoic acid and salicylaldehyde yielded salicylic acid[308]. The reagent also has been employed in the oxidation of ketols to 1,2-diketones. Thus, benzoin yielded benzil and anisoin yielded anisil[308]. In boiling chloroform, cleavage of α-ketols has been reported. Thus the dihydroxyacetonyl side chain of a steroid was cleaved by manganese dioxide to yield a 17-oxo-steroid[309] (equation 121).

$$
\begin{array}{c}
CH_2OH \\
| \\
CO \\
| \\
\text{---OH}
\end{array}
\xrightarrow{MnO_2}
\qquad (121)
$$

Manganese dioxide also has been used to convert Δ^4-3-ketones to $\Delta^{4,6}$-3-ketones[310], but other reagents are preferred for this dehydrogenation.

XIII. GROUP VIII

A. Iron

The oxidation of organic compounds by ferricyanide has been the subject of a review[311].

Simple aldehydes such as acetaldehyde[312], n-butyraldehyde[313] and isobutyraldehyde[259,314] were oxidized by ferricyanide ion, $Fe(CN)_6^{3-}$. Investigation of the products of these oxidations revealed a complex course of reaction in several cases. For example, oxidation of isobutyraldehyde produced 2,2,5,5-tetramethyldihydropyrazine (110) in good yield[259,314] (equation 122). In more concentrated solution 2,2,5,5-tetramethyl-3,6-dicyanopiperazine (111) was also produced.

$$
(CH_3)_2CHCH{=}O \xrightarrow{Fe(CN)_6{}^{3-}}
\qquad + \qquad
\qquad (122)
$$

(110)   (111)

In contrast, furfural was oxidized to furoic acid, and ferricyanide has been proposed as a general oxidant for all furan compounds that are stable to alkali[315]. Benzaldehyde was reported not to be oxidized by ferricyanide[313] and the oxidation of furfural was slow.

Oxidation of ketones such as cyclohexanone[312], acetone[313], ethyl methyl ketone[313], and diisopropyl ketone[313] have been reported. Ferricyanide oxidation again is associated with complex reaction products. Thus, oxidation of phenyl isopropyl ketone yielded a mixture from which phenyl α-aminoisopropyl ketone was isolated[259,314] (equation 123).

$$\text{PhCOCH(CH}_3)_2 \longrightarrow \underset{\underset{NH_2}{|}}{\text{PhCOC(CH}_3)_2} \qquad (123)$$

Furyl methyl ketone (equation 124) and furfuralacetone were oxidized by an excess of ferricyanide to furoic acid[315].

$$\qquad (124)$$

Oxidations of 1,2-diketones such as cyclohexane-1,2-dione and biacetyl have been reported[312], and an interesting reaction was reported involving the formation of a cyclopropane compound (**113**) from 1,3-diketone **112**[316] (equation 125).

(**112**)   (**113**)

$$\qquad (125)$$

The iron(III)-catalysed autoxidation of acetylacetone has been discussed elsewhere in this work[1].

Aldehydes and ketones are reported to form addition compounds with ferricyanide[317], but the nature of these compounds is not known.

Ferric chloride in a boiling ether–water mixture has been used to oxidize acyloins to diketones[230,318]. Thus, acetoin was oxidized to biacetyl[319]. The oxidation of 2-hydroxycyclohexanone by an iron(III)-phenanthroline complex has been studied[312]. Iron(III) has also been used as a catalyst in the air oxidation of ascorbic acid to dehydro-ascorbic acid[233].

Iron(III) complexes have been found to attack α-keto acids such as pyruvic acid[312].

Aldehydes and ketones are oxidized by ferricyanide through their enolic forms[313]. It is believed that the enol anion gives up an electron to the oxidant to yield the resonance stabilized radical **114** which is further oxidized (equation 126). Radical intermediates would explain

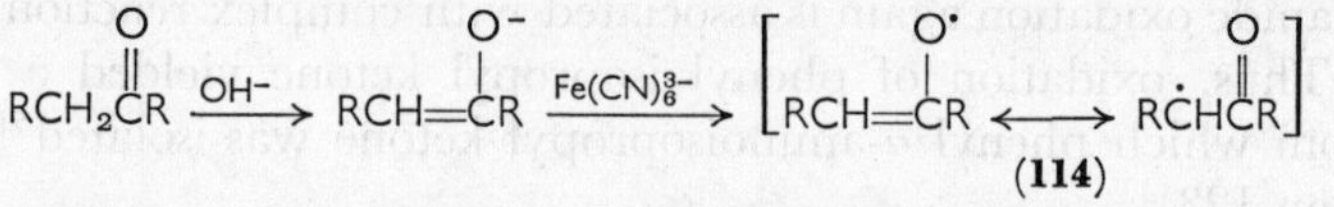

$$RCH_2CR \xrightarrow{OH^-} RCH=CR \xrightarrow{Fe(CN)_6^{3-}} \left[RCH=CR \longleftrightarrow R\dot{C}HCR \right] \quad (126)$$

(114)

the formation of products such as **113**. Nevertheless, vinyl polymerization is not induced by these systems, possibly because the enol radicals are not sufficiently reactive or, alternatively, due to the formation of a radical–oxidant complex[236]. Chelate complexes have been proposed in the iron-catalysed oxidation of ascorbic acid[233]. Formation of the piperazine derivatives **110** and **111** has also been rationalized on the basis of a free-radical intermediate similar to **114**[259,314].

Ferricyanide oxidation of aldehydes and ketones was found to be first order in substrate and hydroxide ion. The reaction order with respect to ferricyanide depended on the compound oxidized. The kinetics of oxidation of various sugars in alkaline ferricyanide has been investigated[320]. The reaction was zero order with respect to ferricyanide and first order in hydroxide and sugar. This supports the theory that the reaction proceeds by a slow enolization step[313].

The kinetics of oxidation of cyclohexanone by tris-1,10-phenanthrolineiron(III) (**115**) has been studied in the presence and in the absence

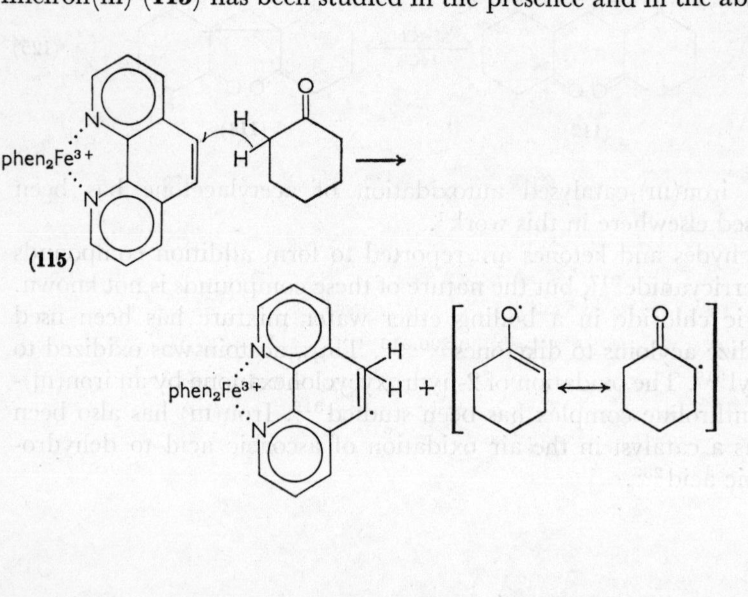

(115)

$$\text{phen}_2\text{Fe}^{3+} \quad + \quad \left[\right] \quad (127a)$$

of oxygen[312]. An electron transfer from the ketone molecule to the oxidant with the generation of 2-oxocyclohexyl radicals has been suggested (equations 127a and 127b).

$$phen_2Fe^{3+} \cdots \longrightarrow phen_3Fe^{2+} + H^+ \qquad (127b)$$

The ferric perchlorate oxidation of acetoin to biacetyl has been investigated[319]. At low concentrations of iron(III) the reaction was first order in acetoin and iron(III). The rate showed a dependence on pH, believed due to equilibria involving various hydrated iron species such as $FeOH^{2+}$ (equation 128). Induced radical polymeri-

$$Fe^{3+} + H_2O \rightleftharpoons FeOH^{2+} + H^+ \qquad (128)$$

zation of methacrylate was observed under the reaction conditions. The authors propose direct attack of iron(III) on the ketol to yield radical **116** which is oxidized in a subsequent fast step to the diketone (equations 129a and 129b).

$$Fe^{3+} + CH_3COCHOHCH_3 \xrightarrow{\text{slow}} Fe^{2+} + H^+ + CH_3CO\dot{C}(OH)CH_3 \quad (129a)$$
$$\text{(116)}$$

$$Fe^{3+} + CH_3CO\dot{C}(OH)CH_3 \longrightarrow Fe^{2+} + H^+ + CH_3COCOCH_3 \qquad (129b)$$

B. Cobalt

A series of papers have appeared on the oxidation of organic compounds by cobalt(III)[321]. Formaldehyde[322,323] and several aromatic aldehydes[324] have been oxidized by cobalt(III) to the corresponding acids. Cobalt(III) has also been used as a catalyst in the autoxidation of aldehydes[46]. It is thought that cobalt initiates radical chains by direct reaction with the aldehyde (equation 130). Electron-withdrawing substituents accelerate the rate of oxidation[324]. The mechanistic aspects of cobalt oxidation of aldehydes have been discussed in Volume 1 of this work[1].

$$CH_3CHO + Co^{3+} \longrightarrow CH_3\dot{C}O + Co^{2+} + H^+ \qquad (130)$$

Oxidation of ketones by cobalt(III) produces extensive decomposition of the organic molecule[325]. The oxidation of diethyl ketone, for example, was so complete that 10 cobaltic ions were required to completely oxidize 1 mole of ketone. The oxidation of alcohols by cobalt(III)

occurs more readily than that of ketones. Cobaltic oxidation of
t-butyl alcohol, for example, yielded acetone, which was highly
resistant to further oxidation[326].

Cobalt is a more powerful oxidant than vanadium, manganese[327],
or cerium. Oxidations are usually run in acid solution and cobalt(II)
is the usual reduction product. The reactive species appears to be
$Co(H_2O)_5OH^{2+}$ in perchloric acid, and a sulphate complex in
sulphuric acid[263].

The kinetics of oxidation of cyclohexanone and 2,2,6,6-tetradeutero-
cyclohexanone have been studied in water and deuterium oxide[263].
It was concluded that cobalt(III) attacked the ketone rather than the
enol. Oxidation by cobalt(III) proceeded much more rapidly than
enolization. The solvent isotope effect observed also points to attack
of cobalt(III) on the ketone in both sulphuric acid and perchloric
acid. In the oxidation of the deuterated ketone, no primary isotope
effect was observed which implies that the α C—H bond is not broken
in the rate-determining step. The authors believe that the kinetics
can be explained by assuming that formation of a cobalt–ketone
complex is slower than its subsequent decomposition. Alternatively,
the slow step may involve direct electron transfer from cyclohexanone
to yield a radical cation **117** (equation 131).

$$Co^{3+} + \text{(cyclohexanone)} \longrightarrow Co^{2+} + \text{(radical cation 117)} \qquad (131)$$

(117)

Studies on diethyl ketone and the α-deuterated ketone lend support
to a mechanism involving direct attack of oxidant on the ketone[325].
No kinetic isotope effect was observed and the rate of oxidation was
greater than the rate of enolization. No 1,2-diketone could be detected.
It was calculated that pentane-2,3-dione, if formed, would have been
detected unless it was oxidized at least 100 times as fast as diethyl
ketone. It was concluded that it was the ketone as such which was
oxidized, and that C—C fission must occur at an early stage, via the
radical cation **118** or the cobalt containing intermediate **119** (equations 132a–d).

$$Co^{3+} + Et_2CO \rightleftharpoons Et_2CO\cdots Co^{3+} \longrightarrow Co^{2+} + Et_2\overset{+}{C}-O^{\cdot} \quad (132a)$$

(118)

$$Et_2\overset{+}{C}-O^{\cdot} \longrightarrow Et^{\cdot} + Et\overset{+}{C}O \qquad (132b)$$

or

$$H_2O + Et_2CO\cdots Co^{3+} \rightleftharpoons Et_2\overset{\overset{\displaystyle OH}{|}}{C}-O\cdots Co^{2+} + H^+ \quad (132c)$$
$$\textbf{(119)}$$

$$Et_2\overset{\overset{\displaystyle OH}{|}}{C}-O\cdots Co^{2+} \longrightarrow Co^{2+} + EtCO_2H + Et^{\cdot} \quad (132d)$$

In sulphuric acid, the reaction was first order in cobalt(III) and in ketone, but the reaction showed an inverse dependence on acidity. This has been attributed to the formation of the inert complex, $CoSO_4^+$.

C. Nickel

Nickel, like palladium, has been used to convert cyclohexanones to phenols[328] (equation 133). The mechanism of catalytic dehydrogenation is discussed in section XIII.E.

$$(133)$$

Addition of a solution of sodium hypochlorite to a solution of a nickel(II) salt yields a black precipitate of nickel peroxide, which after washing and drying can be used in a variety of oxidation reactions.

Aromatic and aliphatic aldehydes are oxidized to the carboxylic acids by nickel peroxide in alkaline solution[329]. Under the alkaline conditions, aliphatic aldehydes with an enolizable α-hydrogen sometime yield aldol by-products. Sorbaldehyde (**120**) and crotonaldehyde were converted to the corresponding acid in good yield by this method[330] (equation 134). Aldehydes such as benzaldehyde and furfural were also smoothly oxidized[329].

$$CH_3CH{=}CHCH{=}CHCHO \xrightarrow{NiO_2} CH_3CH{=}CHCH{=}CHCO_2H \quad (134)$$
$$\textbf{(120)}$$

Nickel peroxide cleaves α-keto alcohols to carboxylic acids and α-keto acids to carboxylic acids and carbon dioxide[331]. Thus acetoin, butyroin and benzoin yielded acetic acid, butyric acid, and benzoic acid, respectively, in good yield (equation 135). Pyruvic acid was oxidized to acetic acid and phenylglyoxalic acid was oxidized to benzoic acid (equation 136).

$$PhCHOHCOPh \xrightarrow[99\%]{NiO_2} 2\,PhCO_2H \quad (135)$$

$$PhCOCO_2H \xrightarrow[98\%]{NiO_2} PhCO_2H + CO_2 \quad (136)$$

D. Ruthenium

Ruthenium tetroxide is similar to osmium tetroxide in its reactions, but is a somewhat more powerful oxidant, and is less toxic, less volatile, and cheaper[332]. Ruthenium tetroxide has been prepared by oxidation of ruthenium chloride with sodium hypobromite[333] and by oxidation of ruthenium sulphate with periodic acid[334]. Rough estimation of the concentration of ruthenium tetroxide can be made by reduction with ethanol to the dioxide which is weighed[335]. Oxidations usually are carried out in carbon tetrachloride at 0°.

Aldehydes are rapidly oxidized to the acids by ruthenium tetroxide[333], in contrast to osmium tetroxide, which normally does not oxidize aldehydes. Thus benzaldehyde and n-heptaldehyde (equation 137) were oxidized to the corresponding acids. Methyl

$$CH_3(CH_2)_5CH{=}O \xrightarrow[30\%]{RuO_4} CH_3(CH_2)_5CO_2H \tag{137}$$

cyclopropyl ketone was unaffected under the reaction conditions. Oxidation of carbohydrate derivatives with ruthenium tetroxide has been investigated[336].

E. Palladium and Platinum

Catalytic oxidation has been the subject of a review[337,338]. Although catalytic oxidation does not have as wide applicability as catalytic hydrogenation, it does offer a control and selectivity not possible with other oxidizing agents.

The apparatus designed for catalytic hydrogenation often serves as a convenient system for catalytic oxidations. Normally, high pressure equipment is not necessary, as the rate-determining step in catalytic oxidation is not dependent on the concentration of oxygen.

Other catalysts besides palladium and platinum have been employed, among them nickel, silver, copper, vanadium oxide, zinc, aluminium, cobalt and manganese[338]. However, Adams catalyst, a hydrated platinum oxide, usually gives higher yields and purer products. Amines and sulphur compounds poison Adams catalyst and must be removed. Catalytic oxidations have been run in water, ethyl acetate, benzene, and dioxan. It is important that both reactant and product be soluble in the inert medium or small drops will tend to coalesce on the surface of the catalyst and stop the reaction.

Aldehydes, such as acetaldehyde, may be oxidized to acids by platinum in the presence of oxygen[339].

Cyclohexanones are converted to phenols[340,341]. Aldoses are readily oxidized to aldonic acids. Thus D-glucose in the presence of alkali gave nearly a quantitative yield of D-gluconic acid[342] (equation 138).

$$
\begin{array}{ccc}
\mathrm{CH{=}O} & & \mathrm{COOH} \\
| & & | \\
\mathrm{H{-}C{-}OH} & & \mathrm{H{-}C{-}OH} \\
| & & | \\
\mathrm{HO{-}C{-}H} & \xrightarrow[\text{NaOH}]{\text{Pt, O}_2} & \mathrm{HO{-}C{-}H} \\
| & & | \\
\mathrm{H{-}C{-}OH} & & \mathrm{H{-}C{-}OH} \\
| & & | \\
\mathrm{H{-}C{-}OH} & & \mathrm{H{-}C{-}OH} \\
| & & | \\
\mathrm{CH_2OH} & & \mathrm{CH_2OH}
\end{array}
\qquad (138)
$$

Corresponding oxidations of the aldehyde group of D-galactose, D-mannose, D-xylose, and L-arabinose have been effected catalytically[343]. Oxidation of fructose occurred almost as readily as oxidation of glucose, the product being 2-keto-D-gluconic acid[344] (equation 139).

$$
\begin{array}{ccc}
\mathrm{CH_2OH} & & \mathrm{COOH} \\
| & & | \\
\mathrm{C{=}O} & & \mathrm{C{=}O} \\
| & & | \\
\mathrm{HO{-}C{-}H} & \xrightarrow{\text{Pt, O}_2} & \mathrm{HO{-}C{-}H} \\
| & & | \\
\mathrm{H{-}C{-}OH} & & \mathrm{H{-}C{-}OH} \\
| & & | \\
\mathrm{H{-}C{-}OH} & & \mathrm{H{-}C{-}OH} \\
| & & | \\
\mathrm{CH_2OH} & & \mathrm{CH_2OH}
\end{array}
\qquad (139)
$$

Similarly, catalytic oxidation of L-sorbose with platinum–carbon gave 2-keto-L-gluconic acid[345] (equation 140).

$$
\begin{array}{ccc}
\mathrm{CH_2OH} & & \mathrm{COOH} \\
| & & | \\
\mathrm{C{=}O} & & \mathrm{C{=}O} \\
| & & | \\
\mathrm{HO{-}C{-}H} & \xrightarrow{\text{Pt–C, O}_2} & \mathrm{HO{-}C{-}H} \\
| & & | \\
\mathrm{H{-}C{-}OH} & & \mathrm{H{-}C{-}OH} \\
| & & | \\
\mathrm{HO{-}C{-}H} & & \mathrm{HO{-}C{-}H} \\
| & & | \\
\mathrm{CH_2OH} & & \mathrm{CH_2OH}
\end{array}
\qquad (140)
$$

Oxidation at the primary hydroxyl adjacent to the keto group is the general rule.

In contrast to permanganate oxidations, it is not necessary to block the other hydroxyl groups when employing catalytic oxidation procedures of this type.

Wieland suggested that the catalyst acts like a dehydrogenation enzyme, and he found that catalytic oxidations could be carried out in the absence of oxygen, using methylene blue as a hydrogen acceptor[346]. Further evidence supporting this theory is to be found in the work of Rottenberg and coworkers on the oxidation of acetaldehyde to acetic acid[339]. When oxidation was run using oxygen containing ^{18}O, only slight incorporation of the label into the acetic acid was observed. On the other hand, if the reaction was run in water containing ^{18}O, using unlabelled molecular oxygen, the acetic acid was found to contain most of the activity. This evidence suggests that the aldehyde is oxidatively dehydrogenated through its hydrate (equation 141).

$$CH_3CH{=}O + H_2O \rightleftharpoons CH_3\overset{\overset{\displaystyle H}{|}}{\underset{\underset{\displaystyle OH}{|}}{C}}{-}OH \xrightarrow{Pt} CH_3COOH \qquad (141)$$

XIV. MICROBIOLOGICAL OXIDATION

Microbiological oxidation has been the subject of a review[347].

It was known as far back as 1919 that bacteria could convert vanillin to vanillic acid[348]. However, only since the importance of steroid derivatives in drug manufacture was realized has microbiological oxidation come into its own as a synthetic method. One of the attractive features of a biochemical method is that there is the possibility that it will take a 'chemically' unlikely course[349]. However, the synthetic procedure suffers from the disadvantage of the large dilutions which have to be employed. Also, at the present state of the art, the method is neither general nor predictable. Finally, the yields are very variable[350-352].

This method has been used to introduce a hydroxyl group in the steroid molecule. Reichstein's substance S (**121**) was oxidized to the 2β-hydroxy isomer by *Rhizoctonia ferrugena*[353] (equation 142).

$$(142)$$

(**121**)

Although oxidation often occurs at a position α to the ketone, it occurs sometimes at a position not activated in the usual sense. Thus, incubation of 5α-androstane-6,17-dione with *Calonectria decora* gave the 1β-hydroxyl derivative in 56% yield[354] (equation 143).

$$(143)$$

Studies on the mechanism of steroid hydroxylation in biological systems indicate that the incorporated oxygen comes from molecular oxygen and not from water[355,356].

Dehydrogenation of steroid molecules has also been accomplished by microorganisms. *Bacillus sphaericus* and *Arthrobacter simplex* have been used to introduce Δ^1-double bonds into Δ^4-3-ketones[357] or $\Delta^{4,6}$-3-ketones[158]. Several ring enlargements have been reported. An unusual Baeyer–Villiger type of oxidation was carried out on progesterone by the action of *Aspergillus orizae*, *Aspergillus chevalieri*, or *Penicillium citrium*[358] (equation 144). A D-homosteroid was produced by the in-

$$(144)$$

cubation of 17α-progesterone with *Aspergillus niger*[359] (equation 145).

$$(145)$$

A similar reaction occurred on incubation of triamcinolone with *Streptomyces roseochromogenus*[360] (equation 146). There is some doubt

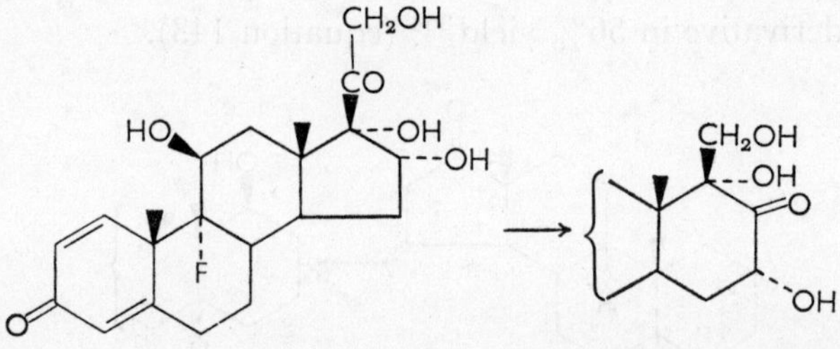

(146)

whether the biological system causes these D-homoisomerizations, or whether the ring expansions are due to metal ion impurities in the reaction medium[360].

Cleavage of the steroid ring system with accompanying aromatization has been observed. Fermentation of 17α-methyl-B-nortestosterone with *Protaminobacter ruber* produced 9α-hydroxy-17α-methyl-B-nortestosterone and two 9,10-secophenols[361] (**122** and **123**) (equation 147).

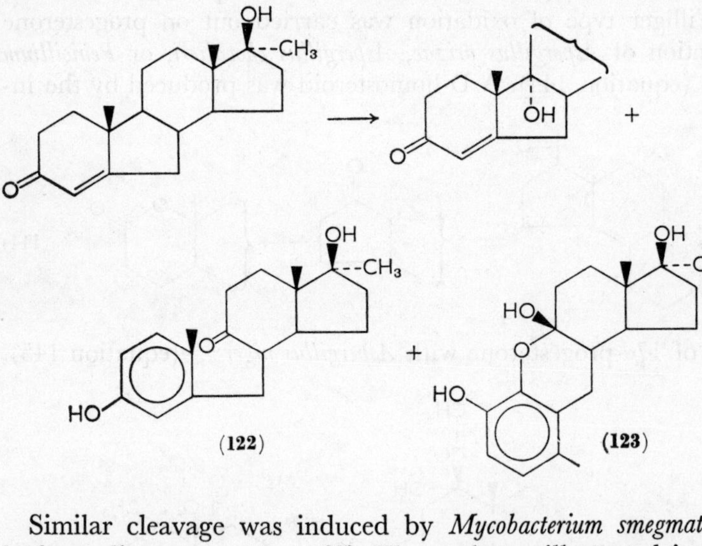

(147)

(**122**)   (**123**)

Similar cleavage was induced by *Mycobacterium smegmatis* in 3β-hydroxyallopregnan-17-one[362]. The pathway illustrated in equation (148) was proposed for the transformation[363,364]. The entire side chain of ring D was cleaved from 19-hydroxy-Δ^4-cholesten-3-one by *Proactinomyces erythropolis* to yield estrone[365].

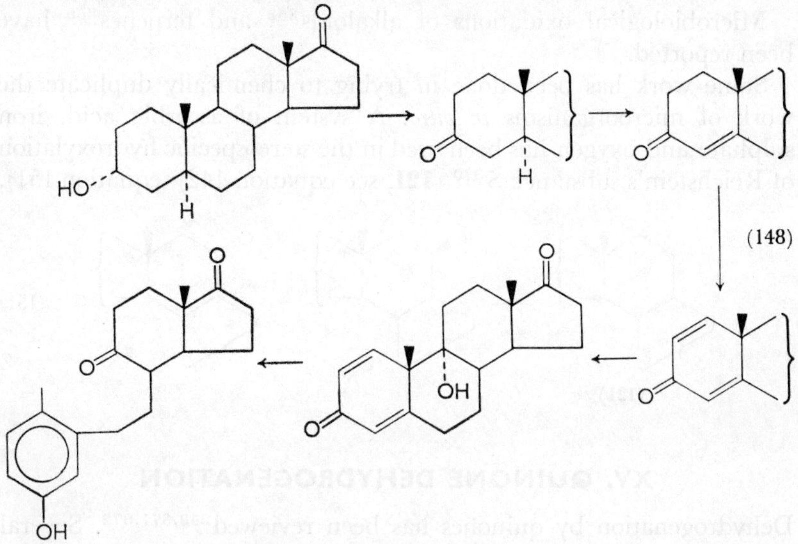

(148)

Dehydrogenation and oxidation of the side chain of a sapogenin diosgenone to a 16-ketosteroid by *Fusarium solani* has been reported[366] (equation 149).

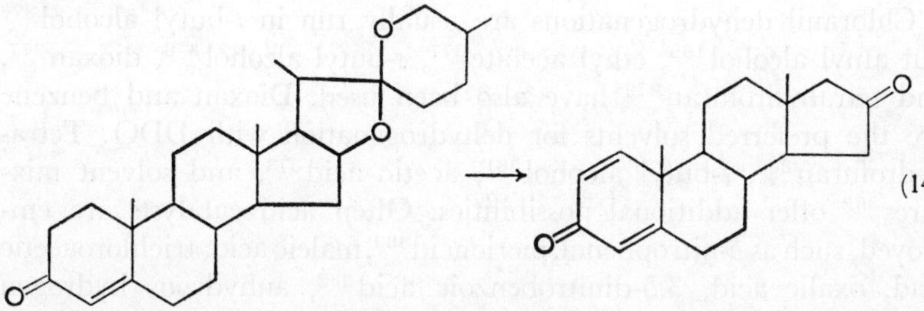

(149)

Corticosterone undergoes an interesting oxidative dimerization when incubated with *Corynespora cassiicola*[367] (equation 150).

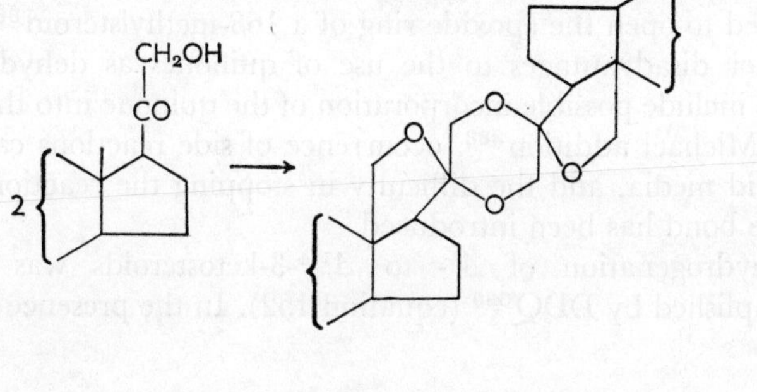

(150)

Microbiological oxidations of alkaloids[368] and terpenes[369] have been reported.

Some work has been done in trying to chemically duplicate the work of microorganisms *in vitro*. A system of ascorbic acid, iron sulphate and oxygen has been used in the stereospecific hydroxylation of Reichstein's substance S[370] (**121**, see equation 142) (equation 151).

$$(151)$$

(**121**)

XV. QUINONE DEHYDROGENATION

Dehydrogenation by quinones has been reviewed[349,371,372]. Several quinones have been used in the dehydrogenation of ketones, such as tetrachloro-*p*-benzoquinone (chloranil), 2,3-dichloro-5,6-dicyanobenzoquinone (DDQ), 2,3-dicyanobenzoquinone[373], 2,3-dibromo-5,6-dicyanobenzoquinone[374], and tetrachloro-*o*-benzoquinone[375].

Chloranil dehydrogenations are usually run in *t*-butyl alcohol[376] but amyl alcohol[155], ethyl acetate[377], *s*-butyl alcohol[378], dioxan[377], and tetrahydrofuran[379] have also been used. Dioxan and benzene are the preferred solvents for dehydrogenation with DDQ. Tetrahydrofuran[380], *t*-butyl alcohol[381], acetic acid[373], and solvent mixtures[382] offer additional possibilities. Often acid catalysts are employed, such as *p*-nitrophenol, picric acid[383], maleic acid, trichloroacetic acid, oxalic acid, 3,5-dinitrobenzoic acid[382], anhydrous hydrogen chloride, and concentrated sulphuric acid[384]. The relative value of various quinones in dehydration reactions has been discussed[349,385]. Chloranil is a milder reagent than DDQ; however, it has been known to undergo a Diels–Alder reaction with the substrate[386] and has been reported to open the epoxide ring of a 16β-methylsteroid[387].

Other disadvantages to the use of quinones as dehydrogenating agents include possible incorporation of the quinone into the molecule via a Michael addition[388], occurrence of side reactions catalysed by the acid media, and the difficulty in stopping the reaction after one double bond has been introduced.

Dehydrogenation of Δ^4- to $\Delta^{1,4}$-3-ketosteroids was effectively accomplished by DDQ[389] (equation 152). In the presence of a strong

$$\text{(152)}$$

acid $\Delta^{4,6}$-3-ketones were formed[390] (equation 153). Saturated

$$\text{(153)}$$

3-ketosteroids can be dehydrogenated to Δ^{1}-, Δ^{4}-, or $\Delta^{1,4}$-3-ketones by DDQ. The stereochemistry of the starting material often determines which hydrogen atoms are removed. Steroidal 3-ketones bearing a 5α-hydrogen atom generally gave Δ^{1}-3-ketones as the major product (equation 154), while a 5β-hydrogen usually yielded Δ^{4}-3-ketones[154,391] (equation 155).

$$\text{(154)}$$

$$\text{(155)}$$

Dehydrogenation of 2-formyl-3-ketosteroids in dioxan yielded 2-formyl-Δ^{1}-3-ketones[392–394] (equation 156). Loss of an angular

$$\text{(156)}$$

methyl group with accompanying aromatization of the A ring has been reported[395] (equation 157).

Chloranil appears to be the reagent of choice for the conversion of Δ^{4}-3-ketones to $\Delta^{4,6}$-3-ketones[396]. Forcing conditions were required for the chloranil dehydrogenation of $\Delta^{4,6}$-3-ketones to $\Delta^{1,4,6}$-3-ketones[397]

however, DDQ smoothly effected this type of dehydrogenation[398]
(equation 158). On the other hand $\Delta^{1,4}$-3-ketones were more easily

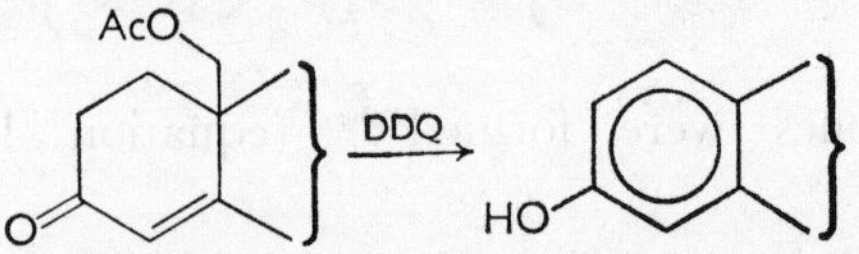

(157)

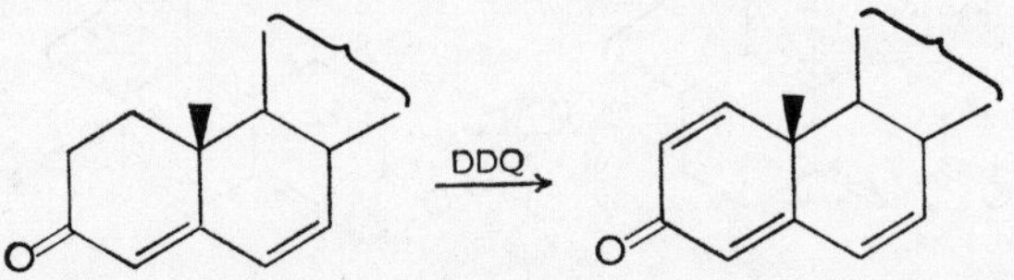

(158)

converted to $\Delta^{1,4,6}$-3-ketones with chloranil[399] than with DDQ[400]
(equation 159).

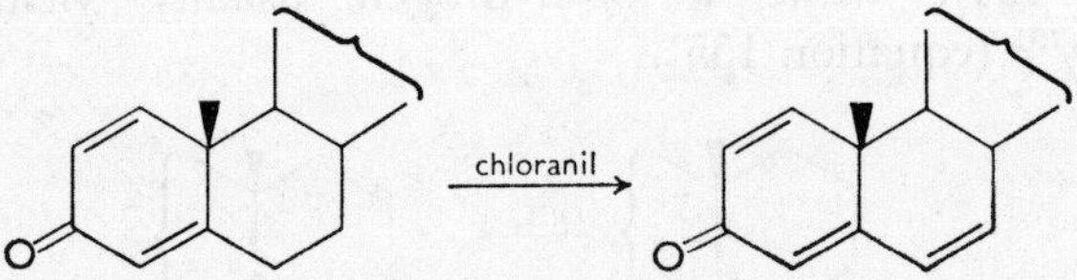

(159)

Steroidal dehydrogenation of Δ^4-3-ketones carrying a substituent
on $C_{(6)}$ to $\Delta^{4,6}$-3-ketones occurred most readily if the substituent
were α[401] (equation 160).

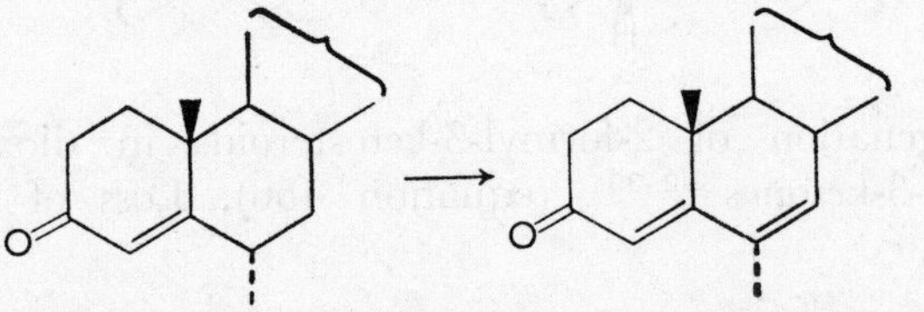

(160)

Molecules other than steroids such as polyhydronapthalene[402]
and cycloheptadiene[153] derivatives have been dehydrogenated by
quinones. A number of interesting oxidative dimerizations effected
by DDQ have been uncovered[403]. Thus 2-arylindane-1,3-diones
were readily converted into 2,2'-diaryl-2,2'-biindane-1,1',3,3'-tetrones
(equation 161). The dimerization is believed to proceed by a one-
electron transfer from the enolate anion to the quinone to yield as an
intermediate the resonance stabilized radical **124**.

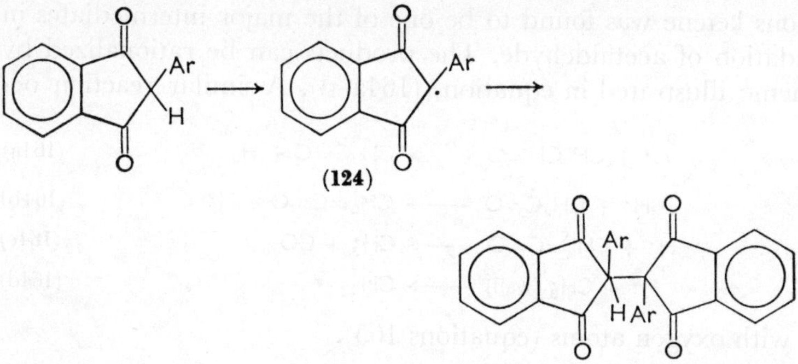

(124)

(161)

The normal course of dehydrogenation of a substrate SH_2 may be rationalized on the basis of a two-step heterolytic mechanism[383,404] (equations 162).

$$SH_2 + Q \xrightarrow{slow} SH^+ + QH^-$$ (162a)

$$SH^+ + QH^- \longrightarrow S + QH_2$$ (162b)

The possibility of hydride transfer via the intermediacy of an addition product such as **125** has been suggested[405] (equation 163).

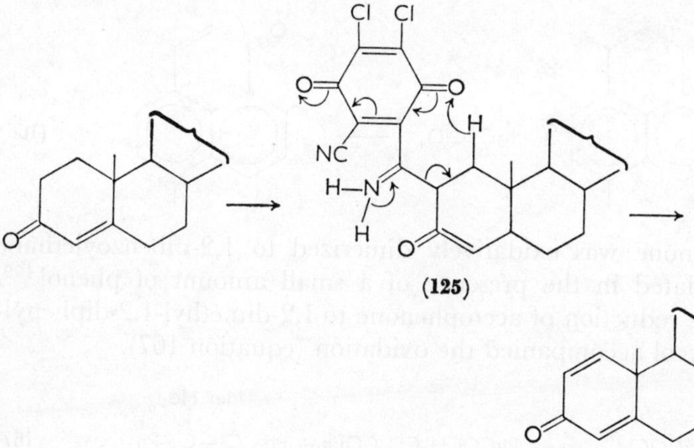

(125)

(163)

XVI. MISCELLANEOUS

Hydrogen has recently been shown to be an oxidizing agent[406]. A stream of hydrogen atoms produced by a microwave discharge was mixed at a nozzle with a gas stream of the compound to be studied. The products were analysed with a mass spectrometer. Under these conditions ketene was found to be one of the major intermediates in the oxidation of acetaldehyde. The products can be rationalized by the scheme illustrated in equations (164a–d). A similar reaction oc-

$$H\cdot + CH_3CH{=}O \xrightarrow{\text{slow}} CH_3\dot{C}{=}O + H_2 \tag{164a}$$

$$H\cdot + CH_3\dot{C}{=}O \longrightarrow CH_2{=}C{=}O + H_2 \tag{164b}$$

$$H\cdot + CH_2{=}C{=}O \longrightarrow CH_3^{\cdot} + CO \tag{164c}$$

$$H\cdot + CH_3^{\cdot} \text{ (wall)} \longrightarrow CH_4 \tag{164d}$$

curred with oxygen atoms (equations 165).

$$O\cdot + CH_3CH{=}O \longrightarrow HO\cdot + CH_3\dot{C}{=}O \tag{165a}$$

$$O\cdot + \dot{C}H_3C{=}O \longrightarrow HO\cdot + CH_2{=}C{=}O \tag{165b}$$

Carbonium ions have also been used as oxidants. Thus 3,6,9-trimethylperinaphthan-1-one was dehydrogenated with triphenylmethyl carbonium perchlorate to 3,6,9-trimethylperinaphthen-1-one[407,408] (equation 166). Carbonium ion dehydrogenation was reported to be the best method for converting perinaphthanones into perinaphthenones.

$$+ Ph_3C^+ClO_4^- \longrightarrow \tag{166}$$

Acetophenone was oxidatively dimerized to 1,2-dibenzoylethane when irradiated in the presence of a small amount of phenol[409]. Bimolecular reduction of acetophenone to 1,2-dimethyl-1,2-diphenylethylene glycol accompanied the oxidation (equation 167).

$$PhCOCH_3 \xrightarrow[\text{phenol}]{h\nu} PhCOCH_2CH_2COPh + Ph{-}\underset{\underset{OH}{|}}{\overset{\overset{Me}{|}}{C}}{-}\underset{\underset{OH}{|}}{\overset{\overset{Me}{|}}{C}}{-}Ph \tag{167}$$

Electrolytic oxidation has not so far found as wide applicability as electrolytic reduction. However, it has been successfully employed in the oxidation of D-glucose to D-gluconic acid[410] in the presence of a catalytic amount of bromide ion. Bromine probably was the actual oxidant which was continually regenerated.

XVII. ACKNOWLEDGEMENTS

I wish to thank my wife, Mona, and Mrs. Alberta Boyko for their help in the preparation of the manuscript. I also wish to thank Dr. Owen H. Wheeler for his suggestions and encouragement.

XVIII. REFERENCES

1. (Ed. S. Patai) *The Chemistry of the Carbonyl Group*, Vol. 1, Interscience, London, 1966.
2. A. G. Sykes, *J. Chem. Soc.*, **1961**, 5549.
3. K. B. Wiberg and W. Koch, *Tetrahedron Letters*, **1966**, 1779.
4. J. S. Littler, *J. Chem. Soc.*, **1962**, 827.
5. G. B. Payne and C. W. Smith, *J. Org. Chem.*, **22**, 1680 (1957).
6. S. M. McElvain and E. J. Eisenbraun, *J. Amer Chem. Soc.*, **77**, 1599 (1955).
7. R. Criegee in *Newer Methods of Preparative Organic Chemistry*, Vol. 1, Interscience, New York, 1948, p. 1.
8. R. Criegee in *Newer Methods of Preparative Organic Chemistry*, Vol. 2 (Ed. W. Foerst), Academic Press, New York, 1963, p. 367.
9. R. Criegee, *Angew. Chem.*, **70**, 173 (1958).
10. A. S. Perlin in *Advances in Carbohydrate Chemistry*, Vol. 14 (Ed. M. L. Wolfrom), Academic Press, New York, 1959, p. 9.
11. Y. Gaoni and E. Wenkert, *J. Org. Chem.*, **31**, 3809 (1966).
12. H. B. Henbest, D. N. Jones, and G. P. Slater, *J. Chem. Soc.*, **1961**, 4472.
13. J. D. Cocker, H. B. Henbest, G. H. Phillips, G. P. Slater, and D. A. Thomas, *J. Chem. Soc.*, **1965**, 6.
14. H. B. Henbest, D. N. Jones, and G. P. Slater, *J. Chem. Soc. (C)*, **1967**, 756.
15. G. W. K. Cavill and D. H. Solomon, *J. Chem. Soc.*, **1955**, 4426.
16. E. Baer, *J. Amer. Chem. Soc.*, **62**, 1597 (1940); **64**, 1416 (1942).
17. L. Vargha, *Nature*, **162**, 927 (1948).
18. L. Vargha and M. Remenyi, *J. Chem. Soc.*, **1951**, 1068.
19. R. E. Partch, *J. Amer. Chem. Soc.*, **89**, 3662 (1967).
20. F. H. Pollard and R. M. H. Wyatt, *Trans. Faraday Soc.*, **45**, 760 (1949).
21. C. Moureu and R. Chaux in *Organic Syntheses*, Coll. Vol. 1 (Eds. H. Gilman and A. H. Blatt), John Wiley and Sons, New York, 1941, p. 166.
22. J. English Jr. and J. E. Dayan in *Organic Syntheses*, Coll. Vol. 4 (Ed. N. Rabjohn), John Wiley and Sons, New York, 1963, p. 499.
23. C. F. Koelsch in *Organic Syntheses*, Coll. Vol. 3 (Ed. E. C. Horning), John Wiley and Sons, New York, 1955, p. 791.
24. A. Hrubesch and O. Schlichting, *Ger. Pat.* 887, 943 (1953).
25. B. A. Ellis in *Organic Syntheses*, Coll. Vol. 1 (Eds. H. Gilman and A. H. Blatt), John Wiley and Sons, New York, 1941, p. 18.
26. Y. Ogata, H. Tezuka, and Y. Sawaki, *Tetrahedron*, **23**, 1007 (1967).
27. W. Rigby, *J. Chem. Soc.*, **1951**, 793.
28. B. Holden and W. Rigby, *J. Chem. Soc.*, **1951**, 1924.
29. W. Rigby, *Nature*, **164**, 185 (1949).
30. W. Rigby, *J. Chem. Soc.*, **1950**, 1903.
31. C. J. W. Brooks and J. K. Norymberski, *Chem. Ind. (London)*, **1952**, 804.

32. J. K. Norymberski, *Nature*, **170**, 1074 (1952).
33. C. J. W. Brooks and J. K. Norymberski, *Biochem. J.* (*London*), **55**, 371 (1953).
34. R. L. Livingston in *Autooxidation and Autooxidants* (Ed. W. O. Lundberg), Interscience, New York, 1961, p. 249.
35. G. O. Schenck, *Angew. Chem.*, **69**, 579 (1957).
36. G. O. Schenck, *Z. Electrochem.*, **64**, 170 (1960).
37. L. Bateman, *Quart. Rev.* (*London*), **8**, 147 (1954).
38. C. E. Frank, *Chem. Rev.*, **46**, 155 (1950).
39. G. H. Twigg, *Chem. Ind.* (*London*), **1962**, 4.
40. G. F. Cullis, *Chem. Ind.* (*London*), **1962**, 23.
41. G. J. Minkoff and C. F. H. Tipper, *Chemistry of Combustion Reactions*, Butterworths, London, 1962.
42. B. V. Suvorov, S. R. Rafikov, and A. D. Kagarlitskii, *Russ. Chem. Rev.* (*Engl. Transl.*), **34**, 657 (1965).
43. K. U. Ingold, *Chem. Rev.*, **61**, 563 (1961).
44. J. W. Wilt, L. L. Maravetz, and J. F. Zawadski, *J. Org. Chem.*, **31**, 3018 (1966).
45. H. L. J. Bäckström, *Z. Physik. Chem.*, **B25**, 99 (1934).
46. C. E. H. Bawn and J. E. Jolley, *Proc. Roy. Soc.* (*London*), **A237**, 297 (1956).
47. H. R. Cooper and H. W. Melville, *J. Chem. Soc.*, **1951**, 1984, 1994.
48. T. A. Ingles and H. W. Melville, *Proc. Roy. Soc.* (*London*), **A218**, 163 (1953).
49. A. Robertson and W. A. Waters, *Trans. Faraday Soc.*, **42**, 201 (1946).
50. W. A. Waters, *Trans. Faraday Soc.*, **42**, 184 (1946).
51. A. M. Markevich and L. F. Filippova, *Zh. Fiz. Khim.*, **33**, 2214 (1959).
52. A. A. Anisonyan, S. Ya. Beider, A. M. Markevich, and A. B. Nalbandyan, *Zh. Fiz. Khim.*, **33**, 1695 (1959).
53. D. E. Cheaney, D. A. Davies, A. Davis, D. E. Hoare, J. Protheroe, and A. D. Walsh, *7 Intern. Symp. Combustion*, Butterworths, London, 1959, p. 183.
54. A. M. Markevich and Yu. I. Pecherskaya, *Zh. Fiz. Khim.*, **35**, 1418 (1961).
55. J. M. Hay, *J. Chem. Soc.*, **1965**, 7388.
56. B. Phillips, F. C. Frostick Jr., and P. S. Starcher, *J. Amer. Chem. Soc.*, **79**, 5982 (1957).
57. P. S. Starcher, B. Phillips, and F. C. Frostick Jr., *J. Org. Chem.*, **26**, 3568 (1961).
58. M. J. Kagan and G. D. Lubarsky, *J. Phys. Chem.*, **39**, 837 (1935).
59. A. Elce, H. M. Stanley, and K. H. W. Tuerck, *Brit. Pat.* 635,054 (1950).
60. M. B. Neiman, V. Y. Efremov, and N. K. Serdyuk, *Kinetika i Kataliz*, **1**, 345 (1960).
61. N. A. Sokoleva, A. M. Markevich, and A. B. Nalbandyan, *Zh. Fiz. Khim.*, **35**, 850 (1961).
62. H. S. Johnston and J. Heicklen, *J. Amer. Chem. Soc.*, **86**, 4249 (1964).
63. P. Thüring and A. Perret, *Helv. Chim. Acta*, **36**, 13 (1953).
64. S. Winstein and F. H. Seubold, *J. Amer. Chem. Soc.*, **69**, 2916 (1947).
65. E. F. P. Harris and W. A. Waters, *J. Chem. Soc.*, **1952**, 3108.

66. A. G. Pinkus, W. C. Servoss and K. K. Lum, *J. Org. Chem.*, **32**, 2649 (1967).
67. W. H. Richardson and R. F. Steed, *J. Org. Chem.*, **32**, 771 (1967).
68. D. B. Sharp, L. W. Patton, and S. E. Whitcomb, *J. Amer. Chem. Soc.* **73**, 5600 (1951).
69. E. J. Bailey, J. Elks, and D. H. R. Barton, *Proc. Chem. Soc. (London)*, **1960**, 214.
70. W. von E. Doering and R. M. Haines, *J. Amer. Chem. Soc.*, **76**, 482 (1954).
71. J. F. Biellmann and M. Rajic, *Bull. Soc. Chim. France*, **1962**, 441.
72. T. J. Wallace, H. Pobiner, and A. Schriesheim, *J. Org. Chem.*, **30**, 3768 (1965).
73. R. Howe and F. J. McQuillin, *J. Chem. Soc.*, **1958**, 1513.
74. I. G. Tishchenko and L. S. Stanishevskii, *Zh. Obshch. Khim.*, **33**, 3751 (1963).
75. K. Crowshaw, R. C. Newstead, and N. A. J. Rogers, *Tetrahedron Letters*, **1964**, 2307.
76. E. L. Shapiro, T. Legatt and E. P. Oliveto, *Tetrahedron Letters*, **1964**, 663.
77. L. F. Fieser, T. W. Greene, F. Bischoff, G. Lopez, and J. J. Rupp, *J. Amer. Chem. Soc.*, **77**, 3928 (1955).
78. D. E. Hoare and D. A. Whytock, *Can. J. Chem.*, **45**, 865 (1967).
79. L. Michaelis and E. S. Fetcher Jr., *J. Amer. Chem. Soc.*, **59**, 1246 (1937).
80. E. Thomas Strom, G. A. Russell, and J. H. Schoeb, *J. Amer. Chem. Soc.*, **88**, 2004 (1966).
81. N. L. Bauld, *J. Amer. Chem. Soc.*, **87**, 4788 (1965).
82. W. G. Dauben, G. A. Boswell, and W. Templeton, *J. Org. Chem.*, **25**, 1853 (1960).
83. A. E. Chichibabin and D. I. Orochko, *Zh. Russ. Fiz.-Khim. Obshch.*, **62**, 1201 (1930).
84. Phillips Petroleum Co., *Brit. Pat.* 609,059 (1948).
85. S. R. Rafikov, D. Kh. Sembaev, and B. V. Suvorov, *Zh. Obshch. Khim.*, **32**, 839 (1962).
86. A. D. Kagarlitskii, B. V. Suvorov, and S. R. Rafikov, *Tr. Inst. Khim. Nauk, Acad. Nauk Kazakh. SSR*, **7**, 57 (1961).
87. D. Kh. Sembaev, B. V. Suvorov, and S. R. Rafikov, in collected papers, *Sintez i Svoistva Monomerov (Synthesis and Properties of Monomers)*, Izd. "Nauka," 1964, p. 276.
88. Union chemique belge S. A., *Belgian Pat.* 580,361 (1960).
89. D. J. Hadley, *Chem. Ind. (London)*, **1961**, 238.
90. B. V. Suvorov, S. R. Rafikov, and M. I. Khmura, *Izv. Akad. Nauk Kazakh. SSR., Ser. Khim.*, **1959**, 80.
91. A. D. Kagarlitskii, B. V. Suvorov, and S. R. Rafikov, *Zh. Obshch. Khim.*, **29**, 157 (1959).
92. P. A. Leermakers, G. F. Vesley, N. J. Turro, and D. C. Neckers, *J. Amer. Chem. Soc.*, **86**, 4213 (1964).
93. N. J. Turro, P. A. Leermakers, H. R. Wilson, D. C. Neckers, G. W. Byers and G. F. Vesley, *J. Amer. Chem. Soc.*, **87**, 2613 (1965).
94. W. B. Hammond and N. J. Turro, *J. Amer. Chem. Soc.*, **88**, 2880 (1966).
95. G. Ciamician and P. Silber, *Ber.*, **41**, 1928 (1908).
96. E. Sernagiotto, *Gazz. Chim. Ital.*, **48**, 52 (1918); **47**, 153 (1917).

97. M. E. Wolff and S.-Y. Cheng, *Tetrahedron Letters*, **1966**, 2507.
98. E. J. Forbes and J. Griffiths, *Chem. Commun.*, **1967**, 427.
99. C. F. Wilcox Jr. and M. P. Stevens, *J. Amer. Chem. Soc.*, **84**, 1258 (1962).
100. G. O. Schenck, *Ann. Chem.*, **584**, 156 (1953).
101. A. P. Dunlop, P. R. Stout, and S. Swadesh, *Ind. Eng. Chem.*, **38**, 705 (1946).
102. A. P. Dunlop, *Ind. Eng. Chem.*, **40**, 204 (1948).
103. E. J. Forbes and J. Griffiths, *Chem. Commun.*, **1966**, 896.
104. E. McKeown and W. A. Waters, *J. Chem. Soc.* (*B*), **1966**, 1040.
105. R. H. Young and H. Hart, *Chem. Commun.*, **1967**, 827.
106. C. S. Foote and W. Wexler, *J. Amer. Chem. Soc.*, **86**, 3879 (1964).
107. E. J. Corey and W. C. Taylor, *J. Amer. Chem. Soc.*, **86**, 3881 (1964).
108. P. S. Bailey, *Chem. Rev.*, **58**, 825 (1958).
109. H. M. White and P. S. Bailey, *J. Org. Chem.*, **30**, 3037 (1965).
110. F. G. Fischer, H. Düll, and J. L. Volz, *Ann. Chem.*, **486**, 80 (1931).
111. R. M. Dorland and H. Hibbert, *Can. J. Res.*, **18B**, 33 (1940).
112. C. R. Dick and R. F. Hanna, *J. Org. Chem.*, **29**, 1218 (1964).
113. Henkel and Cie., G. m. b. H., *Brit. Pat.* 534,525 (1941).
114. M. Stoll and W. Scherrer, *Helv. Chim. Acta*, **13**, 142 (1930).
115. R. E. Erickson, D. Bakalik, C. Richards, M. Scanlon, and G. Huddleston, *J. Org. Chem.*, **31**, 461 (1966).
116. E. Briner and H. Biedermann, *Helv. Chim. Acta*, **15**, 1227 (1932).
117. J. E. Leffler, *Chem. Rev.* **45**, 399 (1949).
118. K. Maruyama, *Bull. Chem. Soc. Japan*, **33**, 1516 (1960); **34**, 102, 105 (1961).
119. N. C. Yang and R. A. Finnegan, *J. Amer. Chem. Soc.*, **80**, 5845 (1958).
120. G. B. Payne, *J. Org. Chem.*, **25**, 275 (1960).
121. G. B. Payne, *J. Amer. Chem. Soc.*, **81**, 4901 (1959).
122. W. A. Waters in *Progress in Organic Chemistry*, Vol. 5 (Ed. J. W. Cook), Butterworths, Washington, 1961, p. 1.
123. P. Gray and A. Williams, *Chem. Rev.*, **59**, 239 (1959).
124. F. H. Seubold, *J. Amer. Chem. Soc.*, **75**, 2532 (1953).
125. F. F. Rust, F. H. Seubold, and W. E. Vaughan, *J. Amer. Chem. Soc.*, **70**, 3258 (1948).
126. M. S. Kharasch, D. Schwartz, M. Zimmermann, and W. Nudenberg, *J. Org. Chem.*, **18**, 1051 (1953).
127. B. S. Middleton and K. U. Ingold, *Can. J. Chem.*, **45**, 191 (1967).
128. H. Kwart, P. S. Starcher, and S. W. Tinsley, *Chem. Commun.* **1967**, 335.
129. R. R. Sauers and R. W. Ubersax, *J. Org. Chem.*, **30**, 3939 (1965).
130. A.v. Wacek and A.v. Bézard, *Ber.*, **74B**, 857 (1941).
131. J. D' Ans and A. Kneip, *Ber.*, **48**, 1143 (1915).
132. M. Gorodetsky, N. Danieli, and Y. Mazur, *J. Org. Chem.*, **32**, 760 (1967).
133. G. Sosnovsky and N. C. Yang, *J. Org. Chem.*, **25**, 899 (1960).
134. M. S. Kharasch, H. C. McBay, and W. H. Urry, *J. Amer. Chem. Soc.*, **70**, 1269 (1948).
135. T. Kuwada, *Japan. Pat.* 3,730 (1951).
136. I. A. Pearl in *Organic Syntheses*, Coll. Vol. 4 (Ed. N. Rabjohn), John Wiley and Sons, New York, 1963, p. 974.
137. R. A. Dytham and B. C. L. Weedon, *Tetrahedron*, **8**, 246 (1960).

138. M. M. Shemyakin and L. A. Shchukina, *Quart. Rev. (London)*, **10**, 261 (1956).
139. L. Ruzicka, *Helv. Chim. Acta.*, **19**, 419 (1936).
140. F. P. Greenspan and H. W. Woodburn, *J. Amer. Chem. Soc.*, **76**, 6345 (1954).
141. R. G. R. Bacon and R. W. Bott, *Chem. Ind. (London)*, **1953**, 897, 1285.
142. D. A. House, *Chem. Rev.*, **62**, 185 (1962).
143. E. Bekier and W. Kijowski, *Roczniki Chem.*, **15**, 136 (1935).
144. K. C. Khulbe and S. P. Srivastava, *Agra Univ. J. Res. (Sci.)*, **9**, 177 (1960).
145. K. Onodera, S. Hirano, and N. Kashimura, *J. Amer. Chem. Soc.*, **87**, 4651 (1965).
146. M. VanDyke and N. D. Pritchard, *J. Org. Chem.*, **32**, 3204 (1967).
147. F. W. Sweat and W. W. Epstein, *J. Org. Chem.*, **32**, 835 (1967).
148. W. Treibs and R. Vogt, *Ber.*, **94**, 1739 (1961).
149. N. Rabjohn in *Organic Reactions*, Vol. 5 (Ed. R. Adams), John Wiley and Sons, New York, 1949, p. 331.
150. E. J. Corey and H. J. Burke, *J. Amer. Chem. Soc.*, **76**, 5257 (1954).
151. E. J. Corey, H. J. Burke, and W. A. Remers, *J. Amer. Chem. Soc.*, **78**, 180 (1956).
152. J. P. Schaefer, *J. Amer. Chem. Soc.*, **84**, 713 (1962).
153. E. E. van Tamelen and G. T. Hildahl, *J. Amer. Chem. Soc.*, **78**, 4405 (1956).
154. Organon, N. V., *Belgian Pat.* 612,592 (1962).
155. Upjohn Co., *Brit. Pat.* 997,167 (1965).
156. A. Bowers and J. A. Edwards, *U.S. Pat.* 3,036,098 (1962).
157. L. Canonica, G. Jommi, F. Pelizzoni, and C. Scolastico, *Gazz. Chim. Ital.*, **95**, 138 (1965).
158. E. Merck A.-G., *Netherlands Pat. Appl.* 295,201 (1965).
159. R. Owyang in *Steroid Reactions* (Ed. C. Djerassi), Holden-Day, San Francisco, 1963, p. 227.
160. M. Amorosa, L. Caglioti, G. Cainelli, H. Immer, J. Keller, H. Wehrli, M. Lj. Mihailovic, K. Schaffner, D. Arigoni, and O. Jeger, *Helv. Chim. Acta*, **45**, 2674 (1962).
161. G. Jommi, P. Manitto, and C. Scolastico, *Gazz. Chim. Ital.* **95**, 138 (1965).
162. J. A. Edwards, H. J. Ringold, and C. Djerassi, *J. Amer. Chem. Soc.*, **82**, 2318 (1960).
163. P. Rona, *J. Chem. Soc.*, **1962**, 3629.
164. E. H. Reerink, P. Westerhof, and H. F. L. Schoeler, *U.S. Pat.* 3,198,792 (1965).
165. K. Florey and A. R. Restivo, *J. Org. Chem.*, **22**, 406 (1957).
166. A. Schubert and S. Schwarz, *Experientia*, **21**, 562 (1965).
167. J. S. Baran, *J. Amer. Chem. Soc.*, **80**, 1687 (1958).
168. C. S. Barnes and D. H. R. Barton, *J. Chem. Soc.*, **1953**, 1419.
169. J. C. Banerji, D. H. R. Barton, and R. C. Cookson, *J. Chem. Soc.*, **1957**, 5041.
170. R. A. Jerussi and D. Speyer, *J. Org. Chem.*, **31**, 3199 (1966).
171. G. Langbein, *J. Prakt. Chem.*, **18**, 244 (1962).
172. C. W. Smith and R. T. Holm, *J. Org. Chem.*, **22**, 746 (1957).
173. H. M. Hellman and R. A. Jerussi, *Tetrahedron*, **20**, 741 (1964).
174. E. Caspi and S. N. Balasubrahmanyam, *Experientia*, **19**, 396 (1963).

175. E. Caspi and S. N. Balasubrahmanyam, *J. Org. Chem.*, **28**, 3383 (1963).
176. I. Barker, *Chem. Ind. (London)*, **1964**, 1936.
177. M. H. Hashmi, *Pakistan J. Sci.*, **10**, 159 (1958).
178. J. W. Green in *Advances in Carbohydrate Chemistry*, Vol. 3 (Eds. W. W. Pigman and M. L. Wolfrom), Academic Press, New York, 1948, p. 129.
179. L. Farkas, B. Perlmutter and O. Schächter, *J. Amer. Chem. Soc.*, **71**, 2827,2829,2833 (1949).
180. P. T. McTigue and J. M. Sime, *J. Chem. Soc.*, **1963**, 1303.
181. G. Romijn, *Z. Anal. Chem.*, **36**, 18 (1897).
182. C. F. Cullis and P. A. Swain, *J. Chem. Soc.*, **1962**, 3348; **1963**, 1139.
183. J. Colonge and L. Pichat, *Bull. Soc. Chim. France*. **1949**, 180.
184. L. T. Sandborn and E. W. Bousquet in *Organic Syntheses*, Coll. Vol. 1 (Eds. H. Gilman and A. H. Blatt), John Wiley and Sons, New York, 1941, p. 526.
185. E. Hardegger and E. Nikles, *Helv. Chim. Acta*, **40**, 1016, (1957).
186. L. Mamlok, A. Horeau, and J. Jacques, *Bull. Soc. Chim. France*, **1965**, 2359.
187. L. A. Freiberg, *J. Amer. Chem. Soc.*, **89**, 5297 (1967).
188. O. Ruff, *Ber.*, **32**, 2273 (1899).
189. J. P. Hart, F. Sheppard, and M. R. Everett, *Oxidation of Carbohydrates, Keturonic Acids and Salt Catalysis*, University of Oklahoma Medical School, 1944, p. 9.
190. L. Kaplan, *J. Amer. Chem. Soc.*, **80**, 2639 (1958).
191. L. C. Gruen and P. T. McTigue, *J. Chem. Soc.*, **1963**, 5217.
192. B. G. Cox and P. T. McTigue, *Australian J. Chem.*, **17**, 1210 (1964).
193. L. S. Levitt, *J. Org. Chem.*, **20**, 1297 (1955).
194. N. N. Lichtin and M. H. Saxe, *J. Amer. Chem. Soc.*, **77**, 1875 (1955).
195. I. Barker, W. G. Overend, and C. W. Rees, *J. Chem. Soc.*, **1964**, 3254.
196. B. Perlmutter-Hayman and Y. Weissman, *J. Amer. Chem. Soc.*, **82**, 2323 (1962).
197. C. G. Swain, R. A. Wiles, and R. F. W. Bader, *J. Amer. Chem. Soc.*, **83**, 1945 (1961).
198. M. W. Farrar and R. Levine, *J. Amer. Chem. Soc.*, **71**, 1496 (1949).
199. R. Levine and J. R. Stephens, *J. Amer. Chem. Soc.*, **72**, 1642 (1950).
200. R. Filler, *Chem. Rev.*, **63**, 21 (1963).
201. C. Djerassi, *Chem. Rev.*, **43**, 271 (1948).
202. L. Horner and E. H. Winkelmann, *Angew. Chem.*, **71**, 349 (1959).
203. M. Yamaguchi and T. Adachi, *Nippon Kagaku Zasshi*, **79**, 487 (1958).
204. M. Yamaguchi, *Nippon Kagaku Zasshi*, **77**, 513 (1956).
205. E. N. Marvell and M. J. Joncich, *J. Amer. Chem. Soc.*, **73**, 973 (1951).
206. R. Bognar and M. Rakosi, *Chem. Ind. (London)*, **1955**, 773.
207. E. E. van Tamelen, T. A. Spencer Jr., D. S. Allen Jr., and R. L. Orvis, *J. Amer. Chem. Soc.*, **81**, 6341 (1959).
208. E. E. van Tamelen, T. A. Spencer Jr., D. S. Allen Jr., and R. L. Orvis, *Tetrahedron*, **14**, 8 (1961).
209. C. Walling, A. L. Rieger, and D. D. Tanner, *J. Amer. Chem. Soc.*, **85**, 3129 (1963).
210. R. E. Pearson and J. C. Martin, *J. Amer. Chem. Soc.*, **85**, 3144 (1963).
211. R. Criegee and H. Beucker, *Ann. Chem.*, **541**, 218 (1939).

212. K. H. Pausacker, *J. Chem. Soc.*, **1953**, 107.
213. W. T. Caraway and L. Hellermann, *J. Amer. Chem. Soc.*, **75**, 5334, 5426 (1953).
214. G. P. Baker, F. G. Mann, N. Sheppard, and A. J. Tetlow, *J. Chem. Soc.*, **1965**, 3721.
215. E. L. Jackson in *Organic Reactions*, Vol. 2 (Ed. R. Adams), John Wiley and Sons, New York, 1944, p. 341.
216. J. M. Bobbitt in *Advances in Carbohydrate Chemistry*, Vol. 11 (Ed. M. L. Wolfrom), Academic Press, New York, 1956, p. 1.
217. W. G. Breck, R. D. Corlett, and G. W. Hay, *Chem. Commun.* **1967**, 604.
218. F. S. H. Head and G. Hughes, *J. Chem. Soc.*, **1952**, 2046.
219. P. W. Clutterbuck and F. Reuter, *J. Chem. Soc.*, **1935**, 1467.
220. J. L. Bose, A. B. Foster and R. W. Stephens, *J. Chem. Soc.*, **1959**, 3314.
221. M. L. Wolfrom and J. M. Bobbitt, *J. Amer. Chem. Soc.*, **78**, 2489 (1956).
222. K. B. Wiberg and K. A. Saegebarth, *J. Amer. Chem. Soc.*, **79**, 2822 (1957).
223. G. King, *J. Chem. Soc.*, **1936**, 1788.
224. D. B. Sprinson and E. Chargaff, *J. Biol. Chem.*, **164**, 433 (1946).
225. C. E. Crouthamel, A. M. Hayes, and D. S. Martin, *J. Amer. Chem. Soc.*, **73**, 82 (1951).
226. V. J. Shiner and C. R. Wasmuth, *J. Amer. Chem. Soc.*, **81**, 37 (1959).
227. C. A. Bunton and V. J. Shiner Jr., *J. Chem. Soc.*, **1960**, 1593.
228. A. H. Webster and J. Halpern, *J. Phys. Chem.*, **61**, 1239 (1957).
229. R. J. Harrisson and M. Moyle in *Organic Syntheses*, Coll. Vol. 4 (Ed. N. Rabjohn), John Wiley and Sons, New York, 1963, p. 493.
230. P. Ruggli, M. Herzog, J. Wegmann, and H. Dahn, *Helv. Chim. Acta*, **29**, 111 (1946).
231. H. T. Clarke and E. E. Dreger in *Organic Syntheses*, Coll. Vol. 1 (Eds. H. Gilman and A. H. Blatt), John Wiley and Sons, New York, 1941, p. 87.
232. M. L. Lewbart and V. R. Mattox, *J. Org. Chem.*, **28**, 2001 (1963).
233. M. M. Taqui Kahn and A. E. Martell, *J. Amer. Chem. Soc.*, **89**, 4176 (1967).
234. M. Weiss and M. Appel, *J. Amer. Chem. Soc.*, **70**, 3666 (1948).
235. B. Klein, *J. Amer. Chem. Soc.*, **63**, 1474 (1941).
236. B. A. Marshall and W. A. Waters, *J. Chem. Soc.*, **1960**, 2392.
237. B. A. Marshall and W. A. Waters, *J. Chem. Soc.*, **1961**, 1579.
238. K. B. Wiberg and W. G. Nigh, *J. Amer. Chem. Soc.*, **87**, 3849 (1965).
239. J. U. Nef, *Ann. Chem.*, **357**, 214 (1907).
240. M. Delépine and P. Bonnet, *Compt. Rend.*, **149**, 39 (1909).
241. E. Campaigne and W. M. LeSuer, *J. Amer. Chem. Soc.*, **70**, 1555 (1948).
242. F. Asinger, *Ber.*, **75B**, 656 (1942).
243. R. R. Burtner and J. W. Cusic, *J. Amer. Chem. Soc.*, **65**, 265 (1943).
244. I. A. Pearl, *J. Amer. Chem. Soc.*, **67**, 1628 (1945).
245. J. B. Lee and T. G. Clarke, *Tetrahedron Letters*, **1967**, 415.
246. I. A. Pearl, *U.S. Pat.* 2,438,789 (1948).
247. I. A. Pearl, *J. Amer. Chem. Soc.*, **70**, 2008 (1948).
248. M. Taniyama, *Toho-Reiyon Kenkyû Hôkoku*, **2**, 51 (1955).
249. C. Liteanu, *Bull. Soc. Stunte Cluj*, **10**, 96 (1948).
250. A. J. Green, T. J. Kemp, J. S. Littler, and W. A. Waters, *J. Chem. Soc.*, **1964**, 2722.

152   Herbert S. Verter

251. S. Patai and I. Shenfeld, *J. Chem. Soc.* (*B*), **1966**, 366.
252. W. Treibs, G. Lucius, H. Kogler, and H. Breslauer, *Ann. Chem.*, **581**, 59 (1953).
253. W. Treibs and H. Bast, *Ann. Chem.*, **561**, 165 (1949).
254. P. R. Jefferies, A. K. MacBeth, and B. Milligan, *J. Chem. Soc.*, **1954**, 705.
255. D. H. R. Barton and W. J. Rosenfelder, *J. Chem. Soc.*, **1951**, 2381.
256. H. D. Hartough, *Anal. Chem.*, **23**, 1128 (1951).
257. G. Hargreaves and L. H. Sutcliffe, *Trans. Faraday Soc.*, **51**, 1105 (1955).
258. K. B. Wiberg and W. H. Richardson, *J. Amer. Chem. Soc.*, **84**, 2800 (1962).
259. J. B. Conant, J. G. Aston, and C. O. Tongsberg, *J. Amer. Chem. Soc.*, **52**, 407 (1930).
260. J. Shorter and C. Hinselwood, *J. Chem. Soc.*, **1950**, 3276.
261. J. Shorter, *J. Chem. Soc.*, **1950**, 3425.
262. T. J. Hardwick and E. Robertson, *Can. J. Chem.*, **29**, 828 (1951).
263. J. S. Littler, *J. Chem. Soc.*, **1962**, 832.
264. J. Shorter, *J. Chem. Soc.*, **1962**, 1868.
265. S. Venkatakrishnan and M. Santappa, *Z. Physik. Chem.* (*Frankfurt*), **16**, 73 (1958).
266. A. Morette and G. Gaudefroy, *Bull. Soc. Chim. France*, **1954**, 956.
267. A. Morette and G. Gaudefroy, *Compt. Rend.*, **237**, 1523 (1953).
268. J. S. Littler and W. A. Waters, *J. Chem. Soc.*, **1959**, 1299.
269. V. S. Kudinova, S. R. Rafikov, K. D. Sagintaeva, and B. V. Suvorov, *Zh. Prikl. Khim.*, **35**, 2313 (1962).
270. J. S. Littler and W. A. Waters, *J. Chem. Soc.*, **1959**, 3014, 4046.
271. J. R. Jones and W. A. Waters, *J. Chem. Soc.*, **1962**, 1629.
272. W. A. Waters, *Quart Rev.* (*London*), **12**, 277 (1958).
273. F. H. Westheimer, *Chem. Rev.*, **45**, 419 (1949).
274. A. Darapsky, J. Germscheid, C. Kreuter, E. Engelmann, W. Engels, and W. Trinius, *J. Prakt. Chem.*, **146**, 238 (1936).
275. C. D. Hurd, J. W. Garrett and E. N. Osborne, *J. Amer. Chem. Soc.*, **55** 1082 (1933).
276. J. Roček and A. Riehl, *Tetrahedron Letters*, **1966**, 1437.
277. J. C. Craig and E. C. Horning, *J. Org. Chem.*, **25**, 2098 (1960).
278. K. Bowden, I. M. Heilbron, E. R. H. Jones, and B. C. L. Weedon, *J. Chem. Soc.*, **1946**, 39.
279. J. R. Schaeffer and A. O. Snoddy in *Organic Syntheses*, Coll. Vol. 4 (Ed. N. Rabjohn), John Wiley and Sons, New York, 1963, p. 19.
280. W. A. Mosher and F. C. Whitmore, *J. Amer. Chem. Soc.*, **70**, 2544 (1948).
281. H. W. Underwood Jr., and E. L. Kochmann, *J. Amer. Chem. Soc.*, **46**, 2071 (1924).
282. D. T. Cropp, J. S. E. Holker, and W. R. Jones, *J. Chem. Soc.* (*C*), **1966**, 1443.
283. L. F. Fieser, *J. Amer. Chem. Soc.*, **75**, 4377, 4386, 4395 (1953).
284. N. J. Leonard and P. M. Mader, *J. Amer. Chem. Soc.*, **72**, 5390 (1950).
285. W. A. Mosher and D. M. Preiss, *J. Amer. Chem. Soc.*, **75**, 5605 (1953).
286. R. Stewart, *Oxidation Mechanisms*, W. A. Benjamin, New York, 1964.
287. P. A. Best, J. S. Littler, and W. A. Waters, *J. Chem. Soc.*, **1962**, 822.
288. E. Wenkert and B. G. Jackson, *J. Amer. Chem. Soc.*, **80**, 211 (1958).

289. J. Roček and A. Riehl, *J. Amer. Chem. Soc.*, **88**, 4749 (1966).
290. J. Hampton, A. Leo, and F. H. Westheimer, *J. Amer. Chem. Soc.*, **78**, 306 (1956).
291. R. M. Evans, *Quart. Rev. (London)*, **13**, 61 (1959).
292. J. W. Ladbury and C. F. Cullis, *Chem. Rev.*, **58**, 403 (1958).
293. J. Kenyon and B. C. Platt, *J. Chem. Soc.*, **1939**, 636.
294. J. R. Ruhoff in *Organic Syntheses*, Coll. Vol. 2 (Ed. A. H. Blatt), John Wiley and Sons, New York, 1943, p. 315.
295. R. L. Shriner and E. C. Kleiderer in *Organic Syntheses*, Coll. Vol. 2 (Ed. A. H. Blatt), John Wiley and Sons, New York, 1943, p. 538.
296. H. Land and W. A. Waters, *J. Chem. Soc.*, **1957**, 4312.
297. T. J. Kemp and W. A. Waters, *J. Chem. Soc.*, **1964**, 339.
298. C. H. Rogers, *J. Amer. Pharm. Assoc.*, **12**, 503 (1923).
299. V. W. Markownikov, *Ber.*, **33**, 1908 (1900).
300. A. Y. Drummond and W. A. Waters, *J. Chem. Soc.*, **1953**, 435.
301. P. Levesley and W. A. Waters, *J. Chem. Soc.*, **1955**, 217.
302. K. B. Wiberg and R. Stewart, *J. Amer. Chem. Soc.*, **77**, 1786 (1955).
303. A. Y. Drummond and W. A. Waters, *J. Chem. Soc.*, **1955**, 497.
304. K. B. Wiberg and R. D. Geer, *J. Amer. Chem. Soc.*, **87**, 5202 (1965).
305. C. F. Cullis and J. W. Ladbury, *J. Chem. Soc.*, **1955**, 2850.
306. E. A. Braude and J. A. Coles, *J. Chem. Soc.*, **1952**, 1430.
307. B. C. L. Weedon and R. J. Woods, *J. Chem. Soc.*, **1951**, 2687.
308. M. Z. Barakat, M. F. Abdel-Wahab, and M. M. El-Sadr, *J. Chem. Soc.*, **1956**, 4685.
309. J. Padilla and J. Herrán, *Bol. Inst. Quim. Univ. Nacl. Auton. Méx.*, **8**, 3 (1956).
310. F. Sondheimer, C. Armendolla, and G. Rosenkranz, *J. Amer. Chem. Soc.*, **75**, 5932 (1953).
311. B. S. Thyagarajan, *Chem. Rev.*, **58**, 439 (1958).
312. J. S. Littler and I. G. Sayce, *J. Chem. Soc.*, **1964**, 2545.
313. P. T. Speakman and W. A. Waters, *J. Chem. Soc.*, **1955**, 40.
314. J. B. Conant and J. G. Aston, *J. Amer. Chem. Soc.*, **50**, 2783 (1928).
315. E. V. Brown, *Iowa State Coll. J. Sci.*, **11**, 227 (1937).
316. O. H. Mattsson and C. A. Wachmeister, *Tetrahedron Letters*, **1967**, 1855.
317. P. Duprat, *Bull. Inst. Pin.*, **17**, 36 (1933).
318. J. Wegmann and H. Dahn, *Helv. Chim. Acta*, **29**, 101 (1946).
319. J. K. Thomas, G. Trudel, and S. Bywater, *J. Phys. Chem.*, **64**, 51 (1960).
320. R. K. Srivastava, N. Nath, and M. P. Singh, *Tetrahedron*, **23**, 1189 (1967).
321. T. A. Cooper and W. A. Waters, *J. Chem. Soc. (B)*, 687 (1967) and preceding papers.
322. G. Hargreaves and L. H. Sutcliffe, *Trans. Faraday Soc.*, **51**, 786 (1955).
323. D. G. Hoare and W. A. Waters, *J. Chem. Soc.*, **1964**, 2560.
324. T. A. Cooper and W. A. Waters, *J. Chem. Soc.*, **1964**, 1538.
325. D. G. Hoare and W. A. Waters, *J. Chem. Soc.*, **1962**, 971.
326. D. G. Hoare and W. A. Waters, *J. Chem. Soc.*, **1964**, 2552.
327. C. E. H. Bawn and A. G. White, *J. Chem. Soc.*, **1951**, 331.
328. H. Adkins, L. M. Richards, and J. W. Davis, *J. Amer. Chem. Soc.*, **63**, 1320 (1941).
329. K. Nakagawa, R. Konaka, and T. Nakata, *J. Org. Chem.*, **27**, 1597 (1962).

330. Farbwerke Hoechst A.-G., *French Pat.* 1,381,867 (1964).
331. K. Nakagawa, K. Igano, and J. Sugita, *Chem. Pharm. Bull. (Tokyo)*, **12**, 403 (1964).
332. J. L. Howe, *J. Amer. Chem. Soc.*, **23**, 775 (1901).
333. L. M. Berkowitz and P. N. Rylander, *J. Amer. Chem. Soc.*, **80**, 6682 (1958).
334. F. S. Martin, *J. Chem. Soc.*, **1952**, 2682.
335. F. D. Dean and J. C. Knight, *J. Chem. Soc.*, **1962**, 4745.
336. P. J. Beynon, P. M. Collins, P. T. Doganges, and W. G. Overend, *J. Chem. Soc.(C)*, **1966**, 1131.
337. K. Heyns and H. Paulsen, *Angew. Chem.*, **69**, 600 (1957).
338. K. Heyns and H. Paulsen in *Advances in Carbohydrate Chemistry*, Vol. 17 (Ed. M. L. Wolfrom), Academic Press, New York, 1962, p. 169.
339. M. Rottenberg and P. Baertschi, *Helv. Chim. Acta*, **34**, 1973 (1956).
340. E. C. Horning, M. G. Horning, and G. N. Walker, *J. Amer. Chem. Soc.*, **71**, 169 (1949).
341. R. P. Linstead and K. O. A. Michaelis, *J. Chem. Soc.*, **1940**, 1134.
342. J. S. Buckley Jr. and H. D. Embree (Chas. Pfizer and Co., Inc.), *Brit. Pat.* 786,288 (1957).
343. K. Heyns and O. Stöckel, *Ann. Chem.*, **558**, 197 (1947).
344. K. Heyns, *Ann. Chem.*, **558**, 177 (1947).
345. T. Reichstein and A. Grussner, *Helv. Chim. Acta*, **17**, 311 (1934).
346. H. Wieland, *Ber.*, **54**, 2353 (1921).
347. L. M. Kogan, *Russ. Chem. Rev. (Engl. Transl.)*, **31**, 294 (1962).
348. W. J. Robbins and E. C. Lathrop, *Soil Sci.*, **7**, 475 (1919).
349. L. M. Jackman in *Advances in Organic Chemistry: Methods and Results*, Vol. 2 (Eds. R. A. Raphael, E. C. Taylor, and H. Wynberg), Interscience, New York, 1960, p. 329.
350. J. Fried, R. W. Thomas, and A. Klingsberg, *J. Amer. Chem. Soc.*, **75**, **1953**, 5764.
351. H. J. Ringold, M. Gut, M. Hayano, and A. Turner, *Tetrahedron Letters*, 835 (1962).
352. H. J. Ringold, M. Hayano, and V. Stefanovic, *J. Biol. Chem.*, **238**, 1960 (1963).
353. G. Greenspan, C. P. Schaffner, W. Charney, H. L. Herzog, and E. B. Hershberg, *J. Amer. Chem. Soc.*, **79**, 3922 (1957).
354. J. E. Bridgeman, P. C. Cherry, E. R. H. Jones, and G. D. Meakins, *Chem. Commun.*, **1967**, 482.
355. B. Kadis, *J. Amer. Chem. Soc.*, **88**, 1846 (1966).
356. P. Talalay, *Ann. Rev. Biochem.*, **34**, 347 (1965).
357. H. J. Brodie, M. Hayano, and M. Gut, *J. Amer. Chem. Soc.*, **84**, 3766 (1962).
358. A. Čapek, O. Hanč, K. Maček, M. Tadra, and Riedl-Tůmowá, *Naturwissenschaften*, **43**, 471 (1956).
359. J. Fried, R. W. Thoma, J. R. Gerke, J. E. Herz, M. N. Donin, and D. Perlman, *J. Amer. Chem. Soc.*, **74**, 3962 (1952).
360. J. J. Goodman and L. L. Smith, *Appl. Microbiol.* **9**, 372 (1961).
361. K. G. Holden, L. R. Fare, and J. R. Valenta, *J. Org. Chem.*, **32**, 960 (1967).
362. K. Schubert, K. H. Böhme, and C. Hörhold, *Naturforschung*, **15b**, 584 (1960).
363. R. M. Dodson and R. D. Muir, *J. Amer. Chem. Soc.*, **83**, 4627 (1961).

364. R. M. Dodson and R. D. Muir, *J. Amer. Chem. Soc.*, **83**, 4631 (1961).
365. C. J. Sih, K. C. Wang, and H. H. Tai, *J. Amer. Chem. Soc.*, **89**, 1956 (1967).
366. E. Kondo and T. Mitsugi, *J. Amer. Chem. Soc.*, **88**, 4737 (1966).
367. E. Kondo, T. Mitsugi, and K. Tori, *J. Amer. Chem. Soc.*, **87**, 4655 (1965).
368. G. Greenspan, R. Rees, L. L. Smith, and H. E. Alburn, *J. Org. Chem.*, **30**, 4215 (1965).
369. P. J. Chapman, G. Meerman, I. C. Gunsalus, R. Srinivasan, and K. L. Rinehart Jr., *J. Amer. Chem. Soc.*, **88**, 618 (1966).
370. A. Revol, C. Nofre, and A. Cier, *Compt. Rend.*, **247**, 2486 (1958).
371. D. Walker and J. D. Hiebert, *Chem. Rev.*, **67**, 153 (1967).
372. P. J. Neustaeder in *Steroid Reactions* (Ed. C. Djerassi), Holden-Day, San Francisco, 1963, pp. 89, 104.
373. D. N. Kirk and V. Petrow, *Brit. Pat.* 852,847 (1960).
374. A. E. Hydorn, *U.S. Pat.* 3,035,050 (1962).
375. Sociedad General de Farmacia, S. A., *Spanish Pat.* 288,343 (1963).
376. M. M. Coombs and H. R. Roderick, *Nature*, **203**, 523 (1964).
377. E. Merck A.-G., *Belgian Pat.* 625,215 (1963).
378. British Drug Houses, Ltd., *Netherlands Pat. Appl.* 6,413,529 (1965).
379. E. Merck A.-G., *Belgian Pat.* 618,731 (1962).
380. Syntex, S. A., *Belgian Pat.* 621,197 (1962).
381. J. A. Edwards, *U.S. Pat.* 3,080,396 (1963).
382. Osokeyhtio Medica, A. B., *Brit. Pat.* 969,558 (1964).
383. E. A. Braude, L. M. Jackman, and R. P. Linstead, *J. Chem. Soc.*, **1954**, 3548.
384. Syntex Corp., *Netherlands Pat. Appl.* 299,968 (1965).
385. E. A. Braude, A. G. Brook, and R. P. Linstead, *J. Chem. Soc.*, **1954**, 3569.
386. R. Gaertner, *J. Amer. Chem. Soc.*, **76**, 6150 (1954).
387. E. Merck A.-G., *Belgian Pat.* 624,885 (1963).
388. S. K. Pradhan and H. J. Ringold, *J. Org. Chem.*, **29**, 601 (1964).
389. D. Burn, V. Petrow, and G. Weston, *J. Chem. Soc.*, **1962**, 29.
390. American Cyanamid Co., *Netherlands Pat. Appl.* 6,402,447 (1964).
391. Organon Laboratories Ltd., *Brit. Pat.* 927,158 (1963).
392. A. Bowers and J. A. Edwards, *U.S. Pat.* 3,101,353 (1963).
393. J. A. Edwards, M. C. Calzada, L. C. Ibanez, M. E. C. Rivera, R. Urquiza, L. Cardona, J. C. Orr, and A. Bowers, *J. Org. Chem.*, **29**, 3481 (1964).
394. J. A. Edwards, J. C. Orr, and A. Bowers, *J. Org. Chem.*, **27**, 3378 (1962).
395. J. F. Bagli, P. F. Morand, K. Wiesner, and R. Gaudry, *Tetrahedron Letters*, 387 (1964).
396. R. E. Schaub and M. J. Weiss, *French Pat.* 1,367,429 (1964).
397. E. J. Agnello and G. D. Laubach, *J. Amer. Chem. Soc.*, **82**, 4293 (1960).
398. R. C. Tweit, *Belgian Pat.* 619,013 (1962).
399. P. Crabbé, *U.S. Pat.* 3,102,126 (1963).
400. H. J. Ringold and A. Turner, *Chem. Ind. (London)*, **1962**, 211.
401. P. Westerhof and J. Hartog, *Rec. Trav. Chim.*, **84**, 918 (1965).
402. B. R. Davis and T. G. Halsall, *J. Chem. Soc.*, **1962**, 1833.
403. H.-D. Becker, *J. Org. Chem.*, **30**, 989 (1965).
404. E. A. Braude, L. M. Jackman, and R. P. Linstead, *J. Chem. Soc.*, **1954**, 3564.

405. G. Lowe, *Quart Rev. (London)*, **57**, 201 (1960).
406. B. Weinstock, H. Niki, C. McKnight, *154 Nat. Meeting Amer. Chem. Soc.* (1967); see *Chem. and Eng. News*, **45**, 42 (25.9.1967).
407. A. Schönberg and G. Schütz, *Ber.*, **93**, 1466 (1960).
408. W. Bonthrone and D. H. Reid, *J. Chem. Soc.*, **1959**, 2773.
409. H.-D. Becker, *J. Org. Chem.*, **32**, 2140 (1967).
410. H. S. Isbell, H. L. Frush, and F. J. Bates, *J. Res. Nat. Bur. Standards*, **8**, 571 (1932).

CHAPTER 3

Enolization

S. Forsén* and M. Nilsson

The Royal Institute of Technology, S–10044 Stockholm, Sweden
Manuscript received March, 1968.

* *Present address: Chemical Centre, Lund Institute of Technology, PO Box 740, S–22007 Lund, Sweden.*

157

I. INTRODUCTION

The enolization reaction (1) is probably the most discussed example of prototropic tautomerism[1-3], bearing close relation to other prototropic rearrangements, e.g. of nitroalkanes to *aci*-nitro compounds and of imines/immonium ions to enamines. The presence of additional carbonyl groups, imino groups, other hetero atoms, double bonds, rings etc., may complicate the simple pattern outlined in equation (1), and sometimes the word 'enolization' will no longer be adequate to describe the process.

$$R^1-\overset{\overset{\textstyle O}{\|}}{C}-\overset{\overset{\textstyle H}{|}}{C}\overset{R^2}{\underset{R^3}{}} \rightleftharpoons R^1-\overset{\overset{\textstyle OH}{|}}{C}=C\overset{R^2}{\underset{R^3}{}} \qquad (1)$$

Enolization affects the physical properties of carbonyl compounds as keto and enol forms often exhibit quite different characteristics, such as solubilities, melting points, and vapour pressures. In organic reactions enolization is important because it is closely connected to deuteration, racemization and halogenation of carbonyl compounds. Functional derivatives of enols, e.g. ethers and esters, can sometimes be used with advantage in certain syntheses, and enol phosphates play important roles in biochemistry. Enolate anions (metal enolates) will not receive much attention in this review, but can be important intermediates, for example in aldol-type reactions and the base-catalysed halogenation of carbonyl compounds.

Enolization generally is a slow process, but is catalysed by acids and bases. The simplest pathways of enolization may be depicted as in equation (2). This scheme also illustrates the close connection existing between the acidities and basicities of ketones and enols and their enolization rates and equilibria. The basicity of ketones and acid-catalysed enolization were discussed previously in the series[4]. Enolization is also closely linked with association phenomena such as solvation and hydrogen bonding.

$$(2)$$

The detailed mechanisms of enolization are presently being investigated with new and improved techniques.

The simple system depicted in equation (1) covers a manifold of structural variations of the groups R^1, R^2 and R^3. Ring formation may also enter the picture, and vinylogues (equation 3) provide other

$$R^1-\overset{\underset{\|}{O}}{C}-(CH=CH)_n-\overset{\underset{R^3}{|}}{C}\overset{\underset{R^2}{H}}{\diagdown} \rightleftharpoons R^1-\overset{\underset{|}{OH}}{C}=CH-(CH=CH)_{n-1}-CH=\overset{\underset{R^3}{\diagdown}}{C}\overset{R^2}{\diagup} \quad (3)$$

generalizations. One can also consider hetero atoms with a mobile proton (equation 4), as they occur in amides. Several structural

$$R^1-\overset{\underset{\|}{O}}{C}-\overset{\overset{H}{\diagup}}{X}\overset{}{\diagdown}_{R^2} \rightleftharpoons R^1-\overset{\underset{|}{OH}}{C}=X-R^2 \quad (4)$$

variations may be exhibited simultaneously and give rise to many enolization possibilities for a given molecule.

II. DETERMINATION OF KETO–ENOL EQUILIBRIUM CONSTANTS

The methods used for the determination of keto–enol equilibrium constants can be classified as chemical methods and physical, particularly spectroscopical, methods.

The former employ reactions which are considered to be specific for enols and are used on the assumption that they are considerably faster than enolization. Obviously neither the reagents nor the reaction products must catalyse enolization. The most important chemical methods employ halogenation, and Meyer's classical bromine titration, after several modifications and improvements, is still outstanding[5]. For markedly acidic systems, particularly β-dicarbonyl compounds, the reactions of enols with diazomethane to form O-methyl compounds has been widely used. The results have sometimes been misleading and have often been criticized. Methods employing reactions specific for the keto forms have apparently not been seriously considered, but would be of great interest for systems where the degree of enolization is very high (say more than 80 or 90%).

Spectroscopic measurements do not interfere with the enolization reaction; however the evaluation of extinction coefficients in u.v. and i.r. spectroscopic methods poses difficult problems. N.m.r. spectroscopy at present offers the best possibilities, but for limitations at low or high enol concentrations.

A. Halogen titrations

In the *direct* method a solution of the substance to be investigated is titrated with a standard solution of bromine[6,7]. The enol present reacts rapidly with bromine, yielding a bromo ketone, whereas the direct bromination of the keto form is very slow. The reaction gives also hydrogen bromide which is an efficient catalyst for the keto–enol interconversion, which might result in too high enolic values and a rather unsharp end point. This is avoided to some extent in a modification often referred to as the *indirect* method[8-10]. Here, a solution of the tautomeric compound is mixed with a solution of bromine estimated to be in excess of the enol form present. The unreacted bromine is then rapidly trapped by β-naphthol. After adding a solution of potassium iodide and heating, the bromo ketone present reacts according to equation (5), and the liberated iodine is titrated. Occasionally, however, also the indirect method fails because reaction (5)

$$\overset{O}{\underset{|}{\overset{\|}{C}}}\overset{Br}{\underset{|}{\overset{|}{C}}} + H^+ + 2I^- \longrightarrow \overset{O}{\underset{|}{\overset{\|}{C}}}\overset{H}{\underset{|}{\overset{|}{C}}} + I_2 + Br^- \tag{5}$$

is slow or incomplete[15,16]. Many of the early results obtained by Meyer[6-9] and others[16] especially for β-dicarbonyl compounds have been corroborated by more recent n.m.r., u.v. and i.r. measurements. In some cases, however, the bromine titration methods have been reported to give absurdly high enolic contents, in excess of 100%[11-14].

A number of modifications and improvements of the original bromine titration methods of Meyer have been described. Schwartzenbach and coworkers[17,18] have introduced an ingenious flow technique, in which a solution of the enol is mixed with a solution of bromine, the concentration of which is continuously increased. The bromine concentration in the mixture is monitored potentiometrically by means of a platinum electrode inserted at a suitable distance downstream from the mixing chamber. A sudden change in the electrode potential is observed when the bromine concentration exceeds that of the enol. This method allows measurements of much smaller enolic concentrations than the original Meyer methods. However, some compounds show poorly defined electrode potential changes.

Gero[19-21] has employed iodine monochloride in place of bromine in a modification of the indirect method described above. Recently, Allinger, Chow and Ford[22] have shown by an n.m.r. technique that some of the values obtained by Gero on the enolic content of cyclic ketones were erroneous, but the reasons for this are not clear.

In the Walisch method[23-25] bromine is produced electrolytically and its concentration followed potentiometrically. The amount of enol form is calculated from the current and time required to reach a certain bromine concentration. The bromine concentration is always very low, thus reducing the danger of interfering side reactions. By varying the time parameter it is possible to estimate the rate of enolization and to correct the determined enolic contents for this effect. Bell and Smith[26] have recently utilized a redox electrode of platinum in combination with a glass electrode to measure the very small bromine uptake of cyclopentanone and cyclohexanone.

B. Acidimetric methods

Measurements of the apparent acid dissociation constant of a compound capable of existing in two tautomeric forms make it possible to calculate the equilibrium constant (K_t), provided the acid dissociation constants of the tautomeric forms are known (see IV.C.1). Such measurements have been frequently used to determine the ratio of oxo form to hydroxy form in aqueous solutions of hydroxy derivatives of pyridines, quinolines, isoquinolines and other heterocyclic compounds[27,28].

C. Spectroscopic methods

Quantitative infrared spectroscopic measurements have been used to study the enolization of simple ketones[31-33], and β-diketones. Ultraviolet spectroscopy has also been widely employed[34-41]. Raman spectra have been used by Kolrausch and coworkers[42,43] and by Shigorin[44].

Nuclear magnetic resonance spectroscopy has rapidly become a powerful tool in the study of keto–enol tautomerism. It was first applied by Jarrett, Sadler and Shoolery[45] who investigated the proton resonance spectra of 2,4-pentanedione and 3-methyl-2,4-pentanedione. Later work has been concerned mainly with proton resonances[46-56]. It has been recently shown, however, that increased knowledge may be gained from [17]O magnetic resonance spectra[57,58]. With the n.m.r. method equilibrium constants $K_t = $ [enol]/[keto] between 0·05 and 20 may at present be determined with reasonably good accuracy, as with the sensitivity of present day spectrometers it is possible to detect a few tenths of a per cent of one tautomeric form in the pure liquid[22]. In solutions the lower limit of detection will be increased correspondingly. With the aid of time-averaging computers the limit of detection of a weak signal can be effectively lowered by one or two powers of ten. It seems highly desirable to use the n.m.r. technique for a reinvestigation of the enolic contents of simple ketones.

III. ENOLIZATION OF SIMPLE CARBONYL COMPOUNDS

A. Equilibrium constants for ketones

In simple aliphatic and alicyclic ketones or aldehydes the equilibrium

$$R^1-\overset{\overset{\displaystyle O}{\|}}{C}-\overset{\overset{\displaystyle H}{|}}{C}\overset{R^2}{\underset{R^3}{<}} \;\rightleftharpoons\; R^1-\overset{\overset{\displaystyle OH}{|}}{C}=C\overset{R^2}{\underset{R^3}{<}}$$

(1a)   (1b)

is greatly in favour of the keto form (1a). Early attempts to determine the extent of enolization in simple ketones by bromine titration did in fact fail to show any evidence for enolization[8]. Attempts to synthesize the enols directly have resulted in the corresponding ketones. For an example of an alleged preparation of enols see references 60 and 61.

Indirect evidence for the occurrence of small amounts of enolic forms in simple ketones came from kinetic studies of acid-catalysed racemization, halogenation and deuterium exchange. It is often assumed that such reactions proceeded via the enol forms.

The enolic content in aqueous solutions of the simple ketones— acetone, cyclopentanone and cyclohexanone—was first determined in 1947 by Schwartzenbach and Wittwer using the flow method[18].

Later, Gero[19-21] has reported enolic contents extrapolated to those of the pure liquids of these and a number of other ketones and esters (Table 1).

Recent work indicates that these earlier values were too high. Dubois and Barbier[62] (cf. also Walisch[25]) found the enolic content

TABLE 1.   Keto–enol equilibrium constants (K_t = [enol]/[keto]) of simple ketones.

Compound	K_t	Method/medium	Reference
Acetone	2.5×10^{-6}	a	18
	1.5×10^{-6}	b	19
	(0.9×10^{-6})	d	26
Ethyl methyl ketone	1.2×10^{-3}	b	19
Methyl propyl ketone	8.6×10^{-5}	b	19
Butyl methyl ketone	1.1×10^{-3}	b	19
Diethyl ketone	6.7×10^{-5}	b	19
Ethyl propyl ketone	4.7×10^{-4}	b	19
Diisopropyl ketone	3.7×10^{-5}	b	19
Camphor	1.4×10^{-3}	b	19
Acetophenone	3.5×10^{-4}	b	19
Propiophenone	1.0×10^{-4}	b	19
Cyclobutanone	0.55×10^{-2}	c	21
Cyclopentanone	4.8×10^{-5}	a	18
	0.9×10^{-3}	c	21
	1.3×10^{-5}	d	26
Cyclohexanone	2.0×10^{-4}	a	18
	1.2×10^{-2}	c	21
	5×10^{-5} to		
	8×10^{-5}	e	62
	4.1×10^{-6}	d	26
Cycloheptanone	0.56×10^{-2}	b	19
	$<0.2 \times 10^{-2}$	f	22
	$<0.4 \times 10^{-2}$	g	62
Cyclooctanone	1.1×10^{-1}	c	21
	$<5 \times 10^{-3}$	f	22
Cyclononanone	4.2×10^{-1}	c	21
Cyclodecanone	6.5×10^{-1}	c	21

The analytical methods used are: in a to e various modifications of the bromine titration method, in f and g n.m.r. spectroscopy.

a In water.
b In 75% methanol extrapolated to the neat liquid.
c In 75% methanol extrapolated to the neat liquid.
d In water at 25°.
e Medium not stated.
f In ca. 70% methanol.
g In chloroform.

in carefully purified cyclohexanone to be a third to a fourth of that given by Schwartzenbach and Wittwer[18]. Careful measurements on aqueous solutions of acetone, cyclopentanone and cyclohexanone have recently been made by Bell and Smith[26]. Under the conditions of their experiments the reaction between enol and bromine was complete within five seconds, whereas the time required for reestablishment of the keto–enol equilibrium was about ten minutes. They also found evidence for the slow continuous production of some bromine-consuming substance other than the enol—possibly an unsaturated ketone. For cyclopentanone and cyclohexanone, the enolic contents found are lower than those found by Schwartzenbach and Wittwer by factors of 4 and 50 respectively (Table 1). For acetone no consistent value of the enolic content could be obtained, but the earlier values (K_t 2 × 10^{-6}) are most likely too high[26]. From the observed rate of reaction of acetone and bromine in dilute solutions a K_t value of 9 × 10^{-7} can be estimated[26].

With n.m.r. spectroscopy Allinger, Chow and Ford[22] have recently shown the enolic contents found by Gero for cycloheptanone and cyclooctanone to be too high (Table 1), and the K_t values given in Table 1 are thus upper limits, estimated from the limit of signal detection under the experimental conditions. Whenever the K_t values found by Gero using the ICl titration method can be compared with values obtained with other methods, Gero's values are consistently seen to be higher. In view of this, it is at present somewhat uncertain what significance should be given to the other K_t values of simple ketones which have been obtained by the ICl method.

B. Direction of enolization

Many unsymmetrical ketones, for example 2-butanone (**2**), may enolize in two ways:

$$\underset{\text{OH}}{\text{CH}_3\text{—CH=C—CH}_3} \rightleftharpoons \underset{\text{O}}{\text{CH}_3\text{—CH}_2\text{—C—CH}_3} \rightleftharpoons \underset{\text{OH}}{\text{CH}_3\text{—CH}_2\text{—C=CH}_2}$$

$$(2)$$

The degree of enolization is very low, however, and only indirect means can be used to estimate the relative importance of the enols. The classical approach would be to investigate orientation by bromination; however it has recently been proposed that halogenation does not necessarily proceed via the enols[63]. Investigation of the rate of exchange of deuterium for the α-hydrogens seems to be more illustrative, though there can be no *a priori* certainty that even deuteration proceeds via the enols. In the acid-catalysed deuteration of 2-butanone, n.m.r. investigations have shown that the orientation is approximately the same as in halogenation and that the ratio 3-deuteration/1-deuteration is approximately 2·5. Base-catalysed deuteration gave lower values, ca. 0·7, and in neutral media intermediate values have been obtained[64,65]. The work on deuteration is being continued[66]. The situation is, however, confused since the essentially kinetic data cannot be used to evaluate the ratio of the competing enols, even if these are intermediates.

C. Enolization of aldehydes

Enolizable aldehydes provide a puzzling subject. Extensive studies

$$
\begin{array}{ccc}
\overset{R^1}{\underset{R^2}{\diagdown}}\text{CH—CH}\overset{OH}{\underset{OH}{\diagup}} & \rightleftharpoons & \overset{R^1}{\underset{R^2}{\diagdown}}\text{CH—CHO} & \rightleftharpoons & \overset{R^1}{\underset{R^2}{\diagdown}}\text{C}{=}\text{C}\overset{OH}{\underset{H}{\diagdown}} \\
(3a) & & (3b) & & (3c)
\end{array}
$$

have been made on their hydration (**3b** → **3a**) which is a dominating reaction in aqueous media[67,68] (see Chapter 1). The facile aldol condensations and rapid halogenation of simple aldehydes would be expected to parallel a marked enolization (**3b** → **3c**), yet very little is found in the literature on this. Gero, using the iodine monochloride method, could not observe any halogen uptake by acetaldehyde (indicating an enol/keto ratio of less than 1×10^{-7}) under conditions where the enol content of acetone was measurable[20]. In a series of papers on vinyl alcohols Fuson and coworkers have described (mainly mesityl-) substituted acetaldehydes, which are largely enolic. The earlier results have been generalized to give the necessary and sufficient conditions for stability of the vinyl alcohol forms (**3c**); both R^1 and R^2 should be groups like mesityl[69]. The generalization was also extended to include α,β-enediols. In these sterically crowded compounds the keto–enol interconversion is sluggish. This is marked even in the less enolic 2-mesityl-2-phenylacetaldehyde[70].

D. Thermochemical considerations

The low degree of enolization of simple carbonyl compounds is consistent with thermochemical data. Klages[71] has estimated the heats of combustion (enthalpy difference) to be in favour of the keto forms by 16 kcal/mole for ketones, 10 kcal/mole for aldehydes and 24 kcal/mole for esters.

Conant and Carlson[72] tried to estimate the free energy change of enolization of acetophenones from measurements on the rates of isomerization of optically active derivatives, under the assumption that the rate of isomerization is a measure of the rate of enolization, as the observed energy of activation (ca. 14 kcal/mole for a number of acetophenone derivatives) is also an upper limit for the enthalpy change of enolization. A lower limit of ca. 6 kcal/mole for the free energy change of enolization was estimated from measurements on equilibria in β-diketones and keto esters. Neglecting entropy changes we thus have for the enolization of acetophenone $6 < \Delta G° < 14$ (kcal/mole) at 25°C, corresponding to K_t between 10^{-5} and 10^{-10}. In an analogous way, values of $\Delta G°$ of enolization in the gaseous phase were estimated to be 14 to 16 kcal/mole for aliphatic esters, 10 to 12 kcal/mole for aliphatic ketones and 7 to 10 kcal/mole for aliphatic aldehydes.

Sunner showed that the heat of enolization of acetone can be calculated from the heats of hydrolysis of isopropenyl acetate (equation 6) and m-cresyl acetate (equation 7)[73]

$$CH_2{=}C(CH_3)OAc(g) + H_2O(g) \longrightarrow CH_3COCH_3(g) + HOAc(g) \qquad (6)$$
$$\Delta H_6 = -13.7 \text{ kcal/mole at } 25°C$$

$$m\text{-}CH_3C_6H_4OAc(g) + H_2O(g) \longrightarrow m\text{-}CH_3C_6H_4OH(g) + HOAc(g) \qquad (7)$$
$$\Delta H_7 = -3.6 \text{ kcal/mole at } 25°C$$

Equation (6) can be split into the two consecutive steps (8) and (9):

$$CH_2{=}C(CH_3)OAc(g) + H_2O(g) \longrightarrow CH_2{=}C(CH_3)OH(g) + HOAc(g) \qquad (8)$$
$$CH_2{=}C(CH_3)OH(g) \longrightarrow CH_3COCH_3(g) \qquad (9)$$

The enthalpy change is not directly accessible but can be estimated from a comparison of the heats of hydrolysis of isopropenyl acetate and m-cresyl acetate. Combination of equation (7) and the reversed equation (8) gives equation (10):

$$m\text{-}CH_3C_6H_4OAc(g) + CH_2{=}C(CH_3)OH(g) \longrightarrow$$
$$CH_3C_6H_4OH(g) + CH_2{=}C(CH_3)OAc(g) \qquad (10)$$

By assuming that $\Delta H_{10} = \Delta H_7 - \Delta H_8 \approx 0$ it follows that $\Delta H_8 = -3.6$ kcal/mole and thus that the heat of enolization of acetone in the gaseous phase at 25°C is $-\Delta H_9 = 10$ kcal/mole. From the estimated and known heats of vaporization of 2-propenol and acetone, respectively, the heat of enolization of acetone in the liquid phase is cycloalkanes in acetic acid.

Keto-enol equilibria	$\Delta G°_{298}{}^a$ (kcal/mole)	Exo-endo C=C equilibria[59]	$\Delta G°_{298}$ (kcal/mole)	$\Delta H°_{298}$ (kcal/mole)	$\Delta S°_{298}$ (e.u.)
(4-membered ring)	(9.7)	(CH$_2$/CH$_3$ 4-membered ring)	−1.05	−0.95	0.0
(5-membered ring)	6.66[26]	(CH$_2$/CH$_3$ 5-membered ring)	−4.15	−3.9	+0.9
(6-membered ring)	7.35[26]	(CH$_2$/CH$_3$ 6-membered ring)	−3.24	−2.4	+2.8
(7-membered ring)	(8.2)	(CH$_2$/CH$_3$ 7-membered ring)	−2.55	−2.3	+0.8
(8-membered ring)	(6.9)	(CH$_2$/CH$_3$ 8-membered ring)	−3.79	—	—

^{a} Values within parentheses are estimated from the corresponding exo-endo values using the average difference in $\Delta G°_{298}$ between the keto-enol and exo-endo equilibria for the 5- and 6-membered rings (ca. 10·7 kcal/mole).

calculated to be 7·0 kcal/mole. The entropy change due to enolization is estimated to be ca. -3 e.u. in the liquid, and the free energy change for enolization in the liquid phase thus becomes 8 kcal/mole at 25°c ($K_t = 2 \times 10^{-6}$). This value is in fair agreement with the lowest experimental values (valid for aqueous solutions) cited in Table 1.

The recent data of Bell and Smith[26] on cyclopentanone and cyclohexanone indicate the enolic content in these compounds to be higher than that in acetone by one and two powers of ten respectively. Furthermore, the variation of the equilibrium constant with ring size is found to be opposite to that found earlier by Schwartzenbach and Wittwer[18]. The values of the latter authors have been frequently quoted in the literature. Whereas the older data conformed nicely to the generalization by Brown, Brewster and Shechter*—and in part served as support for the generalization—the more recent data do not fit this generalization. If the difference in stability of the keto and enol forms in the two cycloalkanones is related to the difference in internal strains (*I*-strains) of the tautomeric forms, one would expect the enol equilibria to be correlated with the *exo–endo* equilibria in methylenecyclopentane and methylenecyclohexane. However, with the older data no such correlation is evident, as pointed out by Shechter and coworkers[75]. With the new data of Bell and Smith[26] the situation is now changed, as shown in Table 2 for 5- and 6- membered rings. In the same table an attempt is made to extrapolate the available data on the enol equilibria in 5- and 6-membered rings to 4-, 7- and 8-membered rings, with no pretension of accuracy, as solvent effects may have a considerable influence, particularly on the keto–enol equilibria.

IV. KETO–ENOL TYPE EQUILIBRIA FOR PHENOLS

A. Mono- and polycyclic phenols

The free energy of ketonization of phenol has been estimated to be 18·6 kcal/mole[76] corresponding to $K_t = [\text{enol}]/[\text{keto}] = 10^{14}$ at 25°c, making it impossible to determine the equilibrium concentration of the keto form in phenol by direct methods. On the other hand it has been noted that in alkaline deuterium oxide solutions of phenol the *ortho* and *para* hydrogen atoms are exchanged[77,78]. A possible

* "... reaction will proceed in such a way as to favour the formation or retention of an *exo* double bond in the 5-ring and to avoid the formation or retention of the *exo* double bond in 6-ring systems"[74].

reason for this is that the exchange proceeds via a phenolate–ketone
equilibrium according to equation (11).

$$(11)$$

In resorcinol and phloroglucinol derivatives one could expect
the keto form to be relatively favoured. In alkaline deuterium oxide
solutions under conditions where phenol, catechol and hydroquinone
do not undergo measurable deuterium exchange during several
hours, the exchange in resorcinol and phloroglucinol reaches equilibrium within one hour or less[79].

There is still no evidence, however, that phloroglucinol should
be able to exist in the triketo form as is often suggested. In fact even
gem-disubstituted phloroglucinol derivatives, e.g. '2,2-dialkylcyclo-
hexane-1,3,5-triones', are still highly enolic.

As the aromatic system becomes larger, the delocalization energy
per ring becomes smaller and the free energy difference between
ketone and enol (phenol) should increasingly favour the keto form.
Deuterium exchange in the 1-position of 2-naphthol takes place
more rapidly than in phenol[79,80]. In dihydroxynaphthalenes certain
substitution patterns favour deuterium exchange. For example, in
2,7-naphthalenediol deuterium exchange in the 1- and 8-positions
takes place rapidly at room temperature, while protons in other
positions are unaffected even on heating. In 1,6-naphthalenediol only
the protons in the 2-, 4- and 5-positions are exchanged. The experimental data are consistent with enolate–ketone equilibria analogous
to those in equation (11), but it is not certain that the keto form
represents a true intermediate. In polyhydroxybenzenes it is observed
that deuterium exchange proceeds most rapidly when the deuterium
oxide solution contains equimolar amounts of the strong base and the
polyphenol[80]. Localization type approximation molecular orbital

calculations, based upon a hypothetical enolate–ketone equilibrium, are in agreement with the observed rates and patterns of deuterium exchange in aromatic hydroxy compounds[81].

More direct and striking evidence for keto–enol tautomerism has

(4a)   (4b)

been found in 1,4-naphthalenediol. Both the dihydroxy form **4a** and the diketo form **4b** are known and can in fact be recrystallized with no structural change[82]. Their rate of interconversion is thus very low under uncatalysed conditions, and gives no direct evidence regarding the relative stability of the two tautomeric forms. As shown by Tomson[82,83] the equilibrium is however established in the melt (at about 210°c), where not less than 10% of the compound is found in the keto form. Further hydroxy substitution in **4a** has been shown to favour the keto form[83]. Thus 1,4,5-trihydroxynaphthalene and 1,4,5,8-tetrahydroxynaphthalene are reported to exist predominantly in the keto forms **5b** and **6b** respectively[83].

(5a)   (5b)

(6a)   (6b)

In this connection it is interesting to note that Garbisch[84] has succeeded in preparing the diketo form **7b** of 1,4-dihydroxybenzene (**7a**) in pure crystalline form. In non-polar aprotic solvents the unstable keto form is slowly converted into the more stable enol form **7a**; in polar protic solvents such as water and alcohols the conversion is very fast.

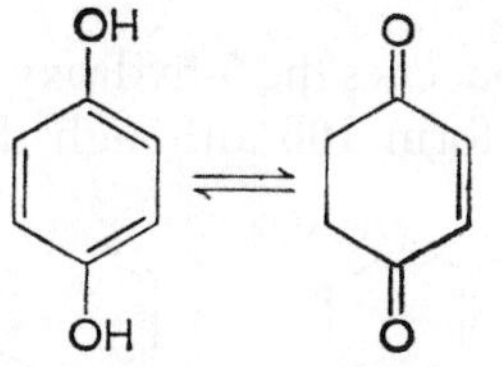

(**7a**)   (**7b**)

Anthrone (**8b**) is perhaps the simplest of those aromatic, tautomeric compounds which exist predominantly as the keto form at room temperature in most solvents. Meyer[85-87] succeeded in preparing both

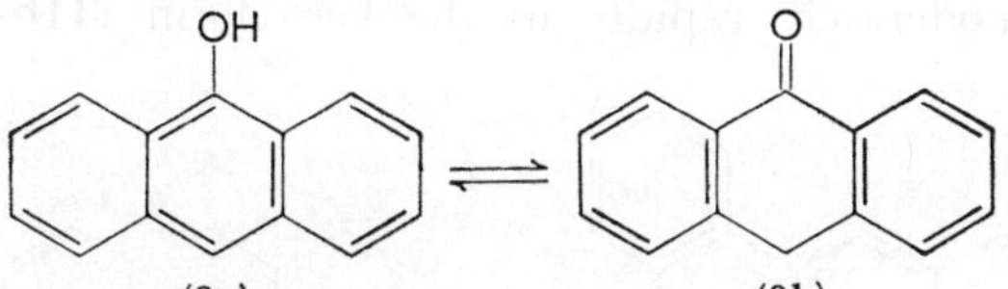

(**8a**)   (**8b**)

the keto and enol form and also determined the equilibrium constant $K_t = $ [enol]/[keto] to be 0·124 (corresponding to 89% keto form) in ethanol at room temperature. The equilibrium constant K_t has also been determined spectroscopically[88] to be $2·1 \times 10^{-3}$ in isooctane at 20°C.

Little is known about the keto–enol equilibrium in polyhydroxy-anthracenes, although for some derivatives both the keto and enol forms are known[83]. As pointed out before, this fact in itself provides little information on the position of the equilibrium. In 9-phenanthrol

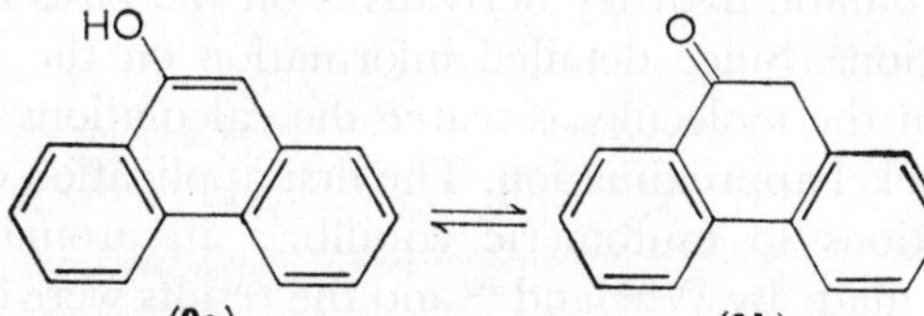

(**9a**)   (**9b**)

the equilibrium is presumably largely in favour of the enol form **9a**[89]. Klages[90] has calculated the heats of combustion in the gas phase at 25°C of **8a** and **8b** to be 1669·4 kcal/mole and 1671·3 kcal/mole respectively. The calculated heats of combustion of **9a** and **9b** are 1665·3 kcal/mole and 1669·6 kcal/mole respectively. Thus in the

case of 9-anthrone the stabilities of the keto and enol forms are calculated to be approximately equal, but in the case of 9-phenanthrol,
the enol form is calculated to be considerably more stable than the
keto form.

Of the hydroxynaphthacenes the 5-hydroxy derivative is presumably
most stable in the keto form **10b** although the equilibrium constant

OH   O

(10a)   **(10b)**

has apparently not been determined. 7-Hydroxybenzanthracene (**11a**)
isomerizes moderately rapidly to the keto form (**11b**) in boiling

OH   O

(11a)   **(11b)**

acetone[92], but the position of the equilibrium is unknown. 6-Hydroxy-
pentacene is only known in the keto form[93]. Only scattered information
is available for other hydroxy derivatives of polycondensed aromatic
hydrocarbons. Rather than discussing this in detail, an attempt will
be made to rationalize the observed variations in the keto–enol
equilibria of aromatic hydroxy derivatives on the basis of molecular
orbital calculations. Since detailed information on the bond angles
and distances in the molecules is scarce the calculations will be confined to the Hückel approximation. The first application of molecular
orbital calculations to tautomeric equilibria in aromatic hydroxy
derivatives was made by Wheland[94] and the results were qualitatively
satisfactory. The present calculation, which has been extended to a
comparatively large number of compounds, is a modification of that
of Wheland. The π-electron energies of the different tautomeric
forms are calculated, but since the number of electrons with π-electron
symmetry in the keto and enol forms differs, it appears more appropriate to compare not the total π-electron energies but the 'total

delocalization energies' (TDE)* introduced by Orgel, Cottrell, Dick and Sutton[95]. The TDE concept has later been used by Del Re[96] and others[97]. In the present calculations no attempt is made to assign a numerical value to the energy difference between the fragments

$$\underset{\overset{\displaystyle |}{\text{—C—CH—}}}{\overset{\overset{\displaystyle O}{\|}}{}} \quad \text{and} \quad \underset{\overset{\displaystyle |}{\text{—C=C—}}}{\overset{\overset{\displaystyle OH}{|}}{}}$$

but rather to estimate this difference approximately in terms of the resonance integral β. The results of the HMO calculations on a number of aromatic hydroxy derivatives are summarized in Table 3. Different strengths in intramolecular hydrogen bonding in the keto and enol forms are neglected. If we further neglect the entropy changes ΔS_t in the keto–enol interconversion or if this difference is approximately constant (little is known about ΔS_t in the case of aromatic compounds, although this quantity should be experimentally accessible in some cases) we may try to correlate the approximate experimental free energy changes on enolization, ΔG_t with the ΔTDE values in Table 3. The TDE values in all cases seem to be greater in the enol forms than in the keto forms. However, when the difference in ΔTDE per keto group formed is less than about $0 \cdot 35\beta$ the experimental data show the keto form to be the most stable tautomer compounds[10,11,19,23,24]. When ΔTDE is in the range $0 \cdot 35\beta$ to about $0 \cdot 50\beta$, the experimental free energy difference between the keto and enol forms is close to zero (solvent effects may shift the equilibrium) (compounds **9b**, **12** and presumably **21**) and when ΔTDE is larger than about $0 \cdot 45\beta$ to $0 \cdot 50\beta$ the enol form is experimentally found to be the most stable tautomer (compounds **1-3**, **5-8**, **14** etc.). In a semiquantitative way the HMO calculations are thus found to reflect the observed relative stabilities of the keto and enol forms of aromatic hydroxy derivatives. If the deuterium exchange of aromatic hydroxy derivatives in alkaline deuterium oxide proceeds via a transition state structurally similar to the keto forms, some of the data in Table 3 may also be used to estimate the relative rates of deuterium exchange at different positions of a given molecule as the TDE of an enolate anion is comparable to that of the parent enol.

* The TDE is defined as the difference between the total π-electron energy and the energy corresponding to the π electrons being localized on the participating atoms. In conjugated molecules involving hetero atoms the TDE concept is of greater utility than the usual 'delocalization energy', defined as the difference between the total calculated π-electron energy and that of a hypothetical reference molecule with the π electrons constrained to isolated double bonds.

TABLE 3. Total delocalization energies (TDE) calculated by the Hückel molecular orbital method of keto and enol forms of aromatic hydroxy derivatives[a].

No.	Enol form	TDE/β	Keto form	TDE/β	ΔTDE[b,c]
1	(phenol structure)	8·197	(a) (keto structure)	7·230	0·967
			(b) (keto structure)	7·296	0·901
2	(catechol structure)	8·386	(a) (keto structure)	7·422	0·964
			(b) (keto structure)	7·485	0·901

3   8.395

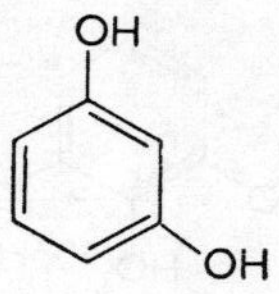

(a)	7.582	0.813
(b)	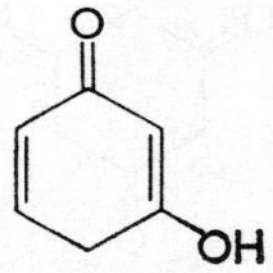 7.518	0.877
(c)	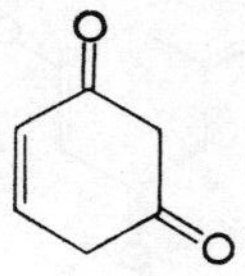 6.995	1.400(0.700)

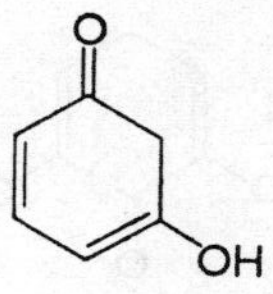

4   8.387

(a)	7.489	0.898
(b)	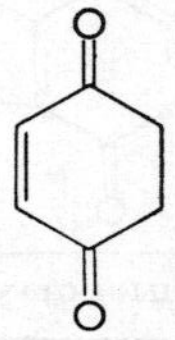 7.464	0.923(0.462)

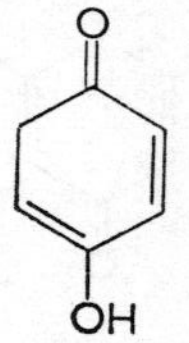

S. Forsén and M. Nilsson

No.	Enol form	TDE/β		Keto form	TDE/β	ΔTDE^{b,c}
5		8.593	(a)		7.841	0.752
			(b)		7.799	0.794
			(c)		7.296	1.297(0.648)
			(d)		6.708	1.885(0.628)
6		8.574	(a)		7.614	0.960

(b) 7·687 0·887

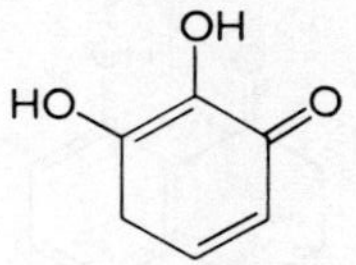

(c) 7·730 0·844

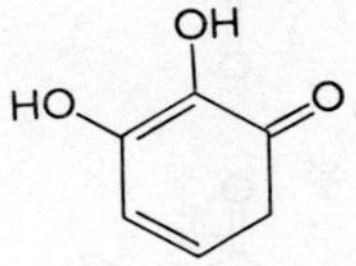

(a) 13·147 0·741

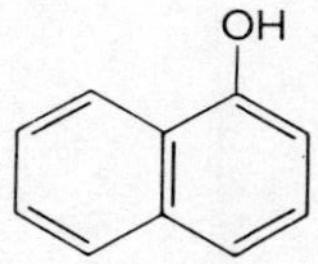

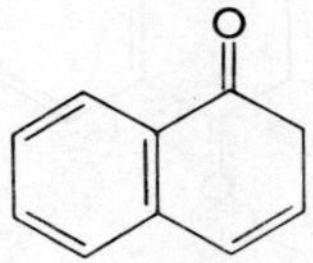

(b) 13·179 0·709

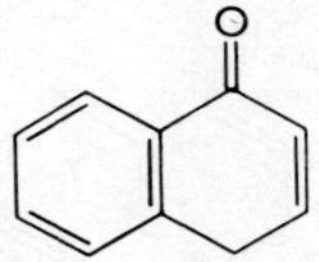

(a) 13·236 0·646

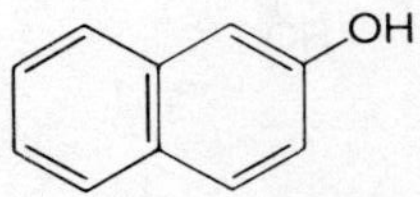

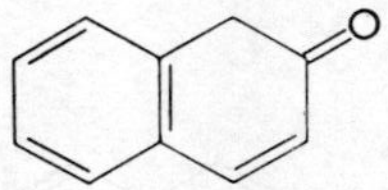

(b) 12·886 0·996

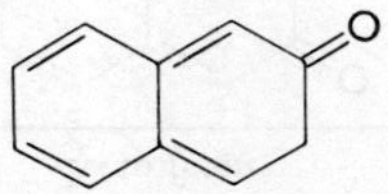

(a) 13·888   7

(b) 13·882   8

No.	Enol from	TDE/β		Keto from	TDE/β	ΔTDE[b,e]
			(c)		12·788	1·094
9	(structure)	14·081	(a)	(structure)	13·345	0·736
				(structure)	13·391	0·690(0·345)
10	(structure)	14·285	(b)	(structure)	13·616	0·669(0·335)

0.642(0.321)

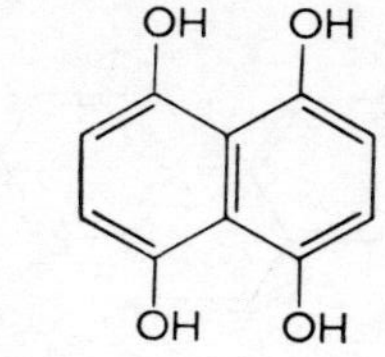

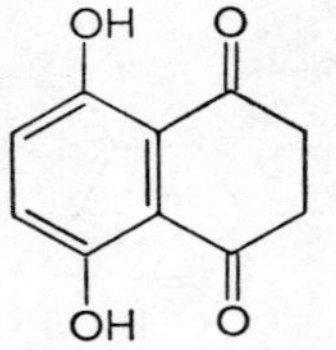

13.837

14.479

11

0.402

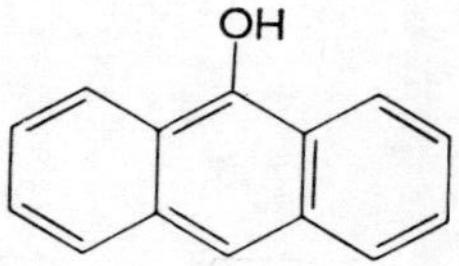

19.127

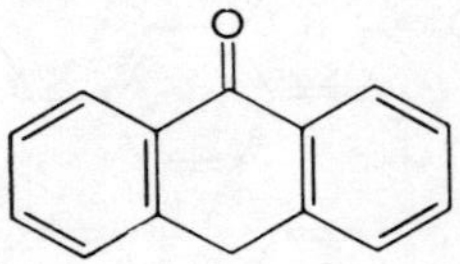

19.529

12

0.652

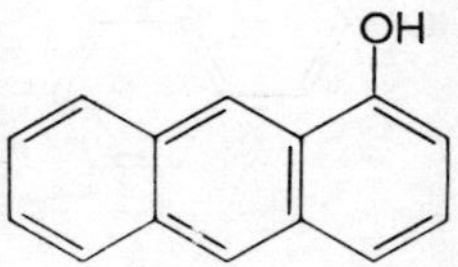

18.868

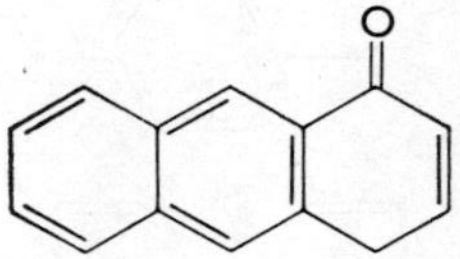

19.520

(a)

13

0.686

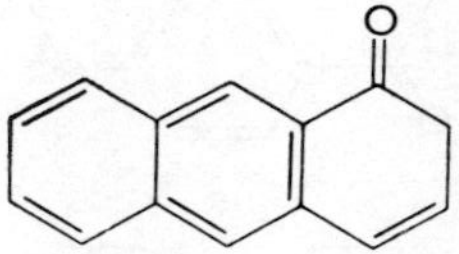

18.834

(b)

0.551

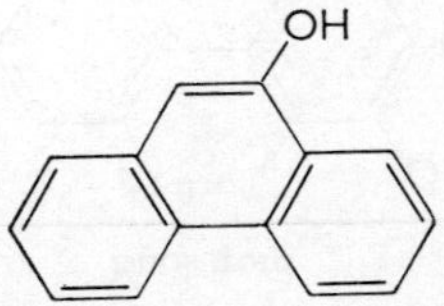

19.102

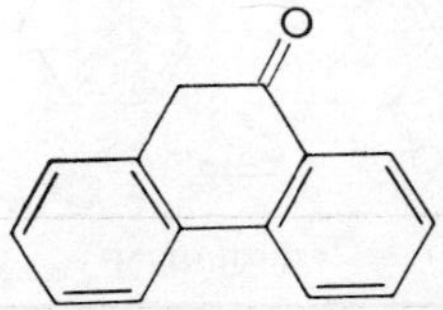

19.653

14

S. Forsén and M. Nilsson

No.	Enol from	TDE/β		Keto from	TDE/β	ΔTDE[b,c]
15		19·652	(a)		18·868	0·784
			(b)		18·864	0·788
16		19·646	(a)		18·954	0·692
			(b)		18·710	0·936
17		19·647	(a)		18·926	0·721

1·058   0·779   0·760   0·333   0·633

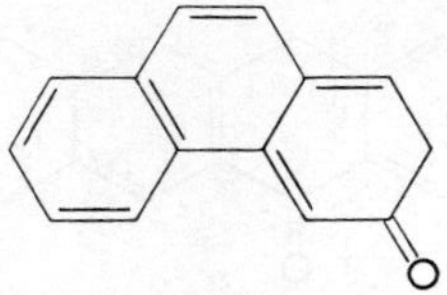

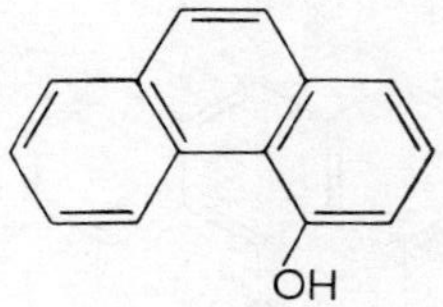

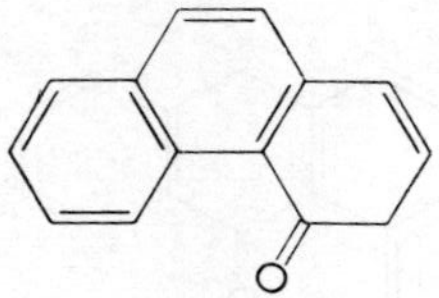

18·689   18·872   18·891   24·816   24·504

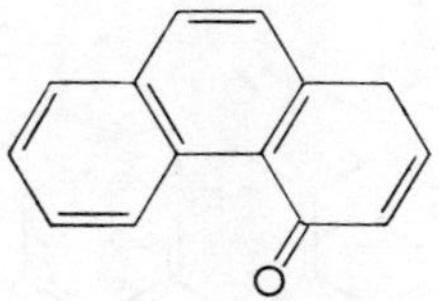

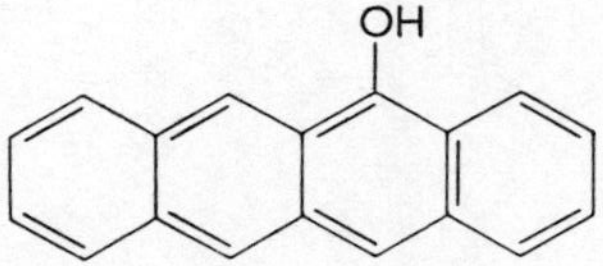

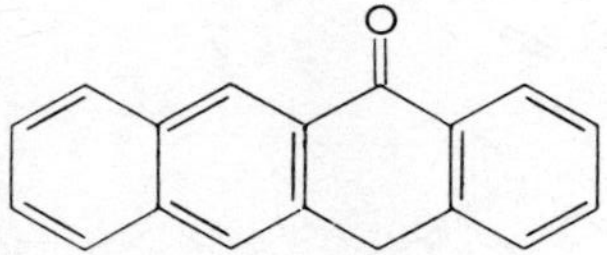

(b)   (a)   (b)   (a)

19·651   25·149   25·137

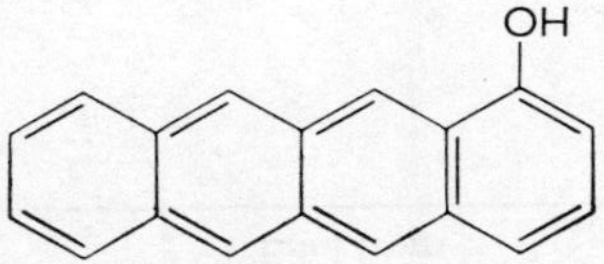

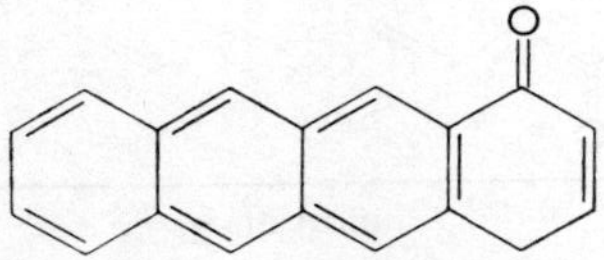

18   19   20

(b)

No.	Enol from	TDE/β	Keto from	TDE/β	ΔTDE[b,c]
				24·472	0·665
21		25·315		24·816	0·499
22		25·313		24·803	0·510
23		30·765		30·505	0·260

30·763

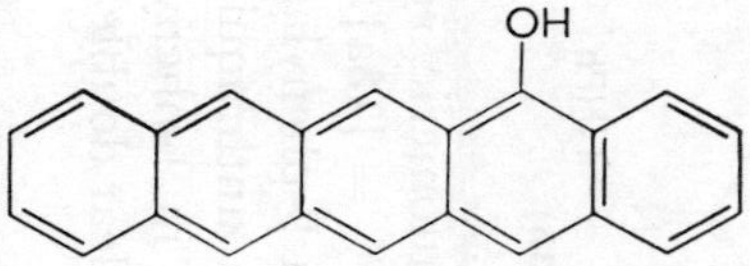

30·452   0·311

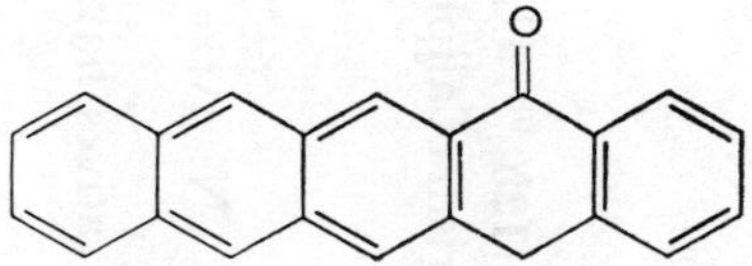

23·242

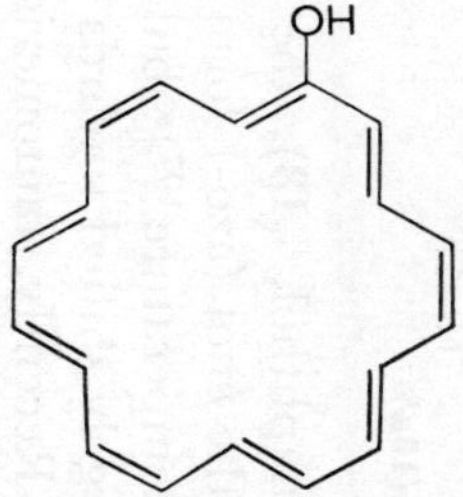

22·562   0·680

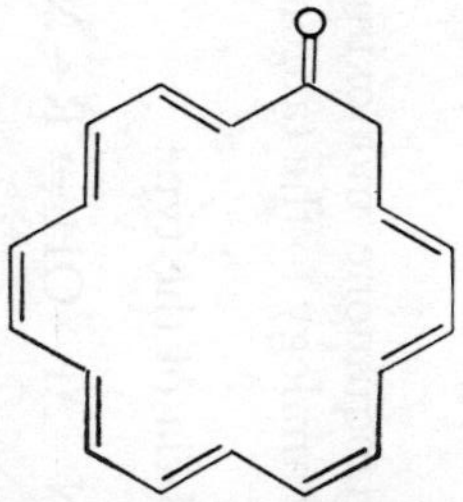

24

25

[a] The parameters employed are: $\alpha_{=O} = \alpha + \beta$, $\alpha_{-O-} = \alpha + 2\beta$, $\beta_{C=O} = \beta$ and $\beta_{C-O} = 0.8\beta$.

[b] $\Delta_{TDE} = (TDE_{enol} - TDE_{keto})/\beta$.

[c] Values within brackets are Δ_{TDE} per keto group formed.

B. Nitroso-, azo- and ketimino-phenols

The observation of keto–enol tautomerism in aromatic systems is not limited only to mono- or polyhydroxy derivatives. A classical case is p-nitrosophenol which has been shown to exist in solution pre-

(12a)   (12b)

dominantly as p-benzoquinone monoxime (**12b**)[98–101]. This system provides an aromatic analogy to the tautomerism of aliphatic α-nitroso ketones.

Tautomeric equilibria of the type

$$R—N{=}N—Ar—OH \rightleftharpoons R—NH—N{=}Ar{=}O$$

in azo derivatives of aromatic hydroxy derivatives have been recognized for a long time[102–106].

(13a)   (13b)

In 4-phenylazo-1-naphthol, (**13**), the tautomeric equilibrium is slightly in favour of the enol (azo-) form $(K_t = [\textbf{13a}]/[\textbf{13b}] = 1.35$ in ethanol at room temperature[107]); but in 9-phenylazo-10-anthrol the equilibrium is largely shifted towards the anthraquinone phenyl-hydrazone form[108a]. Recently, tautomerism in 1-phenylazo-2-naph-thols has been studied by means of heteronuclear double resonance on ^{14}N [108b].

In recent years tautomerism in Schiff bases of aromatic hydroxy aldehydes and hydroxy ketones has been studied in some detail by Dudek and colleagues mainly by n.m.r. spectroscopy[109–113].

The Schiff bases obtained by condensation of β-diketones, β-keto esters or α-formyl ketones with amines may exist in at least three tautomeric forms (*trans*-forms are omitted here): the 'true' Schiff

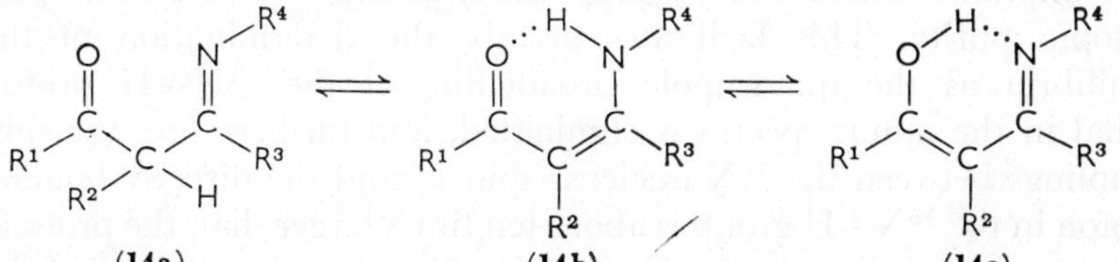

base (**14a**), the enamino ketone (**14b**) and the iminoenol (**14c**). U.v. and i.r. investigations of compounds of this type have provided strong support for the predominance of the enamino ketone form (**14b**)[114-116]. In p.m.r. spectra of aliphatic Schiff bases with R_3 = —CH_2R or —CH_3 and R_4 = H, spin couplings of the N—H proton with the protons on R_3 and R_4 have been observed[109,117]. The values of these spin couplings have been found to be fairly insensitive to variations in temperature or solvent, indicating that tautomer **14b** is appreciably more stable than the other forms. This form may be considered also as a vinylogous amide.

In p.m.r. spectra Schiff bases formed by condensation of methylamine or benzylamine with 1-hydroxy-2-acetonaphthone (**15**) or 2-hydroxy-1-naphthaldehyde (**16**), both the N—CH_3 and the N—CH_2 protons signals appear as doublets due to spin coupling with the N—H proton ($J_{\underline{H}-N-C-\underline{H}}5$ Hz). The spin coupling observed shows clearly

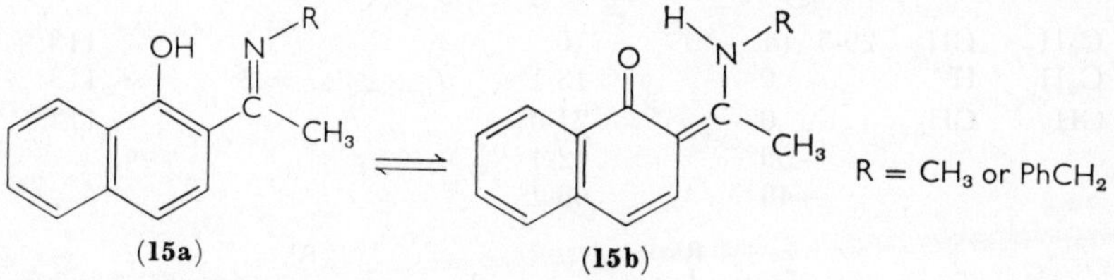

$R = CH_3$ or $PhCH_2$

$R = CH_3$ or $PhCH_2$

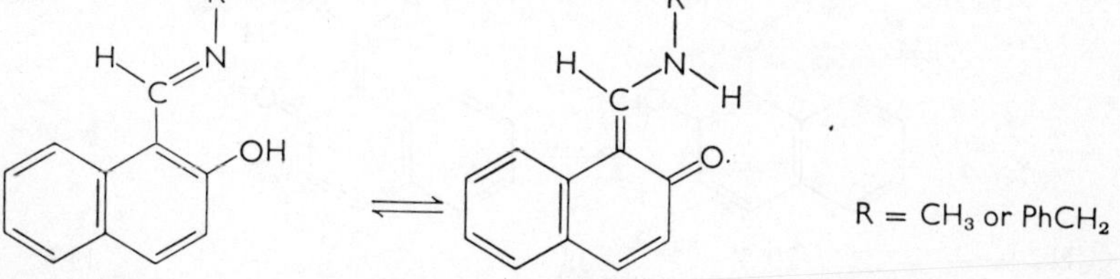

7*

that these aromatic Schiff bases are also present in the enaminoketo forms **15b** and **16b** to a considerable extent.

Later Dudek and Dudek[111–113] synthesized a number of aromatic and aliphatic Schiff bases using methylamine-^{15}N of 99 atom % isotopic purity. This facilitates greatly the determination of the equilibria as the quadrupole broadening of the ^{14}N—H proton signal in the n.m.r. spectra is eliminated, and furthermore, the spin couplings between the ^{15}N nucleus (spin $\frac{1}{2}$) and the directly bonded proton in the ^{15}N—H group is about ten times larger than the proton–proton spin couplings in the H—N—C—H group. A summary of the p.m.r. results obtained by Dudek and coworkers on aromatic ^{15}N-substituted Schiff bases is presented in Table 4. Perhaps the most interesting result is the fairly large ^{15}N—H spin coupling observed in Schiff bases of *o*-hydroxybenzaldehyde and *o*-hydroxyacetophenone, which shows that the free energies of the enamino ketone and iminoenol

TABLE 4.   Spin–spin couplings observed in n.m.r. spectra of ^{15}N-substituted Schiff bases of aromatic hydroxy aldehydes and hydroxy ketones[111–113].

R^1	R^2	Temperature (°c)	$J_{^{15}\text{N—H}}$ (Hz)^{a,b,c}	Other spin couplings (Hz)	Reference
C$_6$H$_5$	CH$_3$	29·5° to −40°	(0)		113
C$_6$H$_5$	H	0°	13·1^b	$J_{\underline{\text{H}}\text{—N—C—}\underline{\text{H}}} = 2$	113
CH$_3$	CH$_3$	0°	21·0		112
		−50°	32·4		
		−46°	68·2^c		
C$_6$H$_5$	—	29·5°	33		113
		−40°	49·4	$J_{\underline{\text{H}}\text{—N—C—}\underline{\text{H}}} = 1·5$	

CH$_3$	—	31°	64·5		111
		−45°	81·4	$J_{H-N-C-H} = 11·2$; $J_{H-N-CH_3} = 5·1$	

(structure: iminoenol ⇌ enamino ketone tautomers, R^1, OH, N, C–CH$_3$)

C$_6$H$_5$	—	31°	27·5		113
		−30°	36·5		
CH$_3$	—	31°	79·1	$J_{H-N-CH_3} = 4·7$	111
		15°	80·3		

(structure: R^1, C=N–CH$_3$, OH ⇌ R^1, C–N–CH$_3$, H, O)

CH$_3$	—	0° to 31°	(0)		111

^{a} In chloroform-d if not otherwise stated.
^{b} In ethanol.
^{c} In methanol.

tautomers are nearly equal. The relative amounts of the enamino ketone and iminoenol forms may be estimated from the value of the observed effective spin coupling constant ($J_{15_{N-H}}^{eff}$). If the 'N—H' or 'O—H' proton in the enamino ketone or iminoenol forms of a Schiff base undergoes a rapid *intramolecular* exchange between the two sites (a condition which is most likely fulfilled under the experimental conditions) the observed spin coupling constant ($J_{15_{N-H}}^{eff}$) should be a weighted average of the spin coupling constants in the pure enamino ketone and iminoenol forms according to equations (12a) and (12b)

$$J_{15_{N-H}}^{eff} = P_{ek}J_{ek}^{0} + P_{ie}J_{ie}^{0} \qquad (12a)$$

$$P_{ek} + P_{ie} = 1 \qquad (12b)$$

where P_{ek} and P_{ie} are the mole fractions of the molecules in the enamino ketone and iminoenol forms respectively. J_{ek}^{0} and J_{ie}^{0} are the spin coupling constants in the true enamino ketone and iminoenol forms, respectively. The value of J_{ie}^0 may be taken as zero* and the value of J_{ek}^0 was estimated by Dudek to be 88–89 Hz. With $J_{ek}^0 = 89$ Hz the relation between the equilibrium constant $K_t(=[\text{iminoenol}]/[\text{enamino ketone}])$ and $J_{15_{N-H}}^{\text{eff}}$ simply becomes $K_t = (89 - J_{15_{N-H}}^{\text{eff}})/J_{15_{N-H}}^{\text{eff}}$. From the temperature variation of K_t the enthalpy change ΔH of the enamino ketone–iminoenol interconversion has been obtained[113]. The ΔH values were found to be solvent dependent and in the range 0 to -3 kcal/mole.

C. Heteroaromatic hydroxy derivatives

For a comprehensive review of tautomerism in aromatic heterocycles see reference 127.

I. Pyridines, quinolines and related compounds

A number of spectroscopic and other physical data show that hydroxy derivatives of N-heteroaromatic compounds may exist in both the hydroxy (hydroxy imine) form and the keto amine (oxoamine) form[127] as exemplified by 2-hydroxypyridine (**17a**) and 2-pyridone (**17b**). Here again, analogies to the tautomerism of amides are apparent.

(**17a**)   (**17b**)

It is possible to imagine also other keto imine (oxo-imine) type of tautomers (**17c** and **17d**) where the interchangeable proton is

(**17c**)   (**17d**)

attached to a sp^3-hybridized carbon atom in the ring. These tautomeric forms have presumably higher energy than the keto amine forms as no evidence for keto imine forms has been found in monohydroxy derivatives of the simplest N-heterocycles. It is however not unlikely

* In o-hydroxybenzaldehydes, o-hydroxyacetophenone and related compounds spin coupling constants of the order of 0·5 Hz have been observed between the hydroxy proton and protons on the aromatic ring[118].

that the difference in free energy between the keto amine and keto imine forms may become quite small in monohydroxy derivatives of certain higher polycondensed *N*-heterocycles, for example in **18**.

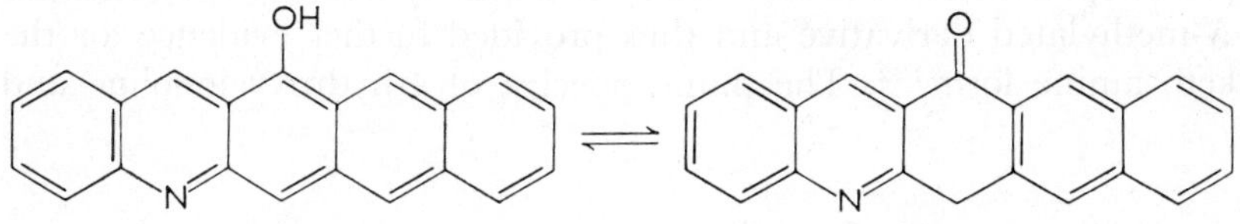

(18a)   (18b)

The position of the keto amine/hydroxy imine equilibrium for hydroxy heterocyclic compounds has been quantitatively or qualitatively investigated with several physical methods, mainly u.v., i.r. and n.m.r. spectroscopy, but also to a considerable extent by potentiometric determinations of ionization constants. In the u.v. spectroscopical approach, the spectrum of the hydroxy compound under investigation is usually compared with those of the *O*-methyl and *N*-methyl derivatives—the latter spectra are assumed to agree closely with those of the 'true' hydroxy imine and keto amine forms respectively. In this way the tautomeric equilibrium constant in a number of *N*-heteroaromatic hydroxy derivatives has been determined[128]. The u.v. spectroscopical method is best applicable for equilibrium constants in the range 10^2 to 10^{-2}. With the u.v. method it was early demonstrated that in **17** and **19** the keto amine forms **17b** and

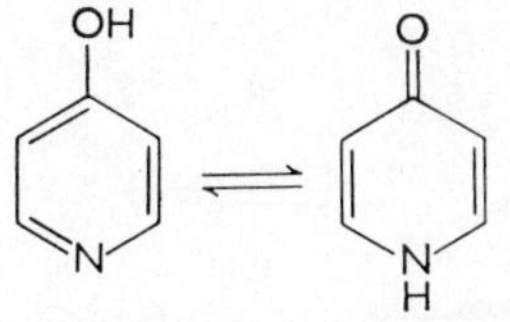

(19a)   (19b)

19b predominate in methanol[129].

Infrared spectroscopic investigations are mainly concerned with the spectral region of the N—H stretching frequency (3360–3500 cm^{-1}) and the C=O stretching frequency (1550–1780 cm^{-1}). Early applications of the i.r. technique were those of Thompson and coworkers[130] who confirmed the predominance of the keto amine form of 2- and 4-hydroxypyridine and of Witkop and coworkers[131] who concluded that 2-hydroxy- and 4-hydroxy-quinoline prevail in the keto amine forms.

N.m.r. studies of tautomeric equilibria have mainly been restricted to proton spectra—it is however conceivable that studies of ^{15}N and ^{17}O resonances should prove to be very informative. The observed p.m.r. spectrum of the **19a–19b** system closely resembled that of the *N*-methylated derivative and thus provided further evidence for the keto amine form[132]. The p.m.r. spectra of 4-hydroxycinnoline and

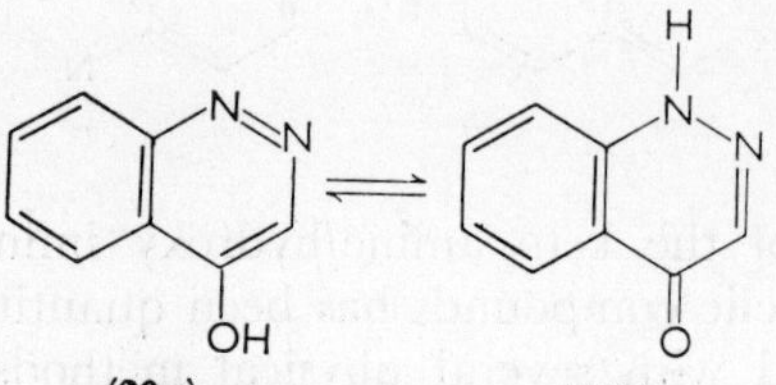

(20a)   (20b)

its 6,7-dimethoxy derivative showed that the most stable tautomer is the keto amine form **20b**[133].

Perhaps the most accurate information regarding the relative stabilities of the tautomeric forms of *N*-heteroaromatic hydroxy derivatives has been obtained by potentiometric determinations of ionization constants. For a more complete discussion see reference 231. The principles of this method may be illustrated by equation (13).

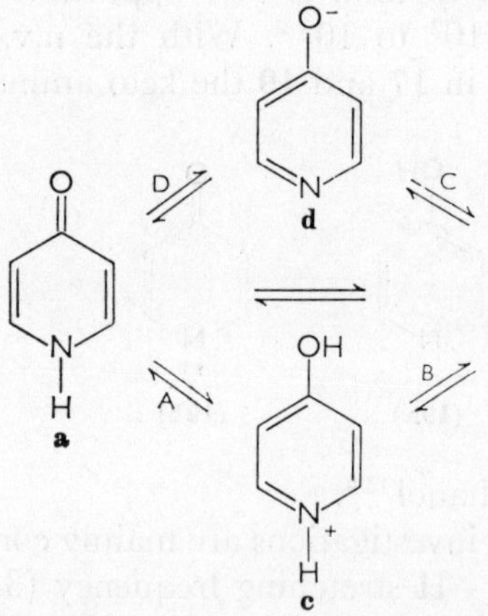

(13)

Anion (**d**) and cation (**c**) are common to both the keto amine (**a**) and the hydroxy form (**b**). The apparent ionization constants $K_1 = [H^+][HA]/[H_2A^+]$ and $K_2 = [H^+][A^-]/[HA]$, where $[HA] = [a] + [b]$, $[H_2A^+] = [c]$ and $[A^-] = [d]$, are related to the true ionization constants of the keto amine form $K_A(= [H^+][a]/[c])$ and $K_D(=$

$[H^+]$ $[\mathbf{d}]/[\mathbf{a}]$) and of the hydroxy form $K_B (= [H^+]$ $[\mathbf{b}]/[\mathbf{c}]$) and $K_C (= [H^+]$ $[\mathbf{d}]/[\mathbf{b}]$) through equations (14) and (15).

$$K_1 = K_A + K_B \tag{14}$$

$$K_2^{-1} = K_C^{-1} + K_D^{-1} \tag{15}$$

The basic ionization constants K_A and K_B of the keto amine and hydroxy forms, respectively, are generally not accessible, but are assumed to be approximately equal to those of the N-methyl (K_{NCH_3}) and the O-methyl (K_{OCH_3}) derivatives respectively. An underlying assumption is that protonation of the keto amine (pyridone) form takes place on the oxygen as indicated in equation (13). Conclusive evidence for predominant O-protonation in 2- and 4-pyridones has been obtained from p.m.r. studies[134,135]. Under these premises the tautomeric equilibrium constant (K_t = [keto amine]/[hydroxy imine]) may be calculated by means of equations (16).

$$K_t = K_A/K_B = K_C/K_D = K_{NCH_3}/K_{OCH_3} \tag{16}$$

Alternatively, if the apparent basic ionization constant K_1 and only *one* of the constants K_{NCH_3} or K_{OCH_3} is known, K_t may be calculated from equations (17) or (18).

$$K_t = (K_1/K_{OCH_3}) - 1 \tag{17}$$

$$K_t = [(K_1/K_{NCH_3}) - 1]^{-1} \tag{18}$$

A summary of reported equilibrium constant in aqueous solutions of the tautomeric forms of a number of N-heteroaromatic hydroxy derivatives is given in Table 5. One of the most surprising results in

TABLE 5.   Tautomerization equilibrium constants for hydroxy derivatives of N-heteroaromatic compounds[a].

Compound	K_t^{-1}	Method[b]	Reference
Pyridines:			
2-hydroxy	3.4×10^2	IC	27
	9.1×10^2	IC	28
3-hydroxy	0.8	IC	28
	1.0	UV	128
4-hydroxy	2.2×10^3	IC	27
	2.0×10^3	IC	28
3-amino-4-hydroxy	4×10^4	IC	c
3-nitro-4-hydroxy	2.5×10^3	IC	c
2-chloro-4-hydroxy	~ 1	—	d
2,6-dihydroxy	$\sim 3^e$	IC (UV)	136

S. Forsén and M. Nilsson

Quinolines:

2-hydroxy	3×10^3	IC	27
	$7\cdot6 \times 10^3$	IC	28
3-hydroxy	0·08	IC	28
	0·06	UV	128
4-hydroxy	$2\cdot4 \times 10^4$	IC	27
	$1\cdot6 \times 10^4$	IC	28
5-hydroxy	0·14	IC	28
	0·05	UV	128
6-hydroxy	0·008	IC	28
	0·014	UV	128
7-hydroxy	4·9	IC	28
	0·42	UV	128
8-hydroxy	0·021	IC	28
	0·035	UV	128

Isoquinolines:

1-hydroxy	$1\cdot8 \times 10^4$	IC	27
	$7\cdot1 \times 10^4$	IC	28
4-hydroxy	2·9	IC	28
	3·8	UV	128
5-hydroxy	0·035	IC	28
	0·038	UV	128
6-hydroxy	2·1	IC	28
	1·9	UV	128
7-hydroxy	0·043	IC	28
	0·038	UV	128
8-hydroxy	2·4	IC	28
	0·87	UV	128

Acridines:

1-hydroxy	0·016^f	IC	28
	0·04	UV	128
2-hydroxy	2·5^g	UV	128
5-hydroxy	10^7	IC	27

---

^{a} Unless otherwise stated the equilibrium constant, $K_t = [\text{'enol'}]/[\text{'keto'}]$, refers to aqueous solutions at 20°.

^{b} IC = ionization constant measurements; UV = u.v. spectroscopy.

^{c} R. A. Jones and B. D. Roney, *J. Chem. Soc.* (*B*), 84 (1967).

^{d} A. R. Katritzky and J. D. Rowe, personal communication.

^{e} About 60% is present as 6-hydroxy-2-pyridone 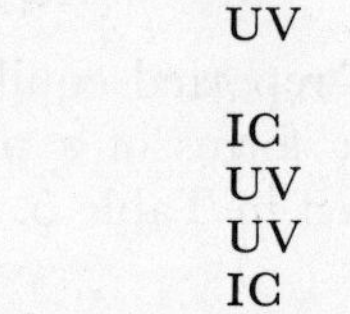 , about 15% in the dioxo form 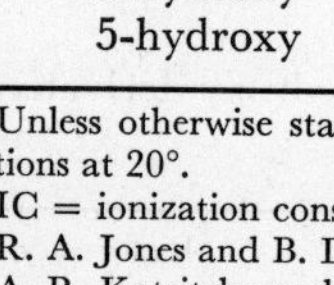 and about 25% in the dihydroxy form .

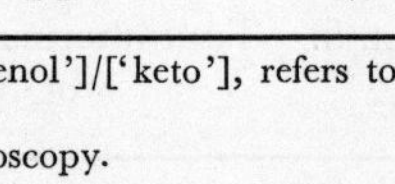

^{f} In 50% ethanol–water.

^{g} In 1:4 ethanol–water.

TABLE 6. Solvent dependence of the tautomeric equilibrium constant (K_t = ['enol']/['keto']) in 3-hydroxyisoquinolines[137].

Solvent	K_t	K_t	K_t
Water	$(0)^a$	$(0)^a$	$(0)^a$
Ethanol	1·5	1	0·5
Chloroform	3	1	0·67
Dimethyl sulphoxide	15	10	4
Benzene	30	10	2
N-Methylpyrrolidone	40	20	10
Carbon tetrachloride	80	20	~4
Diethyl ether	$(\infty)^b$	$(\infty)^b$	$(\infty)^b$

a Only the keto amine form is observed.
b Only the 'enol' form is observed.

Table 5 is the small difference in equilibrium constants between the keto amine and hydroxy forms in the 2-chloro-4-hydroxy- and 2,6-dihydroxypyridines. Decreased base character of the pyridine nitrogen caused by the adjacent substituent is presumably an important factor[136].

That keto amine/hydroxy imine equilibria are greatly affected by solvent interactions has recently been demonstrated by means of u.v. spectroscopy for 3-hydroxyisoquinolines[137]. From the data in Table 6 it appears that the dielectric constant of the solvent as well as the possibilities of hydrogen bond formation influence the value of K_t. That keto imine structures such as 21 are not involved in the equili-

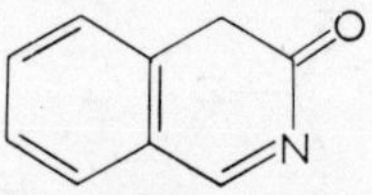

(21)

brium to any significant extent was indicated by the fact that 3-hydroxyisoquinoline does not exchange any C—H hydrogen even after 4 days in benzene solution in the presence of deuterium oxide.

2. Thiophenes, furans and pyrroles

The high resonance stabilization in the benzene ring favours the enolic forms of phenols. In view of the greater stability of the keto fragment relative to the enol fragment in aliphatic and alicyclic ketones one would expect that hydroxy derivatives of unsaturated compounds with low delocalization energy might in some cases exist predominantly in the keto form. In 2-hydroxy- and 3-hydroxy-derivatives of thiophene and furan, three and two tautomeric forms respectively are possible (22 and 23, with Z = O or S).

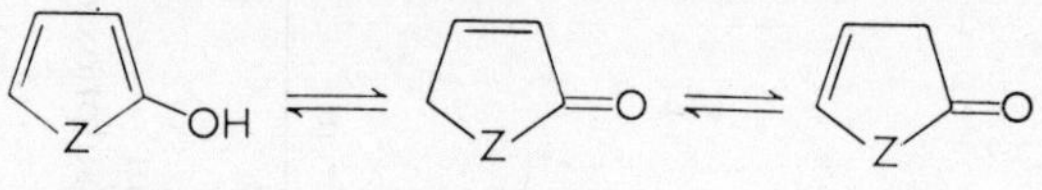

(22a)   (22b)   (22c)

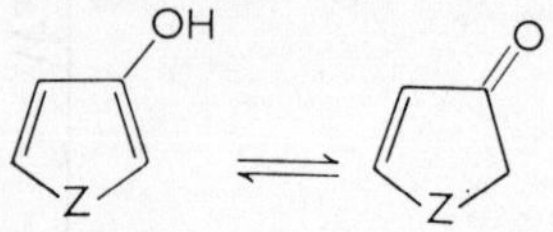

(23a)   (23b)

In the corresponding hydroxypyrroles there are, in addition to the tautomers in **22** and **23**, also possible keto imine type tautomers, for example **24a** and **24b**.

(24a)   (24b)

In recent years a number of potential hydroxy derivatives of thiophenes, furans and pyrroles have been investigated by physical methods. Although these compounds in some cases may have been prepared under conditions where equilibrium between the different tautomeric forms is established, care has not always been taken to ensure that the equilibrium is maintained under the conditions of the physical measurements. For example, in the case of 2-hydroxyfurans it has been observed that rather special conditions are necessary to secure the equilibrium between the tautomeric forms.

It has been shown by n.m.r. spectroscopy that 2-hydroxythiophene exists almost exclusively as 3-thiolene-2-one (**22b**, Z = S)[138]. In the 5-methyl homologue the two possible keto tautomers have been isolated and their equilibrium ratio has been determined[138]. By n.m.r. and i.r. methods it has also been demonstrated that the 3-methyl, 3-methoxy, 3-bromo, 4-methyl and 4-bromo derivatives of 2-hydroxythiophene exist predominantly in the 3-thiolene-2-one form (**22b**, Z = S), in the solid or liquid state as well as in cyclohexane solutions[139]. Hörnfeldt[140] measured the equilibrium ratios of the two predominant keto forms of a number of 5-alkyl-2-hydroxythiophenes in carbon tetrachloride, cyclohexane, acetone and nitromethane. The 3-thiolene-2-one form predominates and constitutes 80–95% of the equilibrium mixture. No spectroscopic evidence for the enol form is found, but the compounds react readily with acyl chlorides to give enol esters.

The free energy difference between the keto and enol forms of 2-hydroxythiophene is presumably not very high, and conjugative interaction with a substituent in the 5-position with the thiophene ring causes this difference to approach zero at room temperature. Thus in methanol solutions of 5-phenyl-2-hydroxythiophene and 5-thienyl-2-hydroxythiophene the equilibrium mixture contains approximately equal amounts of the enol forms **25a** and **26a** and the 4-thiolene-2-one forms **25b** and **26b** ($K_t = $ [enol]/[keto] = 0·43 for

(25a)   (25b)

(26a)   (26b)

25 and 0·33 for 26). Solvent effects such as hydrogen bonding are probably very important, as in chloroform the equilibrium is largely shifted towards the keto form[140].

The i.r. spectrum of 3-hydroxythiophene indicates that this compound exists as a mixture of enol and keto form[141]. This compound is highly unstable, and attempts to study the equilibrium with p.m.r. spectroscopy have as yet failed[140]. On the other hand, 2- and 2,5-alkylsubstituted 3-hydroxythiophenes are considerably more stable, and tautomeric equilibria in CS_2 solution were recently determined[142a]. The results are summarized in Table 7.

TABLE 7. Tautomeric equilibrium constants
(K_t = [enol]/[keto]) in 3-hydroxythiophenes[142,a]

R_1	R_2	K_t
CH_3	H	4
t-Bu	H	1·2
CH_3	CH_3	0·43

a In carbon disulphide at ca. 25°.
Regarding chelated hydroxythiophenes see Jakobsen and Laweson[142b].

In furanoid systems with oxygen attached at position 2 the butenolide forms are greatly favoured and so far no evidence has been presented for the occurrence of 2-hydroxyfurans. The equilibria between the keto forms 27a and 27b of the hypothetical 5-methyl- and 5-t-butyl-2-hydroxyfuran have been studied by p.m.r. spectroscopy[143]. In order to reach equilibrium at 60°c it was found necessary to use either pyridine as solvent or a strong base such as triethylamine in fairly high concentrations in benzene solutions. Under these conditions the form **27a** is somewhat more stable than **27b**. Chemically these furans appear to have little or no enolic character and differ in this respect greatly from the corresponding thiophenes.

$$R \overbrace{\hspace{1cm}}_{O}{=}O \rightleftharpoons R \overbrace{\hspace{1cm}}_{O}{=}O \qquad (R = CH_3 \text{ or } t\text{-Bu})$$

(27a)   (27b)

3-Hydroxyfuran has recently been synthesized via 3-methoxy-furan and studied by p.m.r., i.r. and u.v. spectroscopy[144]. The physical data show that this compound predominantly subsists in the keto form **28** and shows no enolic character in its reactions. The nature of the phenolic furan derivative synthesized by Hodgson and Davies[145] and described by them as true 3-hydroxyfuran remains an open question.

(28)   (29)

A number of 2-, 4- and 5-substituted derivatives of 3-hydroxyfuran have also been studied by spectroscopic methods and with one interesting exception all were found to exist in the keto form[146]. The exception is 2-acetyl-3-hydroxyfuran (isomaltol, **29**). Here the formation of a strong intramolecular hydrogen bond greatly favours the enol form. 2-Hydroxypyrrole as well as a number of substituted 2-hydroxy-pyrroles exists predominantly in the oxo (keto) forms in chloroform and dimethyl sulphoxide solutions[147]. 2-Formyl-3,4-dimethyl-5-hydroxypyrrole has been suggested to exist in the **30a** form in dimethyl sulphoxide solution[147].

(30a)   (30b)

3-Hydroxypyrrole is not known but a number of derivatives of this compound have been studied by Davoll[148]. For example the u.v.

spectrum of ethyl 2-methyl-4-hydroxy-pyrrole-3-carboxylate in ethanol shows that the keto form **31b** is predominant. Conjugative interactions of substituents with the pyrrole ring stabilize the enol form as shown by the behaviour of 1-phenyl derivatives. The u.v. spectra of these

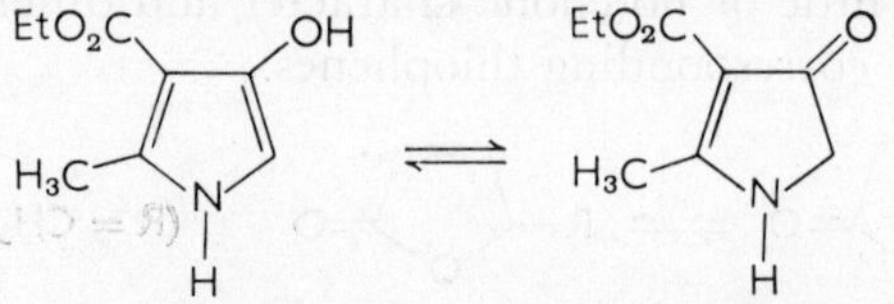

(31a)   (31b)

compounds in ethanol closely resemble those of the corresponding methyl enol ethers[148], suggesting that the hydroxy form **32a** is predominant. When one or two methyl groups are introduced in the

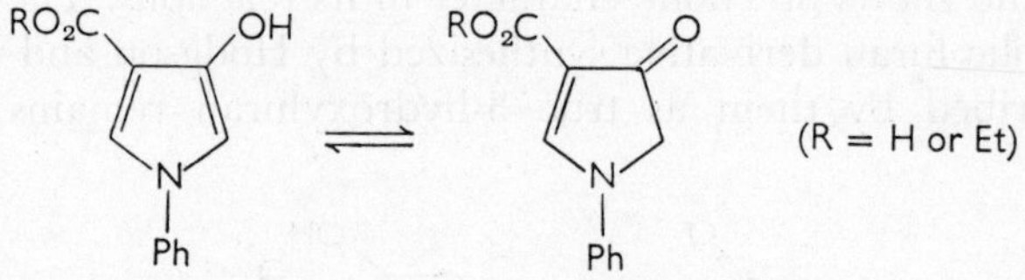

(R = H or Et)

(32a)   (32b)

o-positions of the 1-phenylpyrrole derivatives, steric interactions prevent coplanarity of the phenyl and pyrrole rings. The reduced interannular conjugation decreases the stability of the enol form, shifting the equilibrium markedly towards the keto (oxo) form.

V. KETO–ENOL TAUTOMERISM IN β-DICARBONYL COMPOUNDS

A. Introductory remarks

The present section deals with compounds of structure **33a**, which may enolize in a number of different ways (**33b–33e**).

For a large number of β-dicarbonyl compounds the free energies of the keto and enol forms differ only slightly and the equilibrium constants are fairly easily determined by physical and chemical methods. The number of investigations on β-dicarbonyl compounds are very large[149].

The influence of different substituents R^1, R^2 and R^3 on the keto-enol equilibrium has been the subject of several studies. The interpretation of the substituent effects cannot always be made in a unique manner as the effects on the keto and/or enol forms are difficult to disentangle and solvent effects are often marked. The importance of equilibrium studies on the gaseous phase will be emphasized in the following subsection.

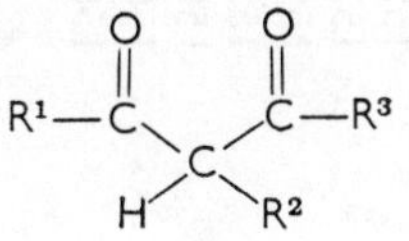

(33a)

cis-Enols:

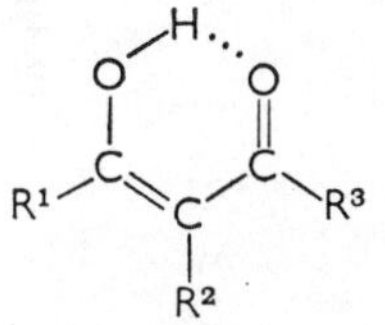

(33b) 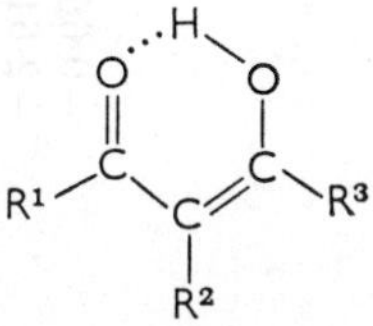   **(33c)**

trans-Enols:

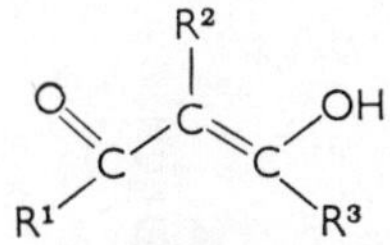

(33d) 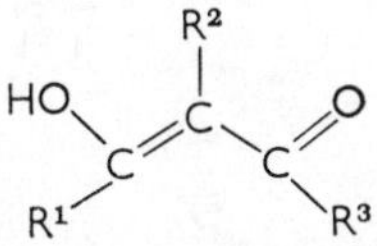   **(33e)**

When $R^1 \neq R^3$ four different enols **33b–33e** may be formed. In recent years information on equilibria between different cis-enol forms has been obtained, mainly from n.m.r. studies of ^1H and ^{17}O nuclei. These results will be discussed in some detail below.

When R^2 is an acyl group the number of possible enolic structures is considerably increased. These will be discussed in section V.F.

B. Thermodynamic data for keto–enol equilibria in the gaseous phase

As direct interconversion of keto and enol forms in the gaseous phase proceeds extremely slowly at ordinary temperatures, Conant and Thompson[150] used the liquid phase as an intermediary. If the keto–enol equilibrium is established in the liquid phase at a given temperature, the composition of the vapour phase in equilibrium with the liquid conforms to the equilibrium condition in the gas phase at the same temperature. By means of a slow isothermal equilibrium distillation at low pressure, the vapour may be removed from the proximity of the liquid surface and rapidly condensed onto a 'cold finger' kept at very low temperature (liquid air or solid carbon

TABLE 8. Thermodynamic data of keto–enol equilibria of β-diketones and β-keto esters in the gaseous phase (K_t = [enol]/[keto]).

R	T(°K)	K_t	ΔG(kcal/mole)	ΔH(kcal/mole)	ΔS(e.u.)	Reference
—H	273.2	19	−1.597	−1.777	−0.66	152
	313.2	11.7	−1.531	−2.847	−2.61	152
	353.2	7.25	−1.392	−2.916	−4.31	152
	413.2	3.66	−1.063	−3.77	−6.55	152
—CH$_3$	298.2	0.77–0.80	0.1	—	—	150
—CH$_2$CH$_3$	298.2	(0.52–0.59)^a	—	—	—	150
—CH$_2$C$_6$H$_5$	298.2	2.1–2.6	−0.5	—	—	150
—H	273.2	1.71	−0.295	−3.705	−12.48	154
	313.2	0.742	0.186	−3.440	−11.47	154
	373.2	0.318	0.850	−3.047	−10.43	154
—CH$_3$	273.2	0.340	0.586	−3.350	−14.41	153
	313.3	0.165	1.120	−2.753	−12.37	153
	373.2	0.0905	1.781	−1.857	−9.75	153
—CH$_2$CH$_3$	273.2	0.332	0.598	−6.261	−25.11	153
	313.2	0.0928	1.479	−4.504	−19.10	153
	373.2	0.0400	2.387	−1.869	−11.40	153

Compound	Temp.					Ref.
—CH₂C₆H₅	273·2	0.158	1·000	−0·616	−5·92	153
	313·2	0.134	1·251	−0·809	−6·58	153
	373·2	0.105	1·672	−1·100	−7·42	153
—C₆H₅	273·2	9·43	−1·217	−5·831	−16·89	153
	313·2	2·66	−0·608	−4·892	−13·68	153
	373·2	0.891	0·086	−3·568	−9·56	153
—CH₃, —C₃H₇-n	273·2	1·44	−0·198	−2·493	−8·40	153
	313·2	0.785	0·151	−2·671	−9·01	153
	373·2	0.381	0·715	−3·652	−0·79	153
—CH₂C₆H₅	273·2	0.471	0·408	−0·332	−2·71	153
	313·2	0.411	0·554	−0·600	−4·54	153
	373·2	0.298	0·898	−1·667	−6·88	153
structure (a) —OCH₂CH₃	298·2	0.38	0·57	—	—	151
structure (b) —OCH₂CH₃	298·2	9·6	−1·35	—	—	151

ᵃ Extrapolated from a higher temperature.

dioxide–acetone), and the composition of the 'quenched' condensate is subsequently investigated.

Selected thermodynamic equilibrium data on the gaseous phase of a few β-diketones and β-keto esters are given in Table 8. In judging the significance of the data in this Table, some of the possible sources of error in the isothermal distillation method should be kept in mind: (a) a true keto-enol equilibrium in the liquid phase was not established, (b) equilibrium between the liquid phase and vapour phase was not established and (c) systematic errors in the analysis (see section II). The following discussion is based upon the assumption that these errors are small.

The degree of enolization of acetylacetone and ethyl acetoacetate in the gaseous phase are higher than that in the neat liquids (in the liquids the enol contents are about 76% and 7·6% respectively at 25°c.). For all compounds studied enolization was found to be an exothermal process. It is also interesting to note that the enthalpy and entropy changes are markedly temperature dependent. The entropy changes accompanying enolization of β-diketones and β-keto esters are observed to be negative. It seems reasonable to assume that —at least at room temperature—the predominant enolic forms are the *cis*-enols which are stabilized by a strong intramolecular hydrogen bond. Thus after enolization the internal rotation of the two acyl groups around the bonds to the central carbon atom will be largely inhibited. The entropy decrease associated with this phenomenon will be dependent on the substituent masses and on the potential barriers to internal rotation of the acyl groups in the keto and enol forms. A crude estimate is 3 to 5 e.u. per group at room temperature.

The entropy decrease due to inhibited internal rotation may partly be compensated for by low frequency skeletal vibrations in the enol form[154b]. If the hydrogen bonded proton in the *cis*-enol form moves in a symmetrical double minimum potential, an additional term of about $R \ln 2$ is also gained[154c]. The observation[153] that the entropy change upon enolization is considerably more negative in CH_3COCH-$(Et)COOEt$ than in $CH_3COCH(CH_3)COOEt$ may possibly be due to restricted rotation of the C-ethyl group in the enol form of the former compound as a consequence of steric interactions with the —$COOEt$ group. Molecular models also show that in the enolic forms internal rotation for a benzyl group is somewhat less constrained than for a phenyl group[153].

The number of substituted β-diketones and β-keto esters so far investigated in the gaseous phase is at present very limited. The substituents have mostly been alkyl groups where steric factors rather than electronic interactions may govern the variations in the tautomeric equilibrium constants. Further studies of keto–enol equilibria in the gaseous state would be of great theoretical interest.

C. Thermodynamic data for keto–enol equilibria in liquids

A large number of investigations on keto–enol equilibria of β-diketones and β-keto esters in solution has been carried out at one particular temperature. Several studies of the temperature dependence of equilibria in solutions were also made[54,55,155–158], and some results are summarized in Table 9. Most data refer to the neat liquid, but acetylacetone was also studied in cyclohexane, carbon tetrachloride and dimethyl sulphoxide (DMSO). The data for the inert solvents do not differ appreciably from those of the neat liquid. The enthalpy changes are found to agree closely in the liquid phase and in the gaseous phase (Table 8), but the entropy change in the gaseous phase is surprisingly somewhat less than in the neat liquid or in the inert solvents cyclohexane and carbon tetrachloride.

No evidence for *trans*-enol forms was found in the p.m.r. spectra obtained by Burdette and Rogers[54]. If present at all, the *trans*-enol forms constitute less than about 3% of the tautomers.

As in the gaseous phase, enolization in the liquid phase is found to be an exothermic process accompanied by large entropy reductions. As for substituent effects, there appears to be a tendency to increase the stability of the enol form relative to that of the keto form with increasing electro-negativity of the substituents (see ΔH in Table 9).

The influence of the medium on the enolization of acetylacetone and ethyl acetoacetate is illustrated in Table 10. In these, as well as in other β-dicarbonyl compounds, the K_t values for highly polar solvents may differ considerably from those for non-polar solvents (cf. section V.G).

It is as yet difficult to rationalize the dependence of K_t on structure. However, it seems from published data, particularly for substituted benzoylacetones[175], that enol substituents R^1 and R^3 (which are electron withdrawing according to resonance theory) tend to increase the stability of the enol form (taken as $\Delta G = - RT \ln K_t$) relative to the keto form. For a series of β-diketones Kabachnik and Ioffe[159] found a Hammett type correlation between $\log K_t$ and the sum of $(\sigma_p - \sigma_I)$ for the two substituents R^1 and R^3 in **33**. σ_p is the Hammett substituent constant for *para*-substitution and σ_I (or σ') is a polar substituent

TABLE 9. Thermodynamic data of keto–enol equilibria in the liquid phase (K_t = [enol]/[keto]) obtained by n.m.r. spectroscopic methods

Compound	Medium	Temperature range (°c)	K_t	ΔG (kcal/mole)	ΔH^c (kcal/mole)	ΔS (e.u.)	Reference
Pentane-2,4-dione	Neat liquid	−11 to 59	3.71[a]	−0.80[a]	−2.84	−6.7[a]	54
		38 to 170	3.65[b]	−0.80[b]	−2.8	−7.3	55
	C_6H_{12} and CCl_4	—	—	—	−3.0	−6	55
	DMSO	—	—	—	−1.6	−5	55
3-Methylpentane-2,4-dione	Neat liquid	−11 to 89	0.52[a]	0.40[a]	−1.33	−5.7[a]	54
		38 to 170	0.39[b]	0.58[b]	−1.3	−6.1	55
3-Allylpentane-2,4-dione	Neat liquid	25 to 125	0.72[b]	0.20[b]	−2.1	−7.3	55
3-Benzylpentane-2,4-dione	Neat liquid	38 to 170	0.86[b]	0.09[b]	−2.5	−8.4	55
3-Chloropentane-2,4-dione	Neat liquid	33 to 80	14[a]	−1.61[a]	−5.92	−14.1[a]	54
		38 to 170	—	—	−4.6	−9.5	55
1-Phenylbutane-1,3-dione	Neat liquid	90 to 170	49[b]	−2.41[b]	−2.8	−3.9	55
Dibenzoylmethane	Neat liquid	90 to 180	—	—	−3.2	−3.4	55
Ethyl 2-chloro-acetoacetate	Neat liquid	−33 to 105	0.65[a]	0.26[a]	−0.88	−3.7[a]	54
Ethyl 4,4,4-tri-fluoroacetoacetate	Neat liquid	5 to 80	6.0[a]	−1.09[a]	−3.91	−9.2[a]	54

[a] At 33°.
[b] At 38°.
[c] Average values over the temperature range studied.

constant essentially equivalent to that determined by Roberts and Mooreland in their classical study of 4-substituted bicyclo[2·2·2]-octane-1-carboxylic acids and esters[160,161]. The difference $\sigma_p - \sigma_I$ may be considered to be a measure of the resonance polar effects of a substituent[162].

TABLE 10. Keto–enol equilibrium constants (K_t = [enol]/[keto]) for acetylacetone and ethyl acetoacetate in some common solvents[a].

Solvent	Acetylacetone		Ethyl acetoacetate	
None	4.3[51]	3.65[55]	0.09[51]	0.08[25]
Hexane or carbon tetra-chloride	19[52]		0.64[52]	1.08[159c]
Cyclohexane	7.3[159b]		0.75[159b]	1.08[159c]
Carbon disulphide	16[52]		0.72[159b]	0.25[52]
Diethyl ether	5.7[159b]	19[52]	0.33[159b]	ca. 0.45[159c]
Benzene	8.1[52]		0.25[159c]	0.19[52]
Dioxan	2.7[159b]	4.6[52]	0.087[159b]	0.12[52]
Methanol	2.3[159b]	2.8[52]	0.064[159b]	0.075[159c]
			0.062[52]	0.064[25]
Ethanol	2.7[159b]	4.6[52]	0.087[159b]	0.11[52]
Water	0.19[159b]		0.005[159a]	—

[a] Temperatures vary generally near 20 to 25°. Various methods of measurement.

For substituents R^2 the situation is less clear. Some substituents which are electron withdrawing in the resonance theory sense, such as —CN or —Ac, increase the stability of the enol form (see section V.F). Bulky R^2 substituents decrease the stability of the enol form relative to that of the keto form, presumably due to steric interactions between the substituents R^1, R^2 and R^3 in the cis-enol forms.

The available data for cyclic β-diketones and β-keto cyclic aldehydes mostly refer to alcoholic or aqueous solutions. In these media one cannot exclude the possibility of interfering hemiacetal formation or hydration of the carbonyl group—especially in the case of aldehydes[163–167]. Reported K_t values are generally based on bromine titrations and may be affected also by by-products such as hydrates and hemiacetals (hemiketals) of the keto and enol forms and may therefore not always represent true equilibrium constants.

The series of alkyl 2-ketocycloalkanecarboxylates (ring size 5 to 17) show an interesting variation in K_t determined in alcoholic solution, with K_t maxima for the 6, 8 and 10-membered rings[16,40,167a]. The ring size dependence of K_t is possibly connected with a similar variation in the difference in the internal ring strains between keto and enol forms. If one assumes that the carbon–carbon double bond in the enols is *endo*-cyclic the difference in K_t between the five- and six-membered rings conforms to the generalizations of Brown, Brewster and Schechter[74]. On the other hand, the data of Campbell and Gilow[173,174] on aryl-substituted 2-benzoylcycloalkanones do not show the same variation with ring size as the cyclic β-keto esters. In the benzoylcycloalkanone series the five-membered compounds show a higher enolic content than the seven- and six-membered ring compounds, in decreasing order. It has recently been demonstrated, however, that in 2-benzoylcycloalkanones the location of the carbon–carbon double bond in the enol fragment is predominantly *endo*-cyclic in the cyclohexanone derivatives and *exo*-cyclic in the cyclopentanes[58] (see discussion in section V.D), and a comparison with cyclic β-keto esters is therefore less valid.

In the extensive series of substituted 2-benzoyl-cyclopentanones and -cyclohexanones summarized in Table 11 it has been found by Campbell and Harmer[168] that log K_t is linearly correlated with the Hammett σ-constants of the substitutents. The enolization of benzoylcyclopentanones is more sensitive to substituent effects than that of benzoylcyclohexanones. This is in accordance with the higher

TABLE 11.   Keto–enol equilibrium constants ($K_t = $ [enol]/[keto]) for 2-benzoyl-cyclopentanones and -cyclohexanones in methanol solution[168].

Substituent in benzoyl group	Cyclopentanones	Cyclohexanones
None	0·630	0·034
p-F	0·630	0·044
p-Cl	0·927	0·043
p-CH$_3$	0·443	0·031
p-OCH$_3$	0·261	0·027
p-Br	1·36	—
p-I	0·976	—
p-NO$_2$	2·97	—
m-Cl	1·15	0·074
m-F	1·13	—
m-CH$_3$	0·435	—
o-F	2·14	0·949
o-Cl	2·99	7·93
o-Br	6·14	13·1
o-CH$_3$	3·53	1·0
o-OCH$_3$	0·942	

degree of coplanarity of the enolized benzoylcyclopentanone molecules. The correlation with the Hammett σ-constants in both series of compounds is in the sense that electron donor substituents ($-CH_3$, $-OCH_3$) decrease the stability (ΔG) of the enol form relative to the keto form, i.e. in the same sense as in acyclic β-diketones.

D. Direction of cis-enolization in β-dicarbonyl compounds

The mechanism of the interconversion between the two possible *cis*-enols (**33b** and **33c**) poses an intriguing problem. The potential energy function characterizing the interconversion will depend in a rather complex way on the various interatomic distances in the molecules. Let us however assume that this structure dependence of the potential energy may be approximated with a single reaction coordinate. Consider first the case when the potential function representing the enol–enol reaction is of a double minimum type with the barrier high above the ground-state of the normal modes of the skeletal vibrations in enols. In this case it would be meaningful to treat the interconversion as an ordinary chemical reaction, and it might be possible to determine its activation energy. However, if the potential barrier becomes very low and comparable to the ground state vibration energies in the enols, quantum-mechanical tunnelling is likely to take place. If the barrier is of a symmetrical or unsymmetrical single minimum type it would be somewhat inadequate to speak of 'interconversion of' and 'equilibrium between' different enol forms. Little experimental evidence is available to differentiate between the representations. Recent x-ray studies[169–171] of the *cis*-enol forms of dibenzoylmethanes have given evidence for both symmetrical [in bis(*m*-chlorobenzoyl)methane and bis(*m*-bromobenzoyl)methane] and unsymmetrical (in dibenzoylmethane) enol forms. From line width observation in the ^{17}O n.m.r. spectra of β-diketones, the lower limit to the rate of exchange between *cis*-enols is found to be at least 10^5–10^6 s^{-1} at room temperature[58].

Several attempts have been made to determine the relative amounts of the different *cis*-enol forms of unsymmetrical β-diketones. Non-chelated fixed *trans*-enols react with diazomethane in ether very rapidly and chelated enols very slowly to form the corresponding methyl enol ethers. The methylation of chelated enols is catalysed by small amounts of methanol.

Chelated enols give primarily the *cis*-enol methyl ethers which, however, may rearrange to the corresponding *trans*-enol ethers[172].

The relative amounts of enol ethers formed have sometimes been assumed to reflect the relative amounts of the corresponding enol forms. The results depend however not only on the relative amounts of enols but also on their reactivity, giving at best a qualitative proof of the occurrence of different enol forms[173]. U.v. and i.r. spectroscopy have also been applied to study equilibria between cis-enols[174,175], but the assignment of absorption bonds and extinction coefficients is uncertain.

The two cis-enols **33b** and **33c** of an unsymmetrical β-diketone should afford different n.m.r. spectra (this applies not only to ¹H but also to ¹³C and ¹⁷O nuclei). If interconversion of the two enol forms takes place at a rate higher than the differences in resonance frequencies between the signals from the two separate cis-enols, a single spectrum will be observed, where the positions of the resonance signals would be weighted averages of the chemical shifts of the separate enol forms. The weight factors would correspond to the relative amounts of the two different cis-enol forms. A similar weighted averaging may also be expected for spin–spin couplings, provided the interconversion takes place strictly intramolecularly—intermolecular exchange might interfere presumably only in the case of the OH proton. In the application of the n.m.r. method to study the relative populations of cis-enol forms, the fundamental problem is to determine or to make a safe estimate of the spectral properties of the individual enol forms.

(34a)   (34b)   (34c)

The series phenylmalondialdehyde (**34**, $R^1 = R^2 = H$), 1-formyl-1-phenylpropan-2-one (**34**, $R^1 = H$, $R^2 = CH_3$) and ethyl 2-formyl-2-phenylacetate (**34**, $R^1 = H$; $R^2 = OEt$) has been studied in carbon tetrachloride solutions at $33°C$[117]*. All three compounds

---

* In this paper the importance of intermolecular proton exchange was not fully realized: The fact that the symmetrical compound phenylmalondialdehyde did not show any spin coupling between the =C—H and —OH protons in the enol form is probably due to intermolecular proton exchange. The value of 6·3 Hz for the spin coupling constant between the —OH and =C—H proton in 1-formyl-1-phenylpropan-2-one given in this paper may also be somewhat too low due to intermolecular proton exchange[176].

were found to be extensively enolized ($K_t \geq 20$) and no *trans*-enols were observed in the solutions. The chemical shifts of the =C—H protons in the enol forms were δ 8·65, 8·15 and 7·33 p.p.m respectively. In ethyl 2-formyl-2-phenylacetate (**34**, R^1 = H, R^2 = OEt) a large H—C—O—H spin coupling constant of 12.5 Hz was observed, which in combination with the chemical shift of the =C—H proton indicates this compound to be predominantly in the hydroxy-methylene form (**34b**). In phenylmalondialdehyde each of the two *cis*-enol forms must have a 50% population, and it is reasonable to assume nearly 100% of the **34b** form for the enol of ethyl 2-formyl-2-phenyl-acetate. The chemical shift of 8·15 p.p.m. for the =C—H proton in

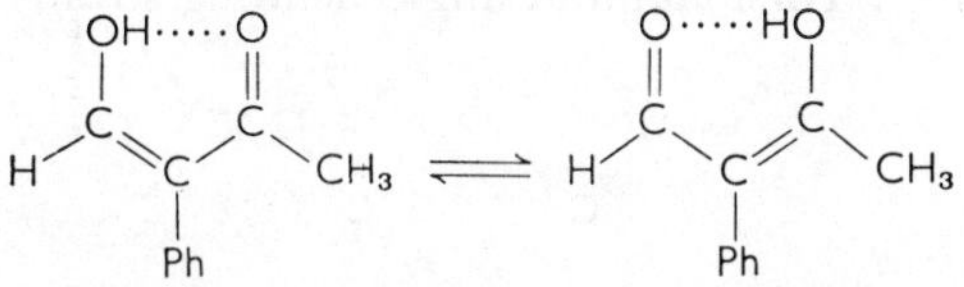

(35a) ⇌ (35b)

the enol of 1-formyl-1-phenylpropan-2-one indicates that the populations of the hydroxymethylene form **35a** and the aldoenol form **35b** are in the ratio 2:1.

Garbisch[56] studied the aldo-enol/hydroxymethylene-ketone equilibria in a number of highly enolized cyclic β-keto aldehydes. A linear

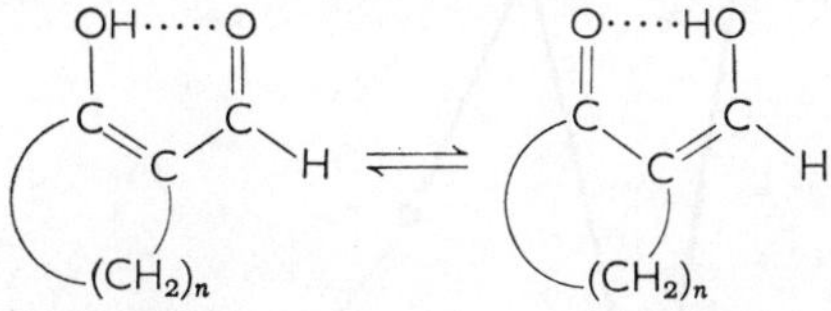

(36a) ⇌ (36b)

correlation was observed between the mole fraction of the hydroxy-methylene-ketone forms (**36b**) as calculated from the observed shifts of the 'aldehydic' proton (which ranged from δ 8·95 to 6·71 p.p.m —the latter value was observed in formylcamphor and was taken to represent the shift in a true hydroxymethylene ketone form) and the effective spin coupling constant $J_{\text{H—C—O—H}}^{\text{eff}}$. This result shows that the mole fractions P_{hk} and P_{ae}, of the hydroxymethyleneketone and aldo-enol forms respectively, may be calculated not only from the observed shifts of the 'aldehydic' proton but also from the observed

8+c.c.g. II

spin coupling constant $J_{\underline{H}-C-O-\underline{H}}^{eff}$ according to equation (19)

$$J_{\underline{H}-C-O-\underline{H}}^{eff} = P_{hk}J_{hk}^0 + P_{ae}J_{ae}^0 \qquad (19)$$

where J_{hk}^0 and J_{ae}^0 are the H—C—O—H spin coupling constant in the hydroxymethylene-ketone and aldo-enol forms, respectively. J_{hk}^0 is presumably about 12·5 Hz and relatively independent of the strength of the intramolecular hydrogen bond, while J_{ae}^0 is virtually zero. The fraction of enol in the **36b** form as a function of ring size in the cyclic β-keto aldehydes is depicted in Figure 1. Later Garbisch was able to rationalize this variation on the basis of strain energy differences between the hydroxymethylene-ketone and aldo–enol forms in the different rings[177]. Torsional and angle bending strains in the

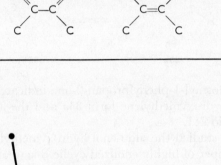

FIGURE 1. Fraction (P_{hk}) of hydroxymethylene-ketone form (**36b**) as a function of ring size (n) for the enols of cyclic β-keto aldehydes.

moieties are shown to have a major influence on the energy. The theoretical equilibrium constants, calculated assuming the entropy changes to be small, are in excellent agreement with the observed values. Compare a recent study of 3-formyl-bornan-2-one by Baker and Bartley on structural effects on the spin–spin coupling constant H—C—O—H[178b].

Recently Russell and Garbisch[178] have studied a large number of formyl ketones over a range of temperatures and obtained the enthalpy and entropy changes for the enol–enol interconversion. The entropy changes were found to be very small. Theoretical calculations by Garbisch[179a] indicate that in planar *exo* conjugated systems the

$$
\begin{array}{cc}
\text{OH}\cdots\text{O} & \text{O}\cdots\text{HO} \\
\text{H}_2\text{C} \diagup\text{C}\diagdown\text{C}\diagup\text{C}\diagdown\text{H} & \text{H}_2\text{C}\diagup\text{C}\diagdown\text{C}\diagup\text{C}\diagdown\text{H} \\
(37a) & (37b)
\end{array}
$$

configuration **37a** has less torsional strain and higher delocalization energy of the π electrons than **37b**, whereas in *endo* conjugated systems, the configuration **38a** has less torsional strain and less delocalization energy than **38b**.

$$
\begin{array}{cc}
\text{O}\cdots\text{HO} & \text{OH}\cdots\text{O} \\
\text{HC}\diagup\text{C}\diagdown\text{C}\diagdown\text{H} & \text{HC}\diagup\text{C}\diagdown\text{C}\diagdown\text{H} \\
(38a) & (38b)
\end{array}
$$

As a consequence, the enol–enol equilibrium in *exo* conjugated α-formyl-ketones is generally shifted more towards the aldo–enol form than in corresponding saturated systems. In *endo* conjugated systems the situation is less clear as the strain effects and delocalization energy tend to cancel each other.

In their most interesting series of communications on hydrogen bonds Musso and coworkers have described, among other things, the tautomerism of cyclopentanediene aldehydes[179b]. The monoaldehyde in nonpolar solvents and in the solid state is present as a mixture of the three possible oxo forms with no evidence for the presence of the 6-hydroxyfulvene form. Acetylcyclopentanedienes behave similarly.

Cyclopentanediene-1,2-dialdehydes, however, are highly enolic and apparently contain a seven-membered 'conjugated chelate' ring with a strong intramolecular hydrogen bond. At low temperatures the n.m.r. signal from the enolic proton is even observed as a triplet ($J = 7.9$ Hz) due to spin coupling with the effectively equivalent 'aldehydic' protons. These compounds accordingly can be described as 6-hydroxyfulvene-2-aldehydes. The studies have also concerned the nitrogen analogues.

The study of equilibria between *cis*-enol forms with proton magnetic resonance is to some extent limited to keto aldehydes since the chemical shifts of protons further removed from the carbonyl groups are barely affected. A comparison of the separate acetyl signals from the ketonic and enolic forms of acetylacetone, ethyl acetoacetate and related compounds indicates that the methyl proton chemical shifts of

$$CH_3-\overset{\overset{\displaystyle O}{\|}}{C}- \quad \text{and} \quad CH_3-\overset{\overset{\displaystyle OH}{|}}{C}=$$

differ by only 0·1 to 0·2 p.p.m.—the signals from the 'enolized' acetyl groups occurring at higher fields. The chemical shifts of the acetyl signal in the enols of 2-acetylcyclopentanone and 2-acetylcyclohexanone in carbon tetrachloride solution occur at δ 1·94 p.p.m. and 2·05 p.p.m. respectively. This difference has been interpreted as an indication that in 2-acetylcyclopentanone the predominant enolic form has an *exo*-cyclic double bond[53]. This result is in accordance with recent [17]O n.m.r. data[58].

The use of [17]O n.m.r. spectroscopy is not limited to the study of keto aldehydes. The [17]O chemical shifts of keto-type oxygen and hydroxy-type oxygen differ by about 500 p.p.m., this difference being larger by more than two powers of ten than that between $-\overset{\overset{\displaystyle O}{\|}}{C}-H$ and $=\overset{\overset{\displaystyle O}{|}}{C}-H$ protons. An inconvenience of the method is the low natural abundance of [17]O, which requires the use of [17]O labelled compounds. The labelling may however be accomplished with relative ease through exchange in [17]O-enriched water[58,180].

The method used to calculate the enol–enol equilibrium constant is analogous to that used by Garbisch for proton spectra[56]. In the spectra of the enol forms of unsymmetrical β-diketones two [17]O resonances are observed. If the rate of interconversion of the two *cis*-enol forms in **39** and **40** is slow one would expect two [17]O signals from each form one (at lower field) from the carbonyl oxygen and one from the hydroxyl oxygen. Since the chemical shifts of the ^{17}O signals are relatively insensitive to variations in the substituents not directly bonded to the oxygen atom, one would expect the two carbonyl resonances to almost coincide, as would be also the case for the hydroxyl resonances. The separation Δ between the hydroxyl and the

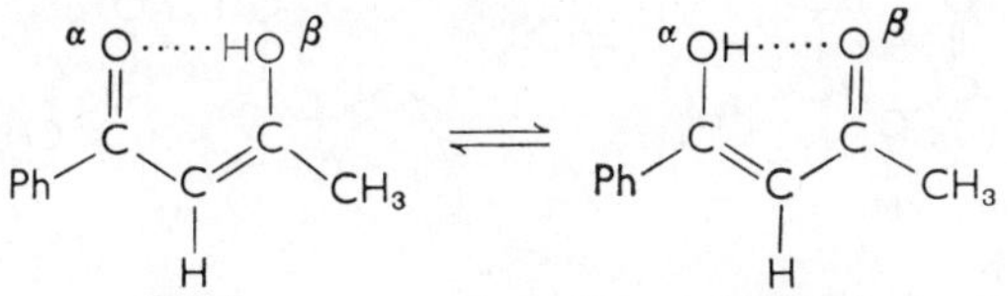

(39a)   (39b)

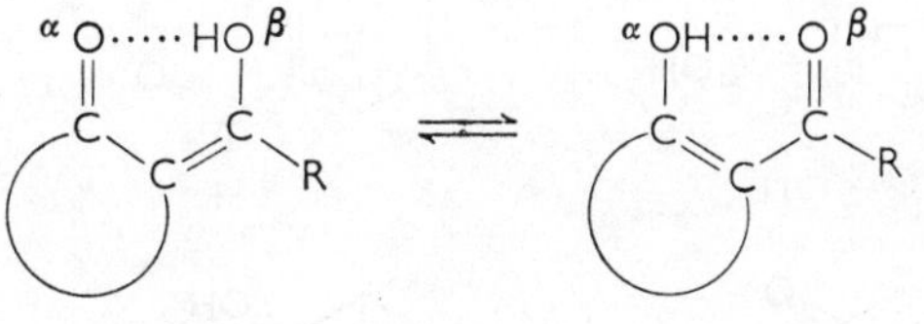

(40a)   (40b)

carbonyl signals was estimated[58] to be 460 p.p.m. If however the rate of interconversion of the two enol forms is faster than the frequency separation of the OH and C=O resonances, one would expect each oxygen (labelled α or β) to show only *one* resonance signal at a shift δ^α or δ^β which is a weighted average of the shifts in the 'true' enol forms ($\delta^{C=O}$ and δ^{OH}) according to

$$\delta^\alpha = P_a \delta^{C=O} + P_b \delta^{OH} \tag{20}$$

$$\delta^\beta = P_b \delta^{C=O} + P_a \delta^{OH} \tag{21}$$

where P_a and P_b are the mole fractions of the enolized molecules in the enol forms (39a or 40a) and (39b or 40b), respectively. This latter behaviour is also experimentally observed. Introducing $\delta = \delta^\alpha - \delta^\beta$ (i.e. the difference between the observed ^{17}O resonances) and $\Delta = \delta^{C=O} - \delta^{OH}$ one can write the equilibrium constant $K = P_b/P_a = P_b/(1 - P_b)$ as[58]

$$K = (\Delta - \delta)/(\Delta + \delta) \tag{22}$$

The results on enol–enol equilibria are summarized in Table 12.

The ^{17}O measurements corroborate the predominance of the hydroxymethylene-ketone form in the enol of formylcamphor (2 in Table 12) found by Garbisch. In the case of α-formylcyclohexanone

TABLE 12. Equilibrium constants of enol–enol equilibria in benzene solution at 27°c as determined by ^{17}O n.m.r. spectroscopy[58].

No.	Enol forms		$K = [B]/[A]$
	A	B	
1			1·29
2			0·10 ± 0·05
3			0·21
4			2·02
5			2·34
6			1·32
7			0·43

(4 in Table 12) the equilibrium constant found by Garbisch on carbon tetrachloride solutions is about 50% higher than the value in benzene solutions obtained from the ^{17}O measurements. This difference may possibly be ascribed to solvent effects and experimental uncertainties but may, however, also be real and an indication of the inadequacy of considering the equilibria between *cis*-enols as ordinary chemical equilibria, as mentioned at the beginning of this section. ^{17}O n.m.r. results on the enols of benzoylacetone (**39**) are in agreement with previous qualitative conclusions drawn from u.v. spectra[175]. The major factors which determine the direction of enolization are, in order of decreasing importance:

(a) *endo*-cyclic double bonds are favoured in cyclohexanone enols whereas *exo*-cyclic double bonds are favoured in cyclopentanone enols;

(b) a phenyl group attached to a carbonylic carbon favours the enol form $C_6H_5C(OH)=CCOR$ relative to the form $C_6H_5COC=C(OH)R$;

(c) a methyl group tends to stabilize the enol form in which the methyl group is adjacent to a carbonyl group better than a hydrogen atom.

The first factor conforms to the generalizations of Brown, Brewster and Schechter[74]. The second factor is in the direction predicted both by Hückel-type MO calculations[97] and by 'extended Hückel' MO calculations[81]. The third factor is in accordance with the p.m.r. results on 1-formyl-1-phenyl-propan-2-one (**35**) reported above and with observations on 1-formyl-1-methylpropan-2-one, 1-formyl-1-methylbutan-2-one[179] and 1-formylpropan-2-one[181]. Extended Hückel calculations on the two *cis*-enol forms of 1-formyl-propan-2-one give the energies $-15652\cdot2$ kcal/mole for enol form **41a** and $-15655\cdot9$

(41a)   (41b)

kcal/mole for enol **41b**[81]. The calculations thus predict the hydroxy-methylene-ketone form **41b** to be the most stable but the difference in stability is exaggerated.

E. Trans-*enolization of* β-*dicarbonyl compounds*

I. Open-chain compounds

In the gaseous phase or in non-polar solvents and in the absence of non-bonded interactions between the substituents one would expect the *cis* forms (**33b** and **33c**) to be energetically (ΔE or ΔH) favoured relatively to the *trans* forms (**33d** and **33e**), due to the presence of an intramolecular hydrogen bond in the *cis* forms. The hydrogen bond however imposes constraints on internal rotation, and entropy effects may tend to destabilize the *cis*-enols at elevated temperatures. Inter-actions between substituents may influence the stability in different ways. When R^1 and R^3 are bulky they may prevent coplanarity of the conjugated system in the *trans*-enol form and thus reduce the delocal-ization energy in the molecule. The internal rotation around the $=$C—C$=$ bond may also become restricted in the *trans*-enol. In the *cis*-enol non-bonded interaction between R^1 and R^3 is less significant. On the other hand, if either R^1 and R^2 or R^2 and R^3 are bulky they may be better accommodated in a *trans*-enol form. If the substituents are polar groups, dipole–dipole interactions will influence the relative stability of the enol forms, and in polar solvents the difference in solvation energy between *cis* and *trans* enol forms may be considerable.

Experimentally, comparatively little is known about equilibria between *cis*- and *trans*-enol forms of β-diketones. Under suitable conditions, the rate of interconversion of *cis*- and *trans*-enol forms may be low as compared to the rate of enol ether formation on reaction with diazomethane. In such cases the relative amounts of *cis*- and *trans*-enol ethers will provide information on the *cis–trans* equilibrium. Using this method Eistert and coworkers[172] concluded that in ethyl ether solutions of acetylacetone the *cis*-enol predominates whereas some *trans*-enol is present in methanol solution. On the other hand Reeves[46] and Burdette and Rogers[54], who studied the p.m.r. spectra of acetylacetone and other β-diketones in alcohols and various polar and non-polar solvents, detected the *cis* form only. Some 3% or less of *trans*-enol might however have escaped detection. From the infrared spectrum of benzoylacetone in carbon tetrachloride it has been concluded that some 2% of *trans*-enol is present in solution[175]. From considerations of the effective spin–spin coupling constant in different

isomers and conformers it was concluded that the enol of malondialdehyde in chloroform-d is predominantly in the *trans-sym-trans* conformation (**42**) [181a].

In the case of acetoacetaldehyde the *cis*-enol (hydroxymethylene-
ketone) form is presumably the most stable tautomer[181]. Recently,
George and Mansell[181b] have shown that polar impurities in the solvent may change the conformation of the enols. Their work indicates
that the enol forms of both acetoacetaldehyde and malondialdehyde
exist in the *s–cis* conformation in carbon tetrachloride, chloroform
and methylene chloride and in the *s–trans* conformation in water.

Ethyl 2-formyl-2-phenylacetate is one of the 'classical' compounds
in the development of the concept of keto–enol tautomerism. This
compound was originally believed to exist in at least four modifications
but the extensive work of Dieckmann showed that only two forms,
one solid and one liquid at room temperature, were present[183]. The
liquid modification showed a colour reaction with $FeCl_3$, whereas
solutions of the solid modification in alcohols did not show such a
colour reaction. It was concluded that the liquid was a mixture of the
keto form and the *cis*-enol and that the solid is the *trans*-enol (**43**).

P.m.r. studies[184] of ethyl 2-formyl-2-phenylacetate strengthen the
conclusions of Dieckmann. The spectrum of the liquid *cis*-enol
modification has already been discussed above. The p.m.r. spectrum
of the solid modification in chloroform solution gives no indication
of an intramolecular hydrogen bond. A singlet at δ 7·95 p.p.m.
must be due to C=C—H, thus excluding the keto form as a possibility.
Rapid intermolecular proton exchange in the *trans*-enol precludes
spin coupling between the =C—H and OH protons. The signals of

8*

the —CH_2— and —CH_3 protons are shifted 0·12 and 0·015 p.p.m. respectively towards higher fields as compared to the spectrum of the cis-enol. In the trans-enol form (43) rotation around the =C—C= bond would no longer be restricted, and therefore the CH_2 and CH_3 protons should on the average be less influenced by the deshielding ring currents of the phenyl group. The spectrum changes with time and characteristic signals of the cis-enol slowly appear. When the liquid cis-enol modification is dissolved in dimethyl sulphoxide the p.m.r. spectrum shows that the cis-form is slowly and almost completely converted into the trans-enol form. The interconversion is catalysed by bases. In ethanol solution (at 33°) the cis-enol is partly converted into the trans-form (20%), but accurate determination of the position of the cis-trans equilibrium is difficult due to the gradual formation of a new compound, possibly an acetal[184].

Substantial amounts of trans-enols in α-alkyl-substituted acetylacetones and ethyl acetoacetates have been reported[185-188]. The α-alkyl derivatives were prepared by alkylation of the sodium salts of acetylacetone or ethyl acetoacetate with alkyl iodides. However, this reaction also results in the formation of O-alkylated derivatives (enol ethers)[189-190], which have apparently been mistaken for trans-enols. After gas chromatographic purification no evidence of trans-enols was found. It is not unlikely that some of the older values of enol contents in α-alkylated β-diketones and keto esters prepared via the sodium salts are also erroneous due to O-alkylated impurities.

2. Cycloalkane-1,3-diones

In this class the cyclohexane-1,3-diones are by far the most investigated ones. Enolization gives 'fixed' trans-enols where intramolecular hydrogen bonding is sterically impossible. Nevertheless these compounds apparently exist largely in the enol forms in the solid state as well as in concentrated solutions. In solvents capable of acting as donors or acceptors of hydrogen bonds the enol form may gain stability whereas in chloroform and other solvents the formation of a dimer (44) has been suggested to stabilize the enol.

(44)   (45)

Recent p.m.r. studies by Cyr and Reeves[119] and ^{17}O n.m.r. and u.v. studies by Yogev and Mazur[57] have shown that the keto–enol tautomerism of cyclohexane-1,3-diones in cyclohexane and chloroform is best described as an equilibrium between keto monomer, enol monomer and enol dimer. At very low total concentration the keto form predominates. At a total concentration of $0·2 \times 10^{-4}$ M in cyclohexane some 5% (and at a concentration of $0·8 \times 10^{-3}$ M some 40%) of cyclohexane-1,3-dione is present in the enol form. At very high concentrations the enol may associate to form higher agglomerates. Probably the enol dimer is stabilized by intermolecular hydrogen bonding as large OH proton shifts are observed for chloroform solutions at concentrations where the dimer form is predominating, δ about 12·3 p.p.m. This value is roughly the same as that observed[120] for the OH proton in the enolized formaldehyde adduct of dimedone (45). (Compare Nakanishi[126].)

The p.m.r. spectrum of cyclobutane-1,3-dione (46)[121] in chloroform shows only one single peak (δ 3·86 p.p.m.) which is consistent with the presence of the pure keto form 46a, but a rapid interconversion of

(46a)   (46b)

the keto and enol forms would also give the same result. The i.r. spectrum indicates that the enol form in chloroform is present in relatively small amounts. In polar media such as DMSO the enol form 46b appears to be present in substantial amounts (For the properties of 2-methylcyclobutane-1,3-dione see reference 122). A characteristic physical property of cyclobutane-1,2-dione and its methyl derivatives is their high acidity, with pK_a about 3 (to be compared with pK_a about 5 for cyclohexane-1,3-diones). Dipolar interactions, as estimated from the Kirkwood–Westheimer theory[123], may roughly account for the difference in acidity[122]. Cyclopentane-1,3-dione seems to be completely enolic in the solid state[124] but has apparently not been more closely investigated. The enolic character of 2-methylcyclopentane-1,3-dione has been described[193].

Cycloheptane-1,3-dione is slightly enolic, though it still shows a markedly acidic character[125]. It is then of some interest to note that 2-acetylcycloheptane-1,3-dione is completely enolic[53].

Stetter and Milbers[194] have shown that decalin-1,8-dione is completely enolic (pK_a 11·1 in aqueous ethanol) and thus provides an example of a 'fixed' *cis*-enol (**47**). The enolization and hydrogen bonding of **47** have been confirmed by n.m.r. studies ($\delta^{OH} = 16·37$)[195].

(47)

F. Enolization in β,β'-tricarbonyl compounds

β,β'-Triketones (**48**) are found in many natural products, especially among the enolic 2-acylcyclohexanediones[196]. Among these, several derivatives of 2-acylcylohexane-1,3,5-trione have been investigated spectroscopically[47,197–200]. The enolization of some similar cyclopentane derivatives from hops has also been investigated[201,202] and the field of (mainly enolic) hop constituents has been reviewed[203].

Six different *cis*-enols are possible in **48** in general. If the acyl

(48)

groups are free to rotate and do not form part of a cyclic system there is no meaningful distinction between *cis* and *trans* enols of a β,β'-triketone. The simplest example, triacetylmethane ($R^1 = R^2 = R^3 = CH_3$), is virtually completely enolized in aqueous solution[204] and in carbon tetrachloride solution[47]. The p.m.r. spectrum in carbon tetrachloride solution shows two methyl signals 0·17 p.p.m. apart with the intensity ratio 2:1 and an OH signal at very low field (δ 17·4 p.p.m.). The occurrence of two separate methyl signals in that intensity ratio shows that the interconversion of enol forms with a β-diketone partial structure such as **49a** and **49b** is, as expected, a rapid process, while interconversion of enol forms involving different β-diketone partial structures, e.g. **49a** + **49b** ⇌ **49c** + **49d** is a slow process in carbon tetrachloride solution but can be accelerated in other solvents. This result appears to be quite general. In order to avoid circumlocutions and facilitate the discussion of complex tautomeric systems, enolic forms which formally can be interconverted by a movement of the OH proton along an intramolecular hydrogen bond, e.g. **49a** and **49b**, or **49c** and **49d**, will be called *internal* tautomers, and enolic forms that cannot be interconverted in this simple way, e.g. **49a** and **49d**, will be called *external* tautomers[48].

(49a)

(49b)

(49c)

(49d)

The interconversion of external tautomers is catalysed by acids and bases and the addition of a trace of triethylamine to a solution of a β,β'-triketone is usually sufficient to cause p.m.r. signals of external tautomers to coalesce into a single averaged spectrum. In unsymmetrical β,β'-tricarbonyl compounds the relative stabilities of the different enol forms may in some cases be determined. The p.m.r. spectrum of the virtually completely enolized compound diacetoacetaldehyde (**48**, $R^1 = R^2 = CH_3$, $R^3 = H$) in carbon tetrachloride solution at 33° shows the presence of two external tautomers in the ratio 4:1 [117]. Two OH signals are observed (δ 17·10 and 18·3 p.p.m.) and the smallest of these (δ 17·10) is a broad doublet (J 6·4 Hz) due to spin coupling with a signal at δ 8·92 p.p.m. A larger 'aldehyde' type' singlet is observed at δ 9·97 p.p.m. and two methyl signals at δ 8·28 and 8·50 p.p.m. with intensity ratio 1:9. The spectrum indicates that the predominant enol form is **50c** and **50d** and the spin coupling of the enolic OH proton in the less stable enol indicates that the tautomer **50b** is slightly favoured over **50a**. The chemical shift of the =C—H proton is not a reliable measure of the position of the equilibrium between **50a** and **50b**, as anisotropy effects from the non-chelated acetyl group may shift this signal downfield. The intensity ratio of the methyl signals is due to overlapping of the signal from the 'chelated' CH_3-group in **50a** and **50b** with that of the CH_3 signal from **50c** and **50d**.

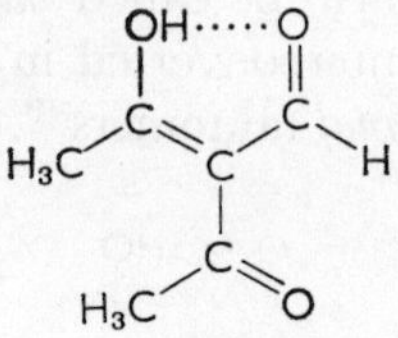

(50a) 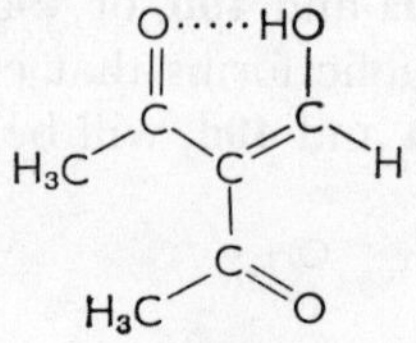   (50b)

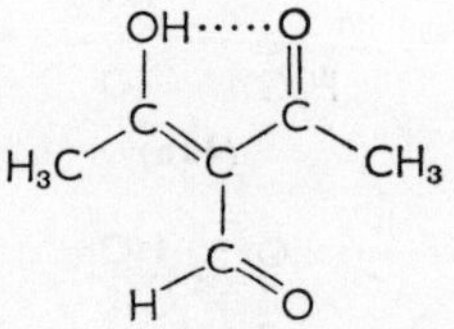

(50c) 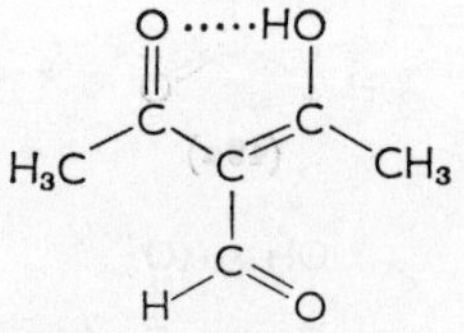   (50d)

The p.m.r. spectrum of ethyl 2-acetyl-2-formylacetate (**48**, R^1 = CH_3, R^2 = OEt and R^3 = H), in which a comparison can be made between three important substituents, shows the presence of two external tautomers in the ratio 9:1. The predominant form is found to be **51a** and **51b**, and the spectra indicate the less stable form to be

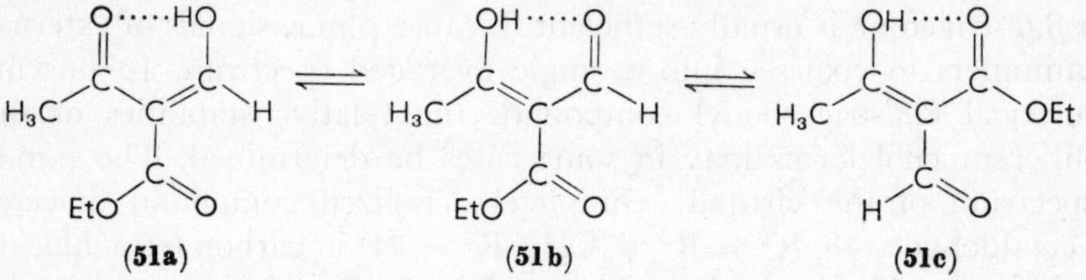

(51a)   (51b)   (51c)

51c. No evidence for the third possible external tautomer is found. In the asymmetric compounds discussed the most stable external tautomer is that with the strongest intramolecular hydrogen bond, as indicated by the chemical shift of the OH proton[117].

In cyclic β,β'-triketones the number of possible tautomeric forms is reduced. In a symmetrical one, such as 2-acetylcyclopent-4-ene-1,3-dione (**52**) the external tautomers are identical and their slow interconversion is manifest in the p.m.r. spectra, which shows an AB-type signal ($J_{H^A H^B}$ = 6·2 Hz) for the olefinic protons. No safe conclusion regarding the equilibrium between **52a** and **52b** could be drawn.

In unsymmetrical cyclic β,β'-triketones the influence of substituents on the equilibrium between external tautomers may be studied. For example the p.m.r. spectrum of 2-acetyl-4-methylcyclopent-4-ene-1,3-dione (53) shows two olefinic quartets ($J_{H_{(5)}CH_3} = 1\cdot7$ Hz) at δ 6·49

(52a)   (52b)

p.p.m. and 6·58 p.p.m. in the intensity ratio 1:1·6. On the basis of anisotropy effects of the C=O group (see however reference 232) one would expect the olefinic proton in 53a and 53b to be at lower fields than in 53c and 53d and thus that [53a and 53b]/[53c and 53d] = 1·6.

(53a)   (53b)

(53c)   (53d)

It may also be demonstrated that *intermolecular* exchange of the OH proton in enolized β,β'-triketones may take place faster than the interconversion of external tautomers. The p.m.r. spectrum of a 1:1 mixture of 2-acetylindane-1,3-dione and 53 in carbon disulphide solution (in which both compounds are virtually completely enolized) shows only one OH signal at a chemical shift intermediate between those of the OH signals of the separate compounds, while the occurrence of two quartets of the olefinic hydrogen in 53 shows that the interconversion of its two external tautomers 53a–53b and 53c–53d is still slow. This indicates too that in the intermolecular transfer of the enolic protons the geometry of the molecules is largely retained.

The p.m.r. spectrum of 2-formyl-5,5-dimethylcyclohexane-1,3-dione[53,117] indicates complete enolization and shows a =C—H signal

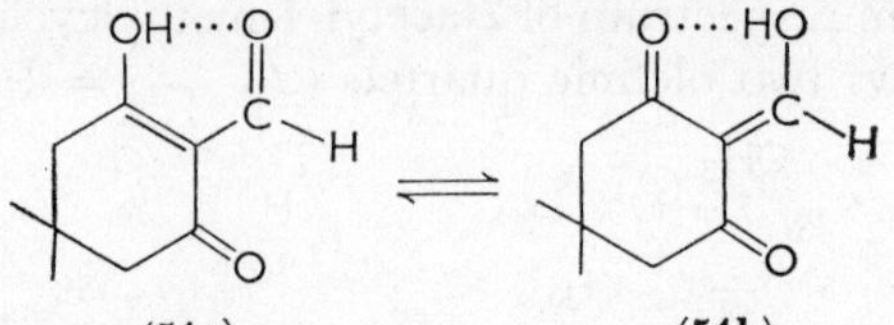

(54a)   (54b)

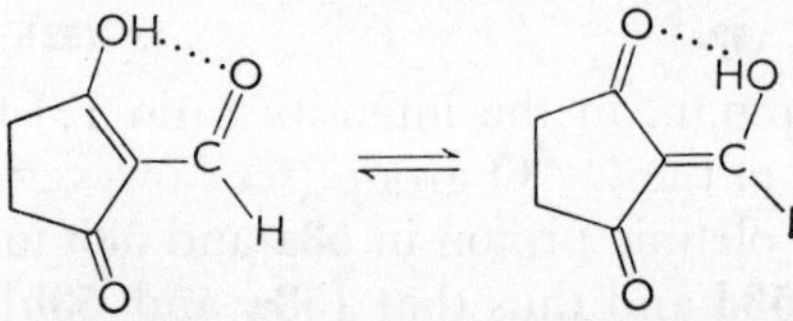

(55a)   (55b)

at δ 9·32 p.p.m. and an OH signal at 15·60 p.p.m. At −75°c a distinct splitting of these signals appears due to mutual spin coupling ($J =$ 1·9 Hz). The magnitude of this spin coupling indicates that the predominant internal tautomer is **54a**. The shift of the =C—H proton also indicates this although the signal is presumably shifted a few tenths of a p.p.m. downfield, due to the deshielding effect of the nonchelated carbonyl group. In the similar five-membered compound 2-formylcyclopentane-1,3-dione (**55**) the =C—H signal is observed at δ 9·59 ppm and the OH signal at δ 12·32 p.p.m. in chloroform-d solution[53]. No H—C—O—H spin coupling was observed even at low temperatures, but the occurrence of intermolecular proton exchange could not be excluded. The shift of the =C—H proton is strongly in favour of **55a** as the predominant enolic form in analogy to the results on **54**. The behaviour of the enols of **55** is thus opposite that of 2-formylcyclopentanone, for which the hydroxymethylene ketone form is the most important, but conforms to the recent data of Russel and Garbisch on *exo*-conjugated formyl ketones (*cf.* section V.D).

Recently it has been shown that triacylmethanes containing a pivaloyl group are wholly ketonic, apparently due to steric effects[205], whereas 2-pivaloyl-1,3-cyclopentanedione is completely enolic[206].

G. Solvent effects

The early work on keto–enol equilibria following the development of the bromine titration method resulted in two 'laws' which comprehended a number of experimental data. The investigations of Dimroth [207] on 3-benzoyl-camphor showed that when the concentrations of the keto (C_k) and enol (C_e) forms in a solvent were expressed relative to the solubilities (S_k) and (S_e) of the keto and enol forms in the same solvent, their quotient was approximately constant (G) and independent of the solvent (equation **23**).

$$G = \frac{C_e/S_e}{C_k/S_k} \tag{23}$$

A relationship for isomer equilibria of this general type had earlier been theoretically derived by Van't Hoff, and equation (23) is now generally referred to as the Van't Hoff–Dimroth relation. Equation (23) is most applicable when the concentrations and solubilities are low; otherwise solute–solute interactions will interfere with the solvent–solute interactions. For an interesting discussion of the Van't Hoff–Dimroth relation in extrathermodynamic terms, see reference 208.

Meyer [209] noted that in a large number of cases the equilibrium constant $K_t = $ [enol]/[keto] could be written as a product of two factors

$$K_t = EL \tag{24}$$

where E is a solvent-independent constant characteristic of a particular tautomeric compound and L is a constant characteristic for a given solvent but independent of the nature of the dissolved tautomeric compound. In the application of equation (24) it has been customary to use ethyl acetoacetate as a 'standard tautomer' $(E = 1)$. The value of L for a given solvent is then equal to K_t of ethyl acetoacetate in the same solvent. If the value of K_t of a tautomeric compound is known in one solvent, equation (24) then permits the calculation of K_t in other solvents.

Meyer's relationship (24) may be given the following interpretation in extrathermodynamic terms: Let us assume that the standard partial molar free energy of an enolizable ketone in a solvent may be expressed for the keto form as

$$\bar{G}_K^0 = g_K + i_K + i_R \tag{25}$$

and for the enol form as

$$\bar{G}_E^0 = g_E + i_E + i_R \tag{26}$$

where g_K and g_E are the intramolecular solvent-independent parts of the partial molar free energies of the keto and enol forms respectively, i_R is that part of the free energy ascribable to interactions between the solvent and all parts of the tautomeric molecule *except* the keto or enol fragments, i_K and i_E finally are the free energies attributable to the interaction between the solvent and the keto and enol fragments respectively in the tautomeric molecules. The difference between the standard partial molar free energy of the enol and keto form, $\bar{G}_E^0 - \bar{G}_K^0 = \Delta \bar{G}_{KE}^0$, may thus be written

$$\Delta \bar{G}_{KE}^0 = (g_E - g_K) + (i_E - i_K) \qquad (27)$$

However, as $\Delta \bar{G}_{KE}^0 = -RT \ln K_t$, where $K_t = [\text{enol}]/[\text{keto}]$, equation (27) leads to equation (24) if $E = -RT \ln (g_E - g_K)$ and $L = -RT \ln (i_E - i_K)$.

The Kirkwood–Onsager theory of solvation of a molecular dipole in a polarizable medium has been applied to keto–enol equilibria by Powling and Bernstein[210].

The following equation may be derived (for dilute solutions)

$$\Delta H_{\text{gas}} = \Delta H_{\text{soln}} + \left(\frac{\epsilon - 1}{2\epsilon + 1} \cdot \frac{1}{a^3} \right) \left(\mu_1^2 - \mu_2^2 \right) \qquad (28)$$

where μ_1 and μ_2 are the dipole moments of the less stable and more stable tautomer, respectively, ϵ the dielectric constant of the solvent and a is the radius of a 'cavity' embodying the solvated dipole in the solvent. Powling and Bernstein equate a^3 with the molar volume of the solvent. It is true that in the original papers by Kirkwood and Onsager[211,212] little was said about how to determine the size of the cavity, but for a large tautomeric molecule in a solvent of relatively small molecules it appears more reasonable to relate the dimensions of the cavity to the molar volume (or partial molar volume) of the solute. The values of ΔH_{gas} of ethyl acetoacetate extrapolated from solution data by equation (28) (-2 kcal/mole) is in fair agreement with the experimental gas phase value of -3.5 kcal/mole found by Briegleb and coworkers at room temperature[153,154]. It has been suggested that equation (28) should be also applicable to free energy changes on enolization. Some correlation is found, but one is willing to agree with the statement of Rogers and Burdette that 'further research will be required for a more complete understanding of the factors influencing tautomeric equilibria'[52].

VI. OTHER ENOLIC CARBONYL COMPOUNDS*

A. α-Diketones

Acyclic α-diketones with at least one adjacent C—H, e.g. biacetyl, are in general less acidic and are enolized to a smaller extent than β-diketones.

Cycloalkane-1,2-diones behave otherwise, and cyclopentane-1,2-dione (56) and derivatives of cyclohexane-1,2-dione are commonly cited examples of highly enolic and acidic α-diketones.

(56)

The difference between cyclic and acyclic α-diketones has been rationalized in terms of repulsion between adjacent carbonyl groups, which leads to 'transoid' conformations of the acyclic compounds. In the simpler cyclic compounds the strain is relieved by enolization[3].

A recent investigation[213] has been made of cycloalkane-1,2-diones with 5- to 11-membered rings by dipole moment measurements. The results, supported by earlier i.r. and u.v. spectral data, confirm the almost complete enolization of cyclopentane-1,2-dione. Cyclohexane-1,2-dione (see also section VI.C) is still largely enolized but for larger rings the enol content is *markedly reduced*. In aqueous and alcoholic media formation of hydrates or hemiacetals may also be important reactions. Conformational factors may affect even the enolization of cyclopentane-1,2-diones as shown by investigations of steroidal 16,17-diketones[214].

B. Enediols

1,1-Enediols (57) can be considered as ketene hydrates or as enols of carboxylic acids.

$$R^1\diagdown C=C=O \underset{}{\overset{+H_2O}{\rightleftharpoons}} R^1\diagdown C=C\diagup OH \rightleftharpoons R^1\diagdown CH-C\diagup O$$
$$R^2\diagup \qquad R^2\diagup \diagdown OH \qquad R^2\diagup \diagdown OH$$

(57)

1,1-Enediols have been little discussed. The ketonization of a transient 1,1-enediol has been investigated mechanistically[215] (see

* The properties of an interesting class are treated in Chapter 4.

section VII.F). These enediols are naturally of interest also in connection with the possible enolization of esters, which has not been discussed in this chapter.

α,β-Enediols (**58**) may be formed by enolization of α-hydroxy-carbonyl compounds. Such enolization has often been discussed in

$$R^1-\underset{\underset{OH}{|}}{CH}-\underset{\underset{O}{\|}}{C}-R^2 \rightleftharpoons R^1-\underset{\underset{OH}{|}}{C}=\underset{\underset{OH}{|}}{C}-R^2 \rightleftharpoons R^1-\underset{\underset{O}{\|}}{C}-\underset{\underset{OH}{|}}{CH}-R^2$$

(**58**)

connection with the isomerization of carbohydrates. Stable α,β-enediols are formed when R^1 and R^2 are bulky substituents, such as mesityl[69,216]. In the most important α,β-enediols, viz. the reductone derivatives, there is (at least) one additional, adjacent carbonyl function. This class of compounds, which includes ascorbic acid and reductinic acid (**59**), is of considerable biological interest and has been reviewed elsewhere[216,217]. Many of these compounds show structural

(**59**)

features which relate them both to α- and to β-dicarbonyl compounds.

C. Oxaloacetic acid

Oxaloacetic acid (**60**) is an important biological intermediate, which is also related to pyruvic acid and acetoacetic acid.

The crystalline oxaloacetic acid is enolic, but it is largely converted into the keto form in aqueous solutions, though in neutral aqueous solution **60** seems to be more enolic than acetoacetic acid[218].

$$\underset{\underset{OH}{|}}{HOCO-C}=CHCOOH \rightleftharpoons HOOC-\underset{\underset{O}{\|}}{C}-CH_2COOH$$

(**60**)

The ketonization has been investigated by ultraviolet spectroscopy and shown to be general acid- and base-catalysed[218].

The degree of enolization of oxaloacetic acid in water and aqueous methanol was also investigated by n.m.r. spectroscopy and was shown to be of the order of 8–20%[219]. Other species were also formed in aqueous solutions, as shown by deuteration experiments; possibly hydration is a competing reaction. As could be expected diethyl oxaloacetate is highly enolic (79% in the neat liquid and 50% in methanol solution). On the other hand, neat diethyl fluorooxaloacetate is virtually completely ketonic[219].

VII. KINETIC AND MECHANISTIC ASPECTS OF ENOLIZATION

A. Current views on the mechanisms

As pointed out in the introduction, enolization is a special case of prototropic rearrangements. It was also considered in that perspective by Ingold[3] and in a certain way it can be seen as a hybrid between allylic rearrangements and the 'tautomerism' of the carboxyl group.

In the introduction to *Fundamentals of Carbanion Chemistry* Cram has pointed out that acid–base theory and concepts of thermodynamic acidity have been developed to a large extent from studies of oxygen acids and bases due to the high rates of proton transfer encountered in these types of molecules, whereas the concept of kinetic acidity in organic chemistry has been applied most frequently to carbon acids (C—H) since the rates of proton transfer are here often low[220].

In enolization–ketonization reactions, involving breaking and formation of C—H and OH bonds, we stand at the cross road between these different ways of treating acidity and basicity and some caution has to be exercised in investigating these reactions[221–224].

Two extremes of the situation may be considered:

Simple carbonyl compounds are so little enolized that 'direct observation' of the enol or the enolization can hardly be achieved. Instead, bromination, deuteration, racemization, etc., have been studied on the assumption that they involve enolization.

On the other extreme stand the completely enolic compounds like phenols and many β,β-triketones, where ketonization is little studied.

Between these cases there are systems where both ketonic and enolic forms can be observed simultaneously by, for example, spectroscopic methods. β-Dicarbonyl compounds provide the most abundant examples. Here it is often possible to observe also the enolization–ketonization reaction directly, by relaxation methods or occasionally also when a 'metastable' keto or enol form, or a non-equilibrium mixture, can be used as starting material.

Two main mechanisms can be proposed for the process.

Consecutive reactions (equation 2, section I) are believed to proceed via cations or anions and are catalysed by acids and bases.

It should be noted that the intermediates in equation (2) may also be intermediates for other types of reactions, e.g. the cations may form hydrates or hemiacetals, while the anions may be more reactive to bromine than the enol itself and may also engage in aldol-type condensations.

A concerted mechanism (equation 29), subject to acid and base catalysis, has sometimes been advanced, but at present it seems to account only for parts of the reaction or for special cases[222].

$$AH + B + R^1\!-\!\overset{\displaystyle O}{\underset{\displaystyle }{C}}\!-\!\overset{\displaystyle H}{\underset{\displaystyle R}{\overset{|}{C}}}{\diagdown}{}^{R^2} \;\rightleftharpoons\; \left[R^1\!-\!\overset{\displaystyle A}{\overset{\vdots}{\underset{\displaystyle}{\overset{\displaystyle H}{\underset{\vdots}{\overset{\displaystyle \dot{O}}{C}}}}}\cdots\overset{\displaystyle B}{\overset{\displaystyle H}{\underset{\displaystyle R^3}{\overset{\vdots}{C}}}}{\diagdown}{}^{R^2} \right] \;\rightleftharpoons\; R^1\!-\!\overset{\displaystyle OH}{\underset{\displaystyle}{C}}\!=\!\overset{}{\underset{\displaystyle R^3}{C}}{\diagdown}{}^{R^2} + A + BH \tag{29}$$

Most mechanistic studies of enolization have been made in protic solvents, where the solvent may play an important role in the transition states. Investigations of enolization in non-polar media are scarce and the possibilities for 'pseudointramolecular' enolization–ketonization reactions analogous to the base-catalysed rearrangements of allylic systems (reference 220, p. 175) have therefore not yet received much attention.

A closer description of the gross mechanisms outlined above needs a deeper understanding of the transition states and the paths of the protons. However, the details of the proton transfer reactions are often obscured by a manifold of other effects. The nature of the catalytic bases and acids, their solvation and the concurrent counterions, affect the reactions and complicate the picture. Thus the metal ions may play a considerable role in base-catalysed enolization and the role of the nucleophile may be critical in acid-catalysed reactions.

No universal mechanisms of enolization have yet been presented, but some fairly recent and typical examples will illustrate both the methods for the mechanistic studies and the results obtained for various types of carbonyl compounds.

B. Relaxation methods

In recent years relaxation measurements have gained large importance for studies of rapidly equilibrating systems and seem particularly promising for proton transfer reactions[225]. Still, only a few relaxation investigations of enolization reactions have been reported and are confined to highly enolic systems. One investigation concerns protolysis and hydrolysis of acetylacetone and some other hydrogen-bonded systems[226], another one dimedone and some of its heterocyclic analogues[227].

The results from the relaxation measurements of rates of protolysis of acetylacetone have recently been summarized[228]. The investigations have as yet been limited to only one temperature and show that the protonation and deprotonation at the carbon atom are slower by a factor of about three powers of ten than those at the oxygen atom, and that the reaction of the chelated enol with OH^- is a thousand-fold slower than the diffusion controlled reaction. Further work in this area may provide a better understanding of this enolic system.

C. Enolization of cyclohexane-1,2-dione

The uncatalysed interconversion between the keto form (**61a**) and the enol form (**61c**) of cyclohexane-1,2-dione is rather slow, therefore allowing their isolation in fairly pure forms. U.v. measurements can be used to determine the concentrations of the enol and the unhydrated ketone.

(61a) ⇌ (61b) ⇌ (61c)

No enol hydrate is formed in aqueous solution but the keto form is rapidly and almost completely hydrated (**61b**) and the kinetics of hydration in aqueous dioxan could be studied. In acidic aqueous media the equilibrium between enol and the hydrated ketone is established slowly and can be easily observed. The equilibrium constant is strongly dependent on electrolyte concentration, apparently due to different influences on the activity coefficients of the tautomeric forms. Temperature studies of the equilibrium gave $\Delta H° = -6$ kcal/mole and $\Delta S° = -21$ e.u. for the formation of the ketone.

A further analysis of the interconversions showed pH regions with acid catalysis and base catalysis, and a pH-independent 'water reaction'. Temperature studies of the rates of ketonization and enolization gave reaction parameters $E_A = 17 \cdot 6$ kcal/mole and $\Delta S^{\ddagger} -26$ e.u. for ketonization; $E_A = 24 \cdot 4$ kcal/mole and $\Delta S^{\ddagger} = -4$ e.u. for enolization. These values, as well as deuterium solvent isotope effects, indicate that the transition state is monohydrated.

D. Racemization of ketones

By studies of D-α-phenylisocaprophenone in strong sulphuric acid (85–93·8%) Swain and Rosenberg[229] showed that the racemization is preceded by a fast equilibrium protonation and that a nucleophile, preferably water, is required in the transition state. The rate constant for water is about a hundred times larger than that for the hydrogen sulphate anion. The kinetic significance of water is further demonstrated by the influence of solvent D_2O. From the racemization experiments and other data, it was concluded that the α-hydrogen in the transition state is nearly symmetrically placed between the O and C atoms in enolizations catalysed by hydroxide ion or hydronium ion, but much closer to the O atom in uncatalysed reactions with water as the nucleophile.

Accordingly, in H_3O^+-catalysed enolizations the conjugate acid of the ketone must be much more reactive and less discriminating (toward nucleophiles) than the free ketone, and the transition state must be nearly symmetrical.

These results and conclusions have been critically discussed by Long and Bakule[167], particularly regarding the relative efficiency of various nucleophiles and the 'ideal' character of the solvent. They found the major aspects of the work by Swain and Rosenberg consistent with their own results on the cyclohexane-1,2-dione system and agree about the need to consider the activity coefficients of all the species present.

E. Ketonization of a 1,1-enediol

In an investigation on the stereochemistry of ketonization, Zimmerman and Cutshall succeeded in generating a transient 1,1-enediol (**62a**) by two different routes[214], and then studied the stereoselectivity of its 'ketonization'. Two diasteromeric carboxylic acids were formed, the *cis*-isomer (**62b**) being favoured. The influence of temperature on the product distribution showed that steric hindrance to C-protonation is important.

F. Rates of bromination

Using refined techniques and very low halogen concentrations Bell and Davis investigated the kinetics of bromination of dimedone, 2-bromodimedone, 3-methyltetronic acid, 3-bromotetronic acid, methylacetylacetone and 2-acetylcyclohexanone in aqueous solution at various pH values[230]. The equilibrium enol concentrations were also determined and it was found that the bimolecular rate constants for bromination of enols and enolate anions were between 5×10^6 and 2×10^7 l/mole s.

For dimedones and tetronic acids there was little difference between the reactivities of the enols and the enolate anions (no systematic variation of rate constant with pH was observed). Otherwise enolate anions have been found to react about 10^5 times faster than enols. The reactions are slower than purely diffusion-controlled reactions by a factor of about 100 and diffusion control alone cannot account for the results. Possibly there is an unusual type of 'stabilization' decreasing the reactivity of the enolate anions.

These puzzling results further illustrate the peculiarities of 'fixed' *trans*-enols and demand some caution in interpretation of other enol–enolate interconversion studies, as well as of halogenation studies.

VIII. CONCLUSION

This review of enolization has perhaps raised more problems than it has provided solutions, explanations or general theories.

We have got the impression that more detailed investigations, particularly concerning the molecular structures and vibrational levels of the chelated enols, would be rewarding. It can be foreseen, for example, that new experimental techniques like e.s.c.a. (i.e.e.) may differentiate between single and double well potentials for intramolecular proton transfer. Also refined *ab initio* molecular orbital calculations may give increased insight into the structures and stabilities of chelated enols.

Further efforts in the tracing of the path of the protons in enolization–ketonization reactions may show more clearly the roles of catalysts and solvents and may also connect these reactions with other prototropic rearrangements.

We hope that this review will stimulate further search for connections to other topics, generalizations and applications.

IX. ACKNOWLEDGEMENTS

We wish to thank Drs. E. W. Garbisch Jr. and J. G. Russell for communicating unpublished material on enol–enol equilibria, Dr. Anna-Britta Hörnfeldt and Dr. Pierre Laszlo for reading and commenting on the manuscript, and Mr. U. Henriksson for carrying out a number of computations.

X. REFERENCES

1. L. P. Hammett, *Physical Organic Chemistry*, McGraw-Hill, New York, 1940, p. 229.
2. C. K. Ingold, *Structure and Mechanism in Organic Chemistry*, G. Bell, London, 1953, p. 529.
3. G. S. Hammond, in *Steric effects in Organic Chemistry* (Ed. M. S. Newman) John Wiley and Sons, New York, 1956, p. 425.
4. V. A. Palm, Ü.L. Halna and A. J. Talvik, *The Chemistry of the Carbonyl Group*, Vol. 1, Interscience, London, 1966, p. 421.
5. H. Henecka and B. Eistert, in *Houben-Weyl, Methoden der organischen Chemie* (Ed. E. Müller), Vol. 2. Thieme, Stuttgart, 1953, p. 380.
6. K. H. Meyer, *Ann. Chem.*, **380**, 212 (1911).
7. K. H. Meyer and P. Kappelmeier, *Ber.*, **44**, 2718 (1911),
8. K. H. Meyer, *Ber.*, **45**, 2843 (1912).
9. K. H. Meyer, *Ber.*, **47**, 826 (1914).
10. K. H. Meyer and V. Schoeller, *Ber.*, **53**, 1412 (1920).
11. J. D. Park, H. A. Brown and J. R. Lacker, *J. Amer. Chem. Soc.*, **75**, 4753 (1953).
12. J. C. Reid and M. Calvin, *J. Amer. Chem. Soc.*, **72**, 2952 (1950).
13. B. Eistert and W. Reiss, *Chem. Ber.*, **87**, 92 (1954).
14. W. Walisch, *Diss. Univ. Saarlandes, Saarbrücken*, 1956.
15. J. B. Conant and A. F. Thompson Jr., *J. Amer. Chem. Soc.*, **54**, 4039 (1932).
16. W. Dieckmann, *Ber.*, **55**, 2470 (1922).
17. G. Schwartzenbach and E. Felder, *Helv. Chim. Acta*, **27**, 1044 (1944).
18. G. Schwartzenbach and C. Wittwer, *Helv. Chim. Acta*, **30**, 656 (1947).
19. A. Gero, *J. Org. Chem.*, **19**, 1960 (1954).
20. A. Gero, *J. Org. Chem.*, **19**, 469 (1954).
21. A. Gero, *J. Org. Chem.*, **26**, 3156 (1961).
22. N. L. Allinger, L. W. Chow and R. A. Ford, *J. Org. Chem.*, **32**, 1994 (1967).
23. W. Walisch, *Ann. Univ. Saraviensis Sci.*, **7**, 289 (1958); *Chem. Abstr.*, **53**, 16607e (1959).

24. W. Walisch and J. E. Dubois, *Chem. Ber.*, **92**, 1028 (1959).
25. W. Walisch, *Chem. Ber.*, **93**, 1481 (1960).
26. R. P. Bell and P. W. Smith, *J. Chem. Soc.* (*B*), **1966**, 241.
27. A. Albert and J. N. Phillips, *J. Chem. Soc.*, **1956**, 1294.
28. S. F. Mason, *J. Chem. Soc.*, **1958**, 674.
29. G. F. Tucher and J. L. Irvin, *J. Amer. Chem. Soc.*, **73**, 1923 (1951).
30. A. M. Bushwell, W. H. Rodebush and R. McL. Whitney, *J. Amer. Chem. Soc.*, **69**, 770 (1947).
31. A. C. Cope and B. D. Tiffany, *J. Amer. Chem. Soc.*, **73**, 4158 (1951).
32. K. Kuratani, *Rep. Instr. Sci. Techn. Univ. Tokyo*, **6**, 217 (1952); *Chem. Abstr.*, **47**, 1493h (1953).
33. R. Mecke in *Hydrogen Bonding*, (Eds. D. Hadži and H. W. Thompson), Pergamon Press, London, 1959.
34. R. A. Morton and W. C. V. Rosney, *J. Chem. Soc.*, **1926**, 706, 713.
35. R. S. Rasmussen, D. D. Tunnicliff and R. R. Brattain, *J. Amer. Chem. Soc.*, **71**, 1068 (1949).
36. B. Eistert and W. Reiss, *Chem. Ber.*, **87**, 108 (1954).
37. G. S. Hammond, W. G. Borduin and G. A. Guter, *J. Amer. Chem. Soc.*, **81**, 4682 (1959).
38. F. Korte and F. Wüsten, *Ann. Chem.*, **647**, 18 (1961).
39. R. D. Campbell and H. M. Gilow, *J. Amer. Chem. Soc.*, **84**, 1440 (1962).
40. S. J. Rhoads and C. Pryde, *J. Org. Chem.*, **30**, 3212 (1965).
41. L. A. Paquette and R. W. Begland, *J. Amer. Chem. Soc.*, **88**, 4685 (1966).
42. K. W. F. Kohlrausch and W. Pongraty, *Ber.*, **67**, 976, 1465 (1934).
43. L. Kahovec and K. W. F. Kohlrausch, *Ber.*, **73**, 1304 (1940).
44. D. N. Shigorin, *Zh. Fiz. Khim.*, **27**, 689 (1953).
45. H. S. Jarrett, M. S. Sadler and J. N. Shoolery, *J. Chem. Phys.*, **21**, 2092 (1953).
46. L. W. Reeves, *Can. J. Chem.*, **35**, 1351 (1957).
47. S. Forsén and M. Nilsson, *Acta Chem. Scand.*, **19**, 1383 (1959).
48. S. Forsén, *Svensk Kem. Tidskr.*, **74**, 439 (1962).
49. C. Giessner-Prettre, *Compt. Rend.*, **250**, 2547 (1960).
50. R. Filler and Saiyid M. Naqvi, *J. Org. Chem.*, **26**, 2571 (1961).
51. J. L. Burdett and M. T. Rogers, *J. Amer. Chem. Soc.*, **86**, 2165 (1964).
52. M. T. Rogers and J. L. Burdett, *Can. J. Chem.*, **43**, 1516 (1965).
53. S. Forsén, F. Merényi and M. Nilsson, *Acta Chem. Scand.*, **18**, 1208 (1964); **21**, 620 (1967).
54. J. L. Burdett and M. T. Rogers, *J. Phys. Chem.*, **70**, 939 (1966).
55. G. Allen and R. A. Dwek, *J. Chem. Soc.* (*B*), **1966**, 161.
56. E. W. Garbisch, Jr., *J. Amer. Chem. Soc.*, **85**, 1696 (1963).
57. A. Yogev and Y. Mazur, *J. Org. Chem.*, **32**, 2162 (1967).
58. M. Gorodetsky, Z. Luz and Y. Mazur, *J. Amer. Chem. Soc.*, **89**, 1183 (1967).
59. R. B. Turner, *J. Amer. Chem. Soc.*, **80**, 1424 (1958).
60. V. Grignard and H. Blanchon, *Bull. Soc. Chim. France*, [4] **49**, 23 (1931).
61. E. P. Kohler and D. Thompson, *J. Amer. Chem. Soc.*, **55**, 3822 (1933).
62. J. E. Dubois and G. Barbier, *Bull. Soc. Chim. France*, **1965**, 682.
63. C. Rappe, *Acta Chem. Scand.*, **20**, 1721 (1966); **21**, 1823 (1967).
64. C. Rappe, *Acta Chem. Scand.*, **20**, 2236, 2305 (1966).

65. A. A. Bothner-By and C. Sun, *J. Org. Chem.*, **32**, 492 (1967).
66. C. Rappe and W. H. Sachs, *Tetrahedron*, **24**, 6287 (1968); W. H. Sachs and C. Rappe, *Acta Chem. Scand.*, **22**, 2031 (1968).
67. R. P. Bell, *Adv. Phys. Org. Chem.*, **4**, 1 (1966).
68. P. Greenzaid, Z. Luz and D. Samuel, *J. Amer. Chem. Soc.*, **89**, 749, 756 (1967).
69. R. C. Fuson, L. J. Armstrong, D. H. Chadwick, J. W. Kneisley, S. P. Rowland, W. J. Shenk and Q. F. Soper, *J. Amer. Chem. Soc.*, **67**, 386 (1945).
70. R. C. Fuson and T.-L. Tan, *J. Amer. Chem. Soc.*, **70**, 603 (1948).
71. F. Klages, *Lehrbuch der Organischen Chemie*, Vol. 2, Walter de Gruyter, Berlin, 1957, p. 376.
72. J. B. Conant and G. H. Carlson, *J. Amer. Chem. Soc.*, **54**, 4048 (1932).
73. S. Sunner, *Acta Chem. Scand.*, **11**, 1757 (1957).
74. H. C. Brown, J. H. Brewster and H. Shechter, *J. Amer. Chem. Soc.*, **76**, 467 (1954).
75. H. Shechter, M. J. Collis, R. Dessy, Y. Okuzumi and A. Chen, *J. Amer. Chem. Soc.*, **84**, 2905 (1962).
76. J. B. Conant and G. B. Kistiakowsky, *Chem. Rev.*, **20**, 181 (1937).
77. C. K. Ingold, C. G. Raisin and C. L. Wilson, *J. Chem. Soc.*, **1936**, 1637.
78. A. P. Best and C. L. Wilson, *J. Chem. Soc.*, **1938**, 28.
79. E. S. Hand and R. M. Horowitz, *J. Amer. Chem. Soc.*, **86**, 2084 (1964).
80. J. Massicot, *Bull. Chim. Soc. France*, **1967**, 2204.
81. S. Forsén and U. Henriksson, unpublished calculations.
82. R. H. Tomson, *J. Chem. Soc.*, **1950**, 1737.
83. R. H. Tomson, *Quart. Rev.*, **10**, 27 (1956).
84. E. W. Garbisch Jr., *J. Amer. Chem. Soc.*, **87**, 4971 (1965).
85. K. H. Meyer, *Ann. Chem.*, **379**, 37 (1911).
86. K. H. Meyer and A. Sander, *Ann. Chem.*, **396**, 133 (1913).
87. K. H. Meyer and A. Sander, *Ann. Chem.*, **420**, 113 (1920).
88. H. Baba and T. Takemura, *Bull Chem. Soc. Japan*, **37**, 124 (1964).
89. P. V. Laakso, R. Robinson and H. P. Vandrewala, *Tetrahedron*, **1**, 103 (1957).
90. F. Klages, *Chem. Ber.*, **82**, 358 (1949).
91. L. F. Fieser, *J. Amer. Chem. Soc.*, **53**, 2329 (1931).
92. L. F. Fieser and E. B. Hershberg, *J. Amer. Chem. Soc.*, **59**, 1028 (1937).
93. C. Marschalk, *Bull. Soc. Chim. France*, **4**, 1547 (1937).
94. G. W. Wheland, *Resonance in Organic Chemistry*, John Wiley and Sons, New York, 1955, pp. 404, 405.
95. L. E. Orgel, T. L. Cottrell, W. Dick and L. E. Sutton, *Trans. Faraday Soc.*, **47**, 113 (1951).
96. G. Del Re, *Tetrahedron*, **10**, 81 (1960).
97. S. Forsén, *Arkiv Kemi*, **20**, 1 (1962).
98. L. C. Anderson and M. B. Geiger, *J. Amer. Chem. Soc.*, **54**, 3064 (1932).
99. L. C. Anderson and R. L. Yanke, *J. Amer. Chem. Soc.*, **56**, 732 (1934).
100. E. Havinga and A. Schors, *Rec. Trav. Chim.*, **69**, 457 (1950).
101. R. K. Norris and S. Sternhell, *Australian J. Chem.*, **19**, 5 (1966).
102. R. Kuhn and F. Bär, *Ann. Chem.*, **516**, 143 (1935).
103. A. Burawoy and A. Tompson, *J. Chem. Soc.*, **1953**, 1443.

104. J. Ospenson, *Acta. Chem. Scand.*, **5**, 491 (1951).
105. G. Badger and R. Buttery, *J. Chem. Soc.*, **1956**, 614.
106. D. Hadži, *J. Chem. Soc.*, **1956**, 2143.
107. E. Sawicki, *J. Org. Chem.*, **22**, 743 (1957).
108a. H. Shingu, *Sci. Papers Inst. Phys. Chem. Res.* (*Tokyo*), **35**, 78 (1939).
108b. A. H. Berrie, P. Hampson, S. W. Longworth and A. Mathias, *J. Chem. Soc.* (*B*), 1308 (1968).
109. G. O. Dudek and R. H. Holm, *J. Amer. Chem. Soc.*, **83**, 2099 (1961); **84**, 2691 (1962).
110. G. O. Dudek, *J. Amer. Chem. Soc.*, **85**, 694 (1963).
111. G. O. Dudek and E. P. Dudek, *J. Amer. Chem. Soc.*, **86**, 4283 (1964).
112. G. O. Dudek and E. P. Dudek, *Chem. Commun.*, **1965**, 464.
113. G. O. Dudek and E. P. Dudek, *J. Amer. Chem. Soc.*, **88**, 2407 (1966).
114. N. H. Cromwell, F. A. Miller, A. R. Johnson, R. L. Frank and D. J. Wallace, *J. Amer. Chem. Soc.*, **71**, 3337 (1949).
115. J. Weinstein and G. M. Wyman, *J. Org. Chem.*, **23**, 1618 (1958).
116. H. F. Holtzclaw, Jr., J. P. Collman and B. Witkop, *J. Amer. Chem. Soc.*, **78**, 2873 (1956).
117. S. Forsén and M. Nilsson, *Arkiv Kemi*, **19**, 569 (1962).
118. S. Forsén, B. Åkermark and T. Alm, *Acta Chem. Scand.*, **18**, 2313 (1964).
119. N. Cyr and L. W. Reeves, *Can. J. Chem.*, **43**, 3057 (1965).
120. S. Forsén, W. E. Frankle, P. Laszlo and J. Lubochinsky, *J. Magn. Res.*, **1**, 327 (1969).
121. H. H. Wasserman and E. V. Dehmlow, *J. Amer. Chem. Soc.*, **84**, 3786 (1962).
122. R. B. Johns and A. B. Kriegler, *Australian J. Chem.*, **17**, 765 (1964).
123. J. G. Kirkwood and F. H. Westheimer, *J. Chem. Phys.*, **6**, 506 (1938).
124. J. H. Boothe, R. G. Wilkinson, S. Kushner and J. M. Williams, *J. Amer. Chem. Soc.*, **75**, 1732 (1953).
125. B. Eistert, F. Haupter and K. Schank, *Ann. Chem.*, **665**, 55 (1963).
126. K. Nakanishi, *Infrared Absorption Spectroscopy*, Holden-Day, San Francisco, 1962, p. 65.
127. A. R. Katritzky and J. M. Lagowski, in *Advances in Heterocyclic Chemistry* (Ed. A. R. Katritzky), Academic Press, New York, 1963, Vol. 1, p. 311.
128. S. F. Mason, *J. Chem. Soc.*, **1957**, 5010.
129. H. Specker and H. Gawrosch, *Ber.*, **75**, 1338 (1942).
130. H. W. Thompson, D. L. Nicholson and L. N. Short, *Disc. Faraday Soc.*, **9**, 222 (1950).
131. B. Witkop, J. B. Patrick and M. Rosenblum, *J. Amer. Chem. Soc.*, **73**, 2641 (1951).
132. R. N. Jones, A. R. Katritzky and J. M. Lagowsky, *Chem. Ind.* (*London*), 870 (1960).
133. J. M. Bruce, P. Knowles and L. S. Besford, *J. Chem. Soc.*, **1964**, 4044.
134. A. R. Katritzky and R. N. Jones, *Proc. Chem. Soc.*, **1960**, 313.
135. A. R. Katritzky and J. Reawilk *J. Chem. Soc.*, **1963**, 753.
136. A. R. Katritzky, F. D. Popp and J. D. Rowe, *J. Chem. Soc.* (*B*), **1966**, 562.
137. D. A. Evans, G. F. Smith and M. A. Wahid, *J. Chem. Soc.* (*B*), **1967**, 590.

138. S. Gronowitz and R. Hoffman, *Arkiv. Kemi*, **15**, 499 (1960).
139. A. B. Hörnfeldt and S. Gronowitz, *Arkiv. Kemi*, **21**, 239 (1963).
140. A. B. Hörnfeldt, *Arkiv Kemi*, **22**, 211 (1963).
141. M. C. Ford and D. Maclay, *J. Chem. Soc.*, **1956**, 4985.
142a.A. B. Hörnfeldt, *Acta Chem. Scand.*, **19**, 1249 (1965).
142b.H. J. Jakobsen and S.-O. Laweson, *Tetrahedron*, **21**, 3331 (1965); **23**, 871 (1967).
143. A. B. Hörnfeldt, private communication.
144. A. Hofmann, W. v. Philipsborn and C. H. Eugster, *Helv. Chim. Acta*, **48**, 1322 (1965).
145. H. H. Hodgson and R. R. Davies, *J. Chem. Soc.*, **1939**, 806, 1013.
146. R. E. Rosenkranz, K. Allner, R. Good, W. v. Philipsborn and C. H. Engster, *Helv. Chim. Acta*, **46**, 1259 (1963).
147. H. Plieninger, H. Bauer and A. R. Katritzky, *Ann. Chem.*, **654**, 165 (1962).
148. J. Davoll, *J. Chem. Soc.*, **1953**, 3802.
149. H. Henecka, *Chemie der Beta-Dicarbonyl-Verbindungen*, Springer, Berlin, 1950.
150. J. B. Conant, *Ind. Eng. Chem.*, **24**, 466 (1932).
151. R. Schreck, *J. Amer. Chem. Soc.*, **71**, 1881 (1949).
152. W. Strohmeier and I. Höhne, *Z. Naturforschung*, **76**, 184 (1952).
153. G. Briegleb, W. Strohmeier and I. Höhne, *Z. Elektrochem.*, **56**, 240 (1952).
154a.G. Briegleb and W. Strohmeier, *Angew. Chem.*, **64**, 409 (1952).
154b.R. Mecke and E. Funk, *Z. Elektrochem.*, **60**, 1124 (1956).
154c.K. S. Pitzer, *Quantum Chemistry*, Prentice-Hall, New York, 1953. p. 247.
155. K. H. Meyer, *Ber.*, **44**, 1147, 2718 (1911).
156. P. Grossman, *Z. Phys. Chem. (Leipzig)*, **109**, 305 (1924).
157. L. W. Reeves, *Can J. Chem.*, **35**, 1351 (1957).
158. E. Funck and R. Mecke, in *Hydrogen Bonding* (Eds. D. Hadži and H. W. Thompson), Pergamon Press, London, 1959, p. 433.
159a.M. I. Kabachnik and S. T. Ioffe, *Dokl. Akad. Nauk SSSR (Engl. Transl.)*, **165**, 1085 (1965).
159b.A. S. N. Murthy, A. Balasubramanian, C. N. R. Ray and T. R. Kasturi, *Can. J. Chem.*, **40**, 2267 (1962).
159c. F. Korte and F. Wüsten, *Ann. Chem.* **647**, 18 (1961).
160. J. D. Roberts and W. T. Moreland Jr., *J. Amer. Chem. Soc.*, **75**, 2167 (1953).
161. R. W. Taft, Jr. and J. C. Lewis, *J. Amer. Chem. Soc.*, **80**, 2436 (1958).
162. R. W. Taft Jr., in *Steric Effects in Organic Chemistry* (Ed. M. S. Newman), John Wiley and Sons, New York, 1956, p. 556.
163. R. P. Bell and A. O. McDougall, *Trans. Faraday Soc.*, **56**, 1281 (1960).
164. Y. Fujiwara and S. Fujiwara, *Bull. Chem. Soc. Japan*, **36**, 574 (1963).
165. L. C. Gruen and P. T. McTique, *J. Chem. Soc.*, **1963**, 5217, 5224.
166. D. L. Hooper, *J. Chem. Soc. (B)* **1967**, 169.
167a. F. A. Long and R. Bakule, *J. Amer. Chem. Soc.*, **85**, 2309, 2313 (1963).
167b.G. Schwarzenback, M. Zimmermann and V. Prelog, *Helv. Chim. Acta*, **34**, 1954 (1951)
168. R. D. Campbell and W. L. Harmer, *J. Org. Chem.*, **28**, 379 (1963).
169. D. E. Williams, W. L. Dumke and R. E. Rundle, *Acta. Cryst.*, **15**, 627 (1962).

170. G. R. Engebretson and R. E. Rundle, *J. Amer. Chem. Soc.*, **86**, 574 (1964).
171. D. E. Williams, *Acta Cryst.*, **21**, 340 (1966).
172. B. Eistert, F. Arndt, L. Loewe and E. Ayca, *Chem. Ber.*, **84**, 156 (1951).
173. R. D. Campbell and H. M. Gilow, *J. Amer. Chem. Soc.*, **82**, 2389 (1960).
174. R. D. Campbell and H. M. Gilow, *J. Amer. Chem. Soc.*, **82**, 5426 (1960).
175. J. U. Lowe, Jr. and L. N. Ferguson, *J. Org. Chem.*, **30**, 3000 (1965).
176. K. I. Dahlqvist and S. Forsén, to be published.
177. E. W. Garbisch, *J. Amer. Chem. Soc.*, **87**, 505 (1965).
178a. J. G. Russell and E. W. Garbisch, personal communication.
178b. K. M. Baker and J. P. Bartley, *Tetrahedron*, **24**, 1651 (1968).
179a. E. W. Garbisch, personal communication.
179b. K. Hafner, H. E. A. Kramer, H. Musso, G. Ploss and G. Schulz, *Chem. Ber.*, **97**, 2066 (1964).
180. D. Samuel and B. L. Silver, *Adv. Phys. Org. Chem.*, **3**, 147 (1965).
181a. A. A. Bothner-By and A. K. Harris, *J. Org. Chem.*, **30**, 254 (1965).
181b. W. O. George and V. G. Mansell, *J. Chem. Soc (B)*, 132 (1968).
182. B. Åkermark and S. Forsén, *Acta Chem. Scand.*, **17**, 1907 (1963).
183. W. Dieckmann, *Ber.*, **50**, 1375 (1917).
184. S. Forsén, to be published.
185. M. I. Kabachnik, S. T. Ioffe and K. W. Vatsuro, *Bull Acad. Sci. USSR*, **1957**, 777.
186. M. I. Kabachnik, S. T. Ioffe and K. W. Vatsuro, *Tetrahedron*, **1**, 317 (1957).
187. M. I. Kabachnik, S. T. Ioffe, E. M. Popov and K. V. Vatsuro, *Zh. Obshch. Khim.*, **31**, 2122 (1961).
188. M. I. Kabachnik, S. T. Ioffe, E. M. Popov and K. V. Vatsuro, *Tetrahedron*, **18**, 923 (1962).
189. S. J. Rhoads, R. W. Hasbrouck, C. Pryde and R. W. Holder, *Tetrahedron Letters*, **1963**, 669.
190. Y. N. Molin, S. T. Ioffe, E. E. Zaev, E. K. Soloveva, E. E. Kugucheva, V. V. Voevodskii and M. I. Kabachnik, *Izv. Akad. Nauk. SSSR, Ser. Khim.*, **9**, 1556 (1965).
191. H. Wassermann and E. V. Dehmlow, *J. Amer. Chem. Soc.*, **84**, 3786 (1962).
192. R. B. Johns and A. B. Kriegler, *Australian J. Chem.*, **17**, 765 (1964).
193. F. A. Long and D. Watson, *J. Chem. Soc.*, **1958**, 2019.
194. H. Stetter and U. Milbers, *Chem. Ber.*, **91**, 977 (1958).
195. I. A. Kaye and R. S. Mattews, *J. Org. Chem.*, **29**, 1341 (1964).
196. C. H. Hassall, *Progr. Org. Chem.*, **4**, 115 (1958).
197. M. Nilsson, *Acta. Chem. Scand.*, **13**, 750 (1959).
198. R. O. Hellyer, I. R. C. Bick, R. C. Nicholls and H. Rottendorff, *Australian J. Chem.*, **16**, 703 (1963).
199. R. O. Hellyer, *Australian J. Chem.*, **17**, 1418 (1963).
200. S. Forsén, M. Nilsson and C. A. Wachtmeister, *Acta. Chem. Scand.*, **16**, 583 (1962).
201. S. Forsén, M. Nilsson, J. A. Elvidge, J. S. Burton and R. Stevens, *Acta Chem. Scand.*, **18**, 513 (1964).
202. J. A. Elvidge and R. Stevens, *J. Chem. Soc.*, **1965**, 2251.
203. P. R. Ashurst, *Prog. Chem. of Org. Nat. Prod.*, **25**, 63 (1967).
204. G. Schwartzenbach and U. Lutz, *Helv. Chim. Acta*, **23**, 1147 (1940).

205. D. C. Nonhebel, *J. Chem. Soc.* (*C*), **1967**, 1716.
206. F. Merényi and M. Nilsson, *Acta Chem. Scand.*, **21**, 1755 (1967).
207. O. Dimroth, *Ann. Chem.*, **377**, 127 (1910); **399**, 110 (1913).
208. J. E. Leffler and E. Grunwald, *Rates and Equilibria of Organic Reactions*, John Wiley and Sons, New York, 1963, p. 22–26.
209. K. H. Meyer, *Ber.*, **47**, 828 (1914).
210. J. Powling and H. J. Bernstein, *J. Amer. Chem. Soc.*, **73**, 4354 (1951).
211. L. Onsager, *J. Chem. Soc.*, **58**, 1486 (1936).
212. L. G. Kirkwood, *J. Chem. Phys.* **2**, 351 (1934); 7, 911 (1939).
213. C. W. N. Cumper, G. B. Leton and A. I. Vogel, *J. Chem. Soc.*, **1965**, 2067.
214. J. Chinn, *J. Org. Chem.*, **29**, 3304 (1964).
215. H. E. Zimmerman and T. W. Cutshall, *J. Amer. Chem. Soc.*, **81**, 4305 (1959).
216. B. Eistert in *Houben-Weyl*, *Methoden der organischen Chemie* (Ed. E. Müller), Vol. 2. Thieme, Stuttgart, 1953, p. 392.
217. H. von Euler and M. Hasselquist, *Reduktone; Sammlung chemischer und chemisch-technischer Vorträge*, Vol. 50. Ferdinand Enke, Stuttgart, 1950.
218. B. E. C. Banks, *J. Chem. Soc.*, **1961**, 5043; **1962**, 63.
219. W. D. Kumler, E. Kun and J. N. Shoolery, *J. Org. Chem.*, **27**, 1165 (1962).
220. D. J. Cram, *Fundamentals of Carbanion Chemistry*, Academic Press, New York, 1965, p. 1.
221. R. P. Bell, *The Proton in Chemistry*, Methuen, London, 1959.
222. R. P. Bell, *Acid-Base Catalysis*, Oxford Univ. Press., 1951.
223. M. Eigen, in *The Kinetics of Proton Transfer*, *Disc. Faraday Soc.* **39**, 7, (1965); R. P. Bell, *Disc. Faraday Soc.*, **39**, 16 (1965).
224. H. E. Zimmerman in *Molecular Rearrangements* (Ed. P. de Mayo), Vol. 1. Interscience, New York, 1963, p. 345.
225. M. Eigen and L. de Maeyer in *Technique of Organic Chemistry*, Vol. 8 (part 2), (Ed. A. Weissberger) Interscience, New York, 1963, p. 895.
226. M. Eigen and W. Kruse, *Z. Naturforsch.*, **18b**, 857 (1963).
227. M. Eigen, G. Ilgenfritz and W. Kruse, *Chem. Ber.*, **98**, 1623 (1965).
228. I. Amdur and G. G. Hammes, *Chemical Kinetics*, McGraw-Hill, New York, 1966, p. 149.
229. C. G. Swain and A. S. Rosenberg, *J. Amer. Chem. Soc.*, **83**, 2154 (1961).
230. R. P. Bell and G. G. Davis, *J. Chem. Soc.*, **1965**, 353.
231. A. Albert in *Physical Methods in Heterocyclic Chemistry* (Ed. A. R. Katritzky), Academic Press, New York, 1963, Vol. 1, p. 1.
232. G. J. Karabatsos, G. C. Sonnichsen, N. Hsi and D. J. Fenoglio, *J. Amer. Chem. Soc.*, **89**, 5067 (1967).

CHAPTER **4**

Oxocarbons and their reactions

*The University of Wisconsin, Madison,
Wisconsin, U.S.A.*

and

JOSEPH NIU

*Wyandotte Chemical Corporation, Wyandotte,
Michigan, U.S.A.*

I. INTRODUCTION

The term 'oxocarbon' designates compounds in which all, or nearly all, of the carbon atoms bear ketonic oxygen functions or their hydrated equivalents[1]. The most important oxocarbons are cyclic compounds, and this review will be limited to the chemistry of the monocyclic oxocarbons.

A few years ago we pointed out that monocyclic oxocarbon anions, $C_nO_n^{m-}$, are aromatic substances, stabilized by electron delocalization of π electrons in the ring[1,2]. At present four such species are known, the squarate (**1**), croconate (**2**) and rhodizonate (**3**) dianions, and the tetraanion of tetrahydroxy-*p*-benzoquinone (**4**). Much of the current

(**1**)   (**2**)   (**3**)   (**4**)

interest in the oxocarbons relates to physical studies of the chemical bonding in these species. However, in this review, in keeping with the purposes of this volume, the chemical reactions of the oxocarbons will be emphasized.

Although the oxocarbon anions have only recently been shown to form a previously unrecognized aromatic series, croconate and rhodizonate ions were actually among the first aromatic species ever synthesized. In 1823 Berzelius, Wöhler and Kindt observed the formation of a powdery black residue in the preparation of potassium by reduction of potassium hydroxide with carbon[3,4]. Dipotassium croconate and croconic acid (named from the Greek *krokos* = yellow, the colour of the acid and its salts) were isolated from this residue in 1825 by Gmelin[4]. A few years later, in 1837, Heller obtained rhodizonic acid (from the Greek *rhodizein* = rose red, the colour of the insoluble alkaline-earth derivatives) and dipotassium rhodizonate from the same material[5].

There is a further point of interest concerning these very early experiments. Both croconic acid and rhodizonic acid are now known to be products of microbiological oxidation of myoinositol (hexahydroxycyclohexane), a compound which is widely distributed in plants[6,7]. Therefore Gmelin's preparation of croconic acid[4] in 1825 represents one of the very first preparations of an 'organic' compound from 'inorganic' starting materials. The classic synthesis of urea by Wöhler did not take place until three years later[8].

II. SYNTHESIS OF OXOCARBONS

A. The Reactions of Carbon Monoxide with Alkali Metals

The earliest preparations of oxocarbons were from potassium hydroxide and carbon[3], but it was soon found by Liebig that the same products could be obtained more conveniently from carbon monoxide and hot potassium metal[9]. The reductive cyclopolymerization of carbon monoxide thus provides a route to both five and six-membered oxocarbons. Although this interesting reaction has been known for more than a century, and the products have been studied repeatedly[10-14], only within the last few years has a clear picture of the nature of the reaction begun to emerge. Because the history of this classical reaction is intimately connected with that of the oxocarbons themselves, it is of interest to trace its development briefly.

1. $K_6C_6O_6$, rhodizonic acid and croconic acid

As early as 1834, it was discovered that hot potassium metal reacts with carbon monoxide[9]. The reaction is exothermic, and heat was usually allowed to develop freely, so that the temperature rose above 150°. Under these conditions, the first product which can be isolated is a grey solid, $K_6C_6O_6$[9]. This compound is very easily oxidized, and upon aqueous work-up in air, dipotassium rhodizonate or dipotassium croconate is obtained.

$K_6C_6O_6$ is now recognized to be the hexapotassium salt of hexahydroxybenzene[12-14]. Formally, it can be regarded as an oxocarbon salt with a $6-$ charge on the anion. It is reasonable that air oxidation could transform $K_6C_6O_6$ into dipotassium rhodizonate:

$$C_6(OH)_6 \xleftarrow{\;H^+\;} K_6C_6O_6 \xrightarrow[H_2O]{\;[O]\;} K_2C_6O_6 + 4\,KOH$$

However, the production of croconate ion from the same reaction, providing an entry into the five-membered ring oxocarbons, is quite surprising. Yet, as early as 1837, in work of amazingly high quality for the time, Heller correctly deduced that rhodizonate is the precursor of croconate ion[5].

The rhodizonate–croconate transformation was first investigated by Nietzki and Benckiser[12,15] and has been studied by several other investigators[6,16,17]. If air or oxygen is bubbled through an alkaline aqueous solution of rhodizonate, complete transformation to croconate takes place:

$$C_6O_6^{2-} \xrightarrow{\text{[O]}} C_5O_5^{2-} + CO_3^{2-}$$

This remarkable reaction, involving contraction from a six to a five-membered ring, provides a most convenient method for the synthesis of croconates. The nature of the transformation has been in doubt, but it now seems clear that the reaction pathway is as follows:

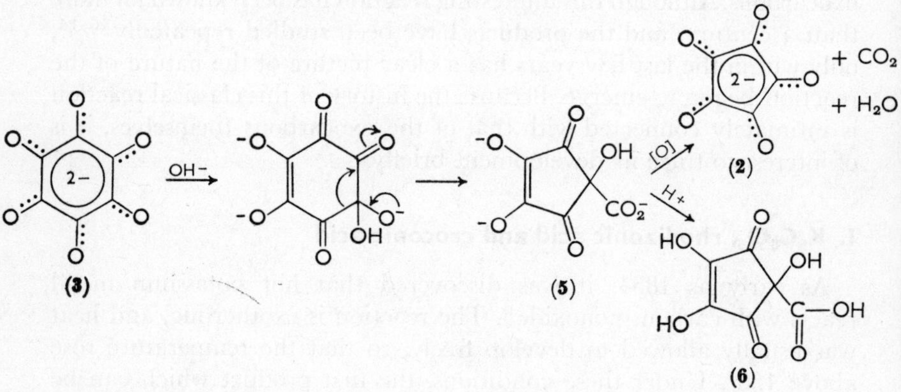

The intermediate anion **5** can be isolated from the solution in the form of the carboxylic acid **6**[15,17]. The ring contraction can be recognized as an example of an α-oxoalcohol rearrangement[18], related to the well-known benzil–benzylic acid rearrangement.

Sodium metal has been reported to react with carbon monoxide at 280–340° to produce the hexasodium salt of hexahydroxybenzene, $Na_6C_6O_6$[19]. This compound is oxidizable to disodium rhodizonate or croconate, just like the potassium compound. A careful recent study shows however that the reaction of sodium with carbon monoxide at temperatures from 200–500°C is complex, leading to a mixture containing sodium acetylide, sodium carbonate, and sodium rhodizonate as well as $Na_6C_6O_6$[19a].

2. The alkali metal salts of dihydroxyacetylene

If care is taken so that the reaction temperature of carbon monoxide with potassium never rises much above the melting point of potassium (63°), an intermediate black solid is formed[13,14,20-22], which is convertible to gray $K_6C_6O_6$ upon heating. In addition, carbon monoxide reacts with all of the alkali metals when they are dissolved in liquid ammonia to give yellowish 'alkali metal carbonyls', with the empirical formula MCO[23-25]. Recent investigations have begun to shed some light on these complex reactions.

In 1963 Sager and coworkers[13] published a study of the reaction between potassium and CO at low temperatures. They present evidence for two intermediates, $(K_3C_2O_2)_x$ and $(KCO)_x$, the latter undergoing conversion to $K_6C_6O_6$ upon heating:

$$K + CO \longrightarrow (K_3C_2O_2)_x \xrightarrow{CO} (KCO)_x \xrightarrow{\Delta > 100°} K_6C_6O_8$$

These investigators alkylated $(KCO)_x$ with methyl iodide and obtained a product with empirical formula $C_4O_4(CH_3)_4$, from which they

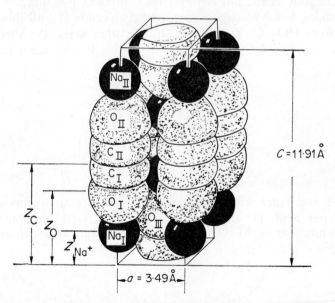

FIGURE 1. Arrangement of Na^+ and $^-O{-}C{\equiv}C{-}O^-$ ions in the crystal structure of the disodium salt of dihydroxyacetylene, $Na_2C_2O_2$, as determined by Weiss and Büchner[28].

deduced the formula $K_4C_4O_4$ for the black $(KCO)_x$, and inferred that the precursor was $K_6C_4O_4$.

The nature of the yellow 'alkali metal carbonyls' obtained in liquid ammonia has been clarified by Weiss and Büchner[26-28]. All of these substances appear to be principally salts of the dihydroxy-acetylene dianion, $M_2^+[OC{\equiv}CO]^{2-}$. The structures of the caesium, rubidium, potassium and sodium salts have been determined by x-ray crystal structure studies (Figure 1). Moreover, the black $(KCO)_x$ obtained by Sager and coworkers, despite its different appearance, is *crystallographically identical with* $K_2C_2O_2$[26]. The black colour must be the result of radical impurities or lattice imperfections.

The alkali metal dihydroxyacetylene salts can be regarded as linear oxocarbons; the anion is formally the first member of the series $C_nO_n^{2-}$. However, the bond lengths in the anion, about 1·20 Å for C—C and 1·27 Å for C—O[26,27], are consistent with the formula ^-O—$C{\equiv}C$—O^-, and suggest that electron delocalization in the system is slight.

The dianion of dihydroxyacetylene is a surprisingly weak base. Protolysis does not take place in acetonitrile solution with water or weak organic acids, but requires HCl, producing a mixture of *aci*-reductones, hexahydroxybenzene, and diglycolide (7), all thought to arise from HO—$C{\equiv}C$—OH[29]. On the other hand, Büchner[28] has found that the $M_2C_2O_2$ salts give a positive test for organometallic

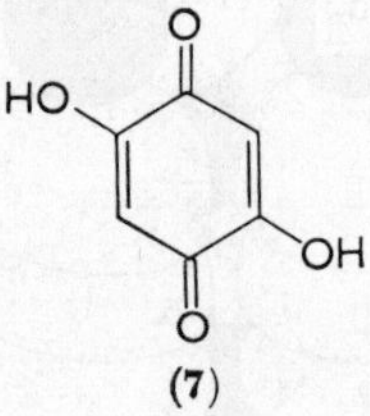

(7)

species and react with *n*-butyl bromide to give small amounts of α-hydroxy acid. He suggests that the 'alkali metal carbonyls' are really mixtures of M^+O^-—$C{\equiv}C$—O^-M^+ and an organometallic salt 8[28]:

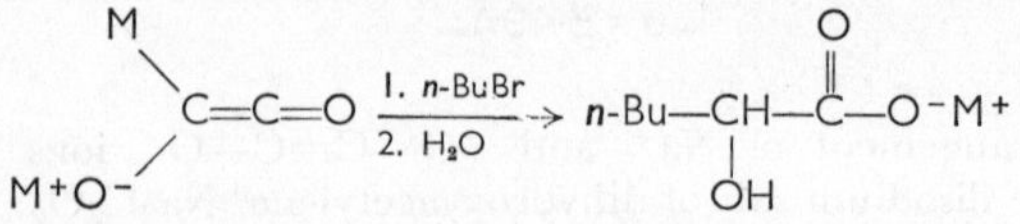

(8)

B. Other Syntheses of Rhodizonic and Croconic Acid and Their Salts

Although the cyclopolymerization of carbon monoxide can be used to synthesize oxocarbons, other methods are often more convenient for laboratory preparations. One of the most useful is the oxidation of the readily available natural product, myoinositol, an isomer of hexahydroxycyclohexane[16,30].

If inositol is oxidized with warm concentrated nitric acid and the solution is then treated with potassium carbonate, the dipotassium salt of tetrahydroxy-p-benzoquinone is produced as green crystals[30]. If the oxidized solution is instead treated with potassium acetate and aerated, dipotassium rhodizonate is the product

Free tetrahydroxyquinone (9) is stable and insoluble in water, so it is easily isolated as purple crystals by acidification of aqueous solutions of the dianion. Rhodizonic acid (10), however, is both sensitive to oxidation and quite soluble in water, so it is difficult to obtain from its salts. Small amounts of 10 can be obtained from basic aqueous solution by ion exchange using a H$^+$-form cation exchange resin[6], followed by low-temperature evaporation[31]; or the aqueous solution can be neutralized, evaporated, and 10 extracted with dioxane and precipitated with pentane[6]. Rhodizonic acid obtained by either of these methods is isolated as the colourless dihydrate, $H_2C_6O_6 \cdot 2H_2O$[32]. Sublimation in vacuum gives the dark orange anhydrous acid, first isolated only in 1958 by Bock and Eistert[32].

Another convenient route to **9** or **10** is the oxidative trimerization of glyoxal[33-35] discovered by Homolka[33]. The reaction is usually carried out in aqueous solution containing sodium sulphite and sodium hydrogen carbonate:

$$\underset{\underset{O}{\parallel}}{\overset{\overset{O}{\parallel}}{HCCH}} \xrightarrow[\text{H}_2\text{O, O}_2]{\text{Na}_2\text{SO}_3,\ \text{NaHCO}_3,} \text{Na}_2\text{C}_6\text{O}_4(\text{OH})_2,\ 5\text{-}10\% \xrightarrow[170°]{[O]} \text{Na}_2\text{C}_6\text{O}_6$$

The yield in this synthesis is low, but according to recent patents considerable improvements in yield result when potassium cyanide is used as a catalyst[35,35a], or when ozone is used in place of oxygen[36]. The disodium salt of tetrahydroxyquinone which is the initial product can be converted to disodium rhodizonate simply by heating in air to 170°[33]. A relatively recent paper by Fatiadi and coworkers gives clear directions for the preparation of rhodizonic acid, croconic acid and their derivatives by this route[6].

Rhodizonic acid has also been prepared by reduction with SO_2 of its oxidation product, triquinoyl octahydrate[12,32] (see section V.A.), and by oxidation of 1,4-diamino-2,3,5,6-tetrahydroxybenzene with potassium persulphate[36a].

Croconic acid (**11**) and the croconates could presumably be made by oxidation of cyclopentane derivatives, but this has apparently never been done. The convenient ring-contraction from rhodizonate has invariably been used. A wide variety of six-membered ring precursors such as hexahydroxybenzene, rhodizonic acid, tetrahydroxyquinone, tetraaminohydroquinone, etc. can be converted to sodium or potassium croconate by oxidation in alkaline solution with air or with manganese dioxide[6,12,15].

(11)

The alkali metal croconates are bright yellow salts, soluble in water to give stable solutions. Free croconic acid can be obtained by ion-exchange on aqueous solutions of the croconate salts, employing H^+-form cation exchange resins, followed by freeze drying of the solution[31]. Alternatively, the insoluble barium salt can be treated with the equivalent amount of sulphuric acid to give an aqueous solution of croconic acid, which can be evaporated[6]. These syntheses give croconic acid as the trihydrate, which can then be recrystallized from ethanol–dioxan by addition of benzene. Drying under vacuum removes the water of crystallization[6].

C. Squaric Acid

Dihydroxycyclobutenedione, or squaric acid (12), was first isolated by Cohen, Lacher and Park by the acid hydrolysis of the halogenated cyclobutene derivatives 14 and 16[37]. Dimerization of 1,1-dichloro-2,2-difluoroethylene followed by dechlorination gives 1,2-dichloro-3,3,4,4-tetrafluorocyclobutene (13)[38]. Ethoxide ion converts the latter to the triether 14[39] which is converted to squaric acid by acid hydrolysis[40,41].

The commercially available hexafluorocyclobutene (15) made by cycloaddition of chlorotrifluoroethylene and dechlorination[38], also serves as a precursor for squaric acid by a similar reaction sequence[41].

Recently, several useful preparations of squaric acid from hexachlorobutadiene have been discovered by Maahs[42-44]. Hexachlorobutadiene reacts with excess morpholine to form a trichlorotrimor-

9*

pholinobutadiene of unknown structure. When this is treated with water at pH 5–7 at 55°, it forms 3-morpholinotrichlorocyclobutene (**17**). This can be further hydrolysed to squaric acid.

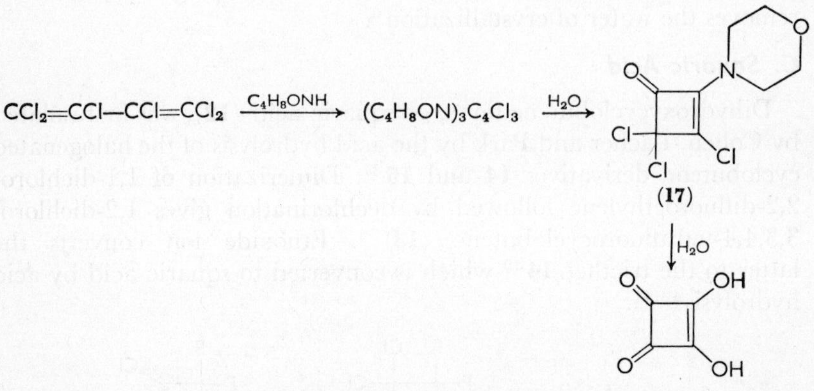

$$CCl_2=CCl-CCl=CCl_2 \xrightarrow{C_4H_8ONH} (C_4H_8ON)_3C_4Cl_3 \xrightarrow{H_2O}$$

(**17**)

An even more convenient synthesis begins with the nucleophilic substitution on hexachlorobutadiene by ethoxide ion to give 1-ethoxy-pentachlorobutane-1,3-diene (**18**), described by Roedig and Bernemann[45]. Compound **18** condenses thermally or catalytically to tetrachlorocyclobutenone (**21**) with elimination of ethyl chloride. Maahs suggests that the reaction proceeds through **20** as an intermediate, but an equally likely pathway appears to be through 1,3 cycloaddition to **19** followed by loss of ethyl chloride. With sulphuric acid, **21** undergoes hydrolysis to squaric acid. The entire reaction sequence can be carried out in a single reaction vessel without isolating intermediates, and it therefore provides a 'one-step' synthesis of squaric

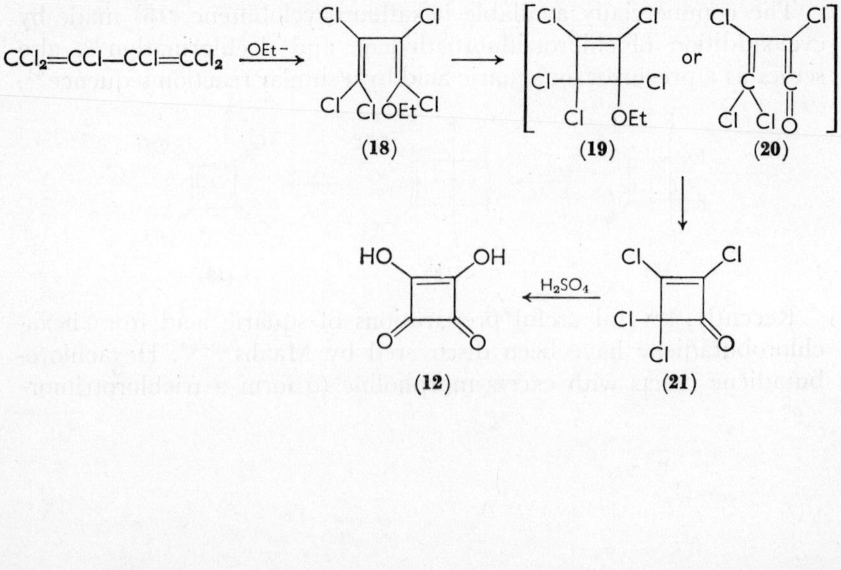

$$CCl_2=CCl-CCl=CCl_2 \xrightarrow{OEt^-}$$

(**18**)   (**19**)   (**20**)

(**12**)   (**21**)

acid from cheap, readily-available starting materials. Squaric acid forms colourless crystals, decomposing only above 293°, and it is now available commercially.

III. STRUCTURES AND PROPERTIES OF THE OXOCARBON ACIDS

Structures of types **22** and **23** for rhodizonic acid (**10**)[46,47], and **24** and **25** for croconic acid (**11**)[12,48] have been proposed in the past, but the α-enediol structures shown for **10** and **11** now seem well established, though x-ray studies have not been done. Neither **10** nor **11** show C—H absorption in their infrared spectra, such as would be present in **22** and **24**. The easy conversion of **10** to **9** argues against structure **23**, as does the normal C=O infrared frequency of 1700 cm^{-1} [31]. Absorption at higher frequency would be expected for the cyclobutanedione structure present in **23**. The same argument applies

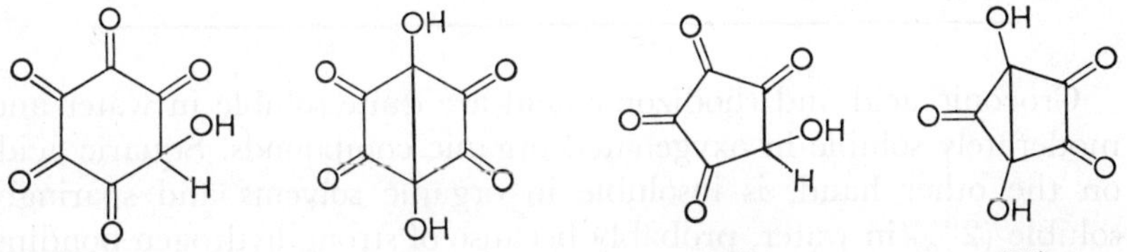

to **11**, which shows $\nu_{C=O}$ at 1755 cm^{-1}, consistent with the assigned structure[49]. Croconic acid has a very high dipole moment in dioxan (9–10 D), as expected for structure **11**[50]. Finally, the crystal structures of rubidium hydrogen croconate and ammonium hydrogen croconate show that these salts are correctly formulated as α-enediolates[51] (**26**). It is difficult to rule out completely the participation of tautomers such as **22** and **24** in solution[52], but there is no firm evidence for their existence.

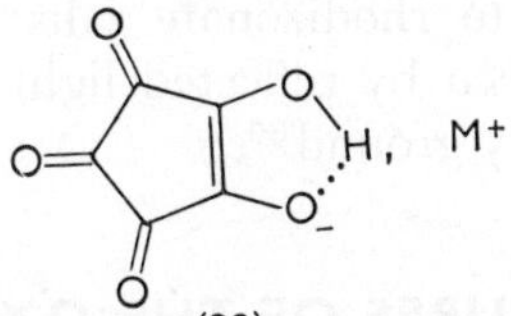

The mass spectra of the oxocarbon acids and of their oxidation products (section V. A) have recently been studied[52a]. The acids all show parent ion peaks in their mass spectra. A characteristic feature of the fragmentation is loss of CO and ring contraction.

The oxocarbon acids are all quite strong. pK values for these substances are given in Table I, which includes only reliable recent data. Croconic acid is decidedly stronger than squaric acid. Rhodizonic acid appears weaker than **11** or **12**, but the species present in water is the dihydrate, $C_6O_4(OH)_2 \cdot 2H_2O$, in which two of the carbonyl groups of **10** are probably hydrated to $\diagdown C(OH)_2$ groups. Unhydrated **10** would be a much stronger acid than the dihydrate[56].

TABLE 1. pK values for oxocarbon acids

Acid	pK_1	pK_2	Reference
$C_4O_2(OH)_2$ (**12**)	1·7, 1·2	3·2, 3·48	53, 53a
$C_5O_3(OH)_2$ (**11**)	0·5	1·5–2·0	54, 55
$C_6O_4(OH)_2 \cdot 2H_2O$ (**10**)	3·1	4·9	56
$C_6O_2(OH)_4$ (**9**)	4·8	6·8	57

Croconic acid and rhodizonic acid are quite soluble in water and moderately soluble in oxygenated organic compounds. Squaric acid, on the other hand, is insoluble in organic solvents and sparingly soluble (2%) in water, probably because of strong hydrogen bonding in the solid. The infrared spectra of all of the oxocarbon acids provides evidence for extensive hydrogen bonding. Croconic and rhodizonic acids give normal tests for the carbonyl group, i.e. with phenylhydrazines. However in squaric acid, both carbonyl groups are 'vinylogous carboxyl' carbonyls, and hence they do not react with carbonyl reagents[37]. All of the oxocarbon acids give colours with ferric chloride characteristic of enols.

The alkali metal salts of the oxocarbon acids are all stable substances soluble only in water. The colour deepens from squarate (colourless) to croconate (yellow) to rhodizonate salts; the latter are strongly dichroic, appearing green by reflected light but red by transmitted light or when very finely ground[30].

IV. STRUCTURES OF THE OXOCARBON ANIONS: AROMATICITY

A. Squarate, Croconate and Rhodizonate Anions

A resonance-stabilized structure for the croconate anion was first suggested in 1958 by Yamada and Hirata[49,58], reasoning from the high acid strength of croconic acid and the lack of infrared absorption in the usual carbonyl region in croconate ion. A delocalized formulation for the squarate ion was proposed on the basis of the same sort of evidence by Cohen, Lacher and Park[37]. These observations led to the suggestion that the anions in question were aromatic species, and to the generalization of the oxocarbons as an aromatic series[1,2].

Aromaticity implies a symmetrical structure for the oxocarbon anions. The first definite evidence for symmetry came from vibrational spectroscopic studies, which indicated that squarate and croconate ions, in their alkali metal salts, are planar species in which all the carbon atoms and all the oxygen atoms are equivalent[2,59]. Detailed normal coordinate analysis of the infrared and Raman spectra of dipotassium squarate and croconate showed, moreover, that there must be substantial π bonding between the ring carbon atoms[59]. Table 2 lists C—O and C—C stretching force constants for the oxocarbon anions and for some related species. The data show that the C—C stretching force constants in the oxocarbons are distinctly greater than for C—C single bonds, though less than for benzene. Moderate multiple bond character (π bonding) between carbon atoms in the oxocarbon rings is indicated.

More recently, the detailed structure of dipotassium squarate monohydrate[60], diammonium croconate[61], and dirubidium rhodizonate[62] have been determined by single-crystal x-ray investigations. Results show that in all of these salts the anions have symmetrical planar structures. The observed C—C bond lengths (Table 3) confirm the existence of substantial delocalized multiple bonding between ring carbon atoms. Differences among the three anions are small,

TABLE 2.   Stretching force constants[a].

Species	K_{CC} (mdyne/Å)	K_{CO} (mdyne/Å)
$C_4O_4^{2-}$	3.95	5.60
$C_5O_5^{2-}$	3.50	6.72
C—C	2.0–2.3	
Benzene	5.17	
C=C	~7	
C—O		2.8
Acetate ion	2.46	7.20
C=O		10.8

[a] Urey–Bradley force field.

and may result from crystal forces rather than differences in the free anions. It is interesting that the C—O bond lengths are very short, particularly in rhodizonate ion, indicating a C—O bond order of nearly 2 for this species.

TABLE 3. Bond lengths in oxocarbon salts

Species	r_{CC}(Å)	r_{CO}(Å)	Reference
$K_2C_4O_4 \cdot H_2O$	1·469	1·259	60
$(NH_4)_2C_5O_5$	1·457	1·262	61
$Rb_2C_6O_6$	1·488	1·213	62

Molecular orbital calculations have been carried out on a variety of oxocarbon anions, including more complex types as well as monocyclics, using the simple Hückel method[1]. Some results are given in Table 4. Even the simplest calculations give remarkably good agreement with experiment in most respects. Thus the bond orders are fairly close to those predicted, and the electronic spectra of the oxocarbon anions, which show bathochromic shifts as ring size increases, are well correlated by the MO calculations. More elaborate SCF calculations have also been attempted, particularly for $C_6O_6^{n-}$ ions[63-65]. Hückel and Pariser-Parr-Pople MO calculations have also been carried out for the oxocarbon acids[65a].

TABLE 4. Results of simple Hückel MO calculations for oxocarbon anions

Species	$D\epsilon/\pi^a$	Mobile Bond Order		Charge Density	
		C—C	C—O	C	O
$C_3O_3^{2-}$	0·280	0·594	0·416	+0·187	−0·854
$C_4O_4^{2-}$	0·240	0·471	0·542	+0·236	−0·736
$C_5O_5^{2-}$	0·231	0·447	0·598	+0·234	−0·634
$C_6O_6^{2-}$	0·227	0·444	0·626	+0·241	−0·571
$C_7O_7^{2-}$	0·223	0·436	0·636	+0·253	−0·539
$C_8O_8^{2-}$	0·220	0·427	0·651	+0·264	−0·513
$C_6O_6^{4-}$	0·209	0·527	0·455	+0·074	−0·741

[a] Delocalization energy (resonance energy) per π electron.

The alternation of aromatic stabilization with ring size, found for the cyclic polyenes, is not predicted (or observed) for oxocarbons. The unknown monocyclic oxocarbons $C_3O_3^{2-}$, $C_7O_7^{2-}$ and $C_8O_8^{2-}$

should all be aromatic[1]. Moreover, most polycyclic oxocarbon anions, regardless of symmetry or ring size, are predicted to have substantial aromatic stabilization[1] (Table 4).

B. The Tetraanion of Tetrahydroxyquinone

Molecular orbital calculations also suggest aromatic properties for oxocarbon anions with charges other than $2-$. However, the lowest-lying unfilled π energy level in all of the monocyclic oxocarbon dianions is *degenerate*, according to the simple Hückel calculations. Therefore, species $C_nO_n^{4-}$ should have two unpaired electrons occupying the degenerate orbitals and should be a triplet state, paramagnetic species[1].

These molecular orbital considerations led directly to the preparation of the only known oxocarbon anion with charge other than $2-$, $C_6O_6^{4-}$, from tetrahydroxyquinone[66]. When tetrahydroxyquinone is treated with base such as potassium hydroxide, the insoluble dipotassium salt $K_2C_6O_4(OH)_2$ is usually produced. $K_4C_6O_6$ was finally obtained by working with dilute anhydrous methanolic solutions of $KOCH_3$ and tetrahydroxyquinone, with rigid exclusion of oxygen[66]. The product is a black solid with absorption throughout the visible region. It is possible that earlier workers had this material in hand, but in quite impure form[12]. In air, it undergoes rapid oxidation to dipotassium rhodizonate. $K_4C_6O_6$ is a probable intermediate in the oxidation of $K_6C_6O_6$[12].

The black solid $K_4C_6O_6$ is diamagnetic, in contradiction to the prediction from Hückel MO theory. The diamagnetism could be explained by distortion of the anion, or by electron correlation neglected in the simple MO approach. The simple infrared spectrum of $K_4C_6O_6$ is consistent with a symmetrical formulation for the anion[31,66] but there is as yet no firm evidence for the structure of the compound.

C. Oxocarbon Radicals

Upon oxidation in air, the diamagnetic $K_4C_6O_6$ gave a dark green, strongly paramagnetic substance giving a single e.s.r. line at

$g = 2 \cdot 003$[66]. This intermediate could not be isolated pure, but underwent rapid quantitative oxidation to give dipotassium rhodizonate. The formulation $C_6O_6^{3-}$ was suggested for the intermediate radical, which would thus be a delocalized oxocarbon anion radical.

Further evidence for oxocarbon anion radicals is given by an e.s.r. study of the products from the potassium–carbon monoxide reaction. When the products were heated to 250°, a single-line e.s.r. spectrum was observed[67], attributed to $C_6O_6^{5-}$. Other such delocalized oxocarbon anion radicals are likely to be discovered in the future.

V. REACTIONS OF THE OXOCARBONS

A. Oxidation Products: Triquinoyl, Leuconic Acid and Octahydroxycyclobutane

It has been known for more than a century that rhodizonic and croconic acids undergo oxidation to give triquinoyl octahydrate $(C_6H_{16}O_{14})$ and leuconic acid $(C_5H_{10}O_{10})$ respectively. The analogous four-membered ring compound, $C_4(OH)_8$ was prepared only in 1963[40].

Triquinoyl octahydrate was first prepared by the oxidation of hexahydroxybenzene with nitric acid[11,12]. It can be made by the bromine oxidation of rhodizonic acid, or, somewhat more conveniently, by oxidation of tetrahydroxyquinone with bromine or nitric acid[6]. The compound forms colourless prisms, decomposing at 99°c. In water, it behaves as a weak acid, but basic solutions undergo rapid degradation[68]. Reduction with sulphur dioxide gives rhodizonic acid[12].

Leuconic acid is easily obtained by oxidation of a solution of croconic acid or an alkali metal croconate with chlorine or nitric acid[6,6a,69]. With SO_2, it is reduced back to croconic acid. Leuconic acid crystallizes as colourless needles decomposing at 158–160°[6]. Like triquinoyl, it is a weak acid which undergoes decomposition to open-chain compounds in basic solution[70,71]. A sodium leuconate $NaC_5H_9O_5$, thought to be cyclic, has been prepared[70].

The oxidation product of squaric acid, $C_4(OH)_8$, was obtained by treatment of squaric acid with bromine or nitric acid at 0°[40]. The product can be precipitated from aqueous solution by addition of acetone. It is a colourless solid decomposing at 140°, and reducible to squaric acid with sulphur dioxide.

The structure of $C_4(OH)_8$ is definitely known to be that of octa-hydroxycyclobutane (**27**). This structure was proposed from study of the infrared and Raman spectra of this compound and its octa-deutero analogue $C_4(OD)_8$, which strongly suggest that the molecule has a fourfold axis of symmetry (D_{4h} structure)[40]. Recently, Bock has carried out a single-crystal x-ray determination on $C_4(OH)_8$ and has confirmed the hydroxycyclobutane structure[72]. The cyclo-butane ring is planar, with bond lengths 1·358 Å (C—O) and 1·562 Å (C—C); all of the hydroxyls are involved in a network of hydrogen bonding.

The infrared spectra of octahydroxycyclobutane, leuconic acid, and triquinoyl octahydrate are all quite similar, and all three com-pounds are now believed to have the fully hydroxylated structures **27**, **28** and **29**[40]. The fact that triquinoyl octahydrate and leuconic

(**27**)   (**28**)   (**29**)

acid show no strong absorption in the carbonyl region of the infrared spectrum is consistent with this assignment[40,73]. The slow rate of reduction of triquinoyl observed in polarographic studies suggests that triquinoyl, but not its reduction product rhodizonic acid, is fully hydrated in solution also[74].

It would be of special interest to obtain unhydrated forms of these polyhydroxy compounds, which would then be neutral polymers of carbon monoxide, $(CO)_n$. Several attempts have been made to prepare these, but with inconclusive results. An anhydrous triquinoyl was claimed by Henle[75], who obtained a yellow syrup by shaking the octahydrate with ether and phosphorus pentoxide, or by treating silver rhodizonate with bromine in dry ether. The yellow solutions gave reactions typical of triquinoyl, but no pure compound was isolated. Later, Bergel prepared a dehydrated form of triquinoyl by warming the octahydrate over P_2O_5 in a vacuum[76]. The material so obtained analysed for $C_6O_6 \cdot 1–2\ H_2O$, but could *not* be converted back to triquinoyl octahydrate on addition of water. A partially dehydrated form of leuconic acid analysing for $C_5H_8O_9$ has also been prepared[12].

Several reactions of the polyhydrated oxocarbons are of interest. Leuconic acid was reported to give a tetra and a pentaoxime with hydroxylamine[12,77]. Leuconic acid[78] and triquinoyl[79] both undergo condensation reactions with o-toluenediamine to give derivatives formulated, on the basis of analysis only, as **30** and **31** respectively.

(30)   **(31)**

Preparation of the condensed ring compounds **32** and **32a** by ultraviolet photolysis of triquinoyl is claimed in a patent[80], but no structural evidence is given. Recently leuconic acid has been photolysed to a dimer believed to have structure **33**[80a].

(32)   **(32a)**

(33)

B. Reduction Products

The six-membered ring oxocarbons are the only ones whose reduction has been carefully studied. Rhodizonate ion or rhodizonic acid can be reduced, for instance with SO_2[12], to tetrahydroxyquinone and ultimately to hexahydroxybenzene (Scheme 1). Potentiometric oxidation-reduction titrations have been carried out by Priesler and his coworkers[30,57] and by Tatibouet and Souchay[74] which show that the reduction takes place in two 2-electron steps and is largely reversible. The reduction potentials, E^0 at pH = 0, for the systems rhodizonic acid–tetrahydroxyquinone, tetrahydroxyquinone–hexahydroxybenzene and rhodizonic acid–hexahydroxybenzene were estimated to be +0·426, +0·350 and +0·388 volt respectively[57]. The electrochemistry was also studied as a function of pH. The measurements explain the

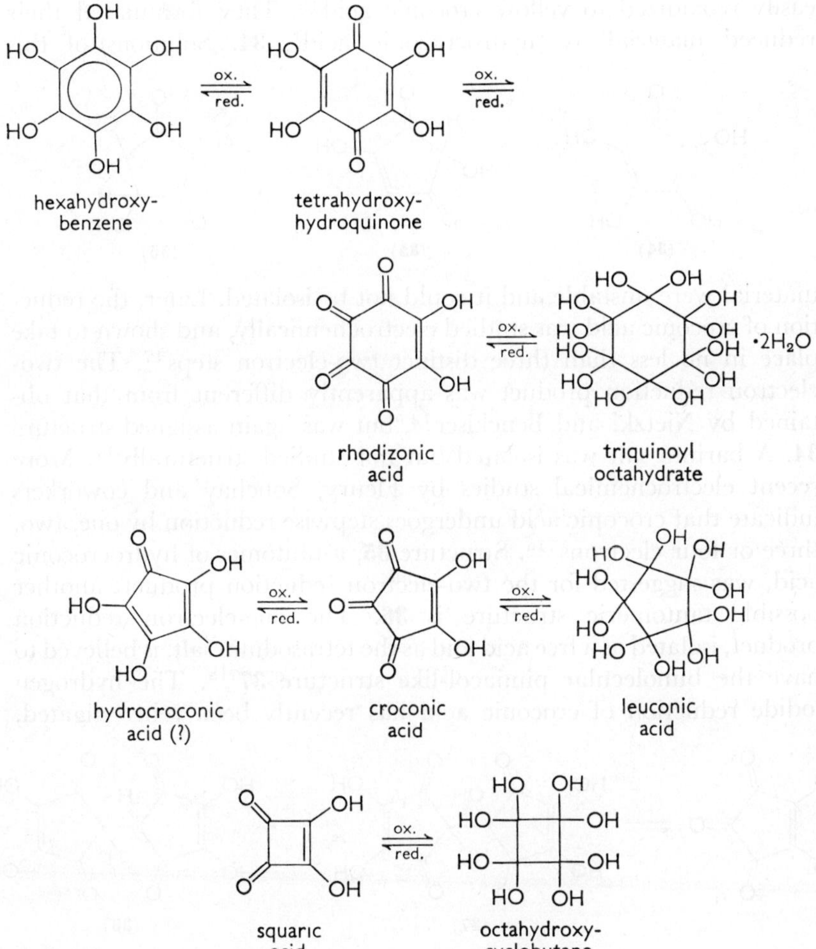

SCHEME 1. Oxidation–Reduction Equilibra Among the Oxocarbons.

exclusive precipitation of dipotassium rhodizonate in an acetate buffer, which results from the dismutation of tetrahydroxyquinone to rhodizonate and $C_6(OH)_6$.

The reduction of croconic acid is not nearly so well understood. The reduction of dipotassium croconate with HI was studied as early as 1862 by Lerch[11], who isolated a yellow-brown reduction product which gave red salts formulated as $K_2C_5H_2O_5$ and $BaC_5H_2O_5$. Later, Nietzki and Benckiser[12] showed however that reduction of croconic acid with SO_2, Zn, or $SnCl_2$ gives colourless solutions which are easily reoxidized to yellow croconic acid[12]. They formulated their reduced material as 'hydrocroconic acid', 34. Solutions of this

(34)　　　　　(35)　　　　　(36)

material were unstable and it could not be isolated. Later, the reduction of croconic acid was studied electrochemically, and shown to take place in no less than three distinct two-electron steps[17]. The two-electron reduction product was apparently different from that obtained by Nietzki and Benckiser[12], but was again assigned structure 34. A barium salt was isolated but not studied structurally[17]. More recent electrochemical studies by Fleury, Souchay and coworkers indicate that croconic acid undergoes stepwise reduction by one, two, three or four electrons[71a]. Structure 35, a tautomer of hydrocroconic acid, was suggested for the two-electron reduction product; another possible tautomeric structure is 36. The one-electron reduction product, isolated as a free acid and as the tetrasodium salt, is believed to have the bimolecular pinnacol-like structure 37[71a]. The hydrogen iodide reduction of croconic acid has recently been reinvestigated,

(37)　　　　　　　　　　(38)

bimolecular products 37 and 38 being reported[81,82]. Both compounds are reported to have four enolic protons; tetramethyl enol ethers of

37 and **38** and a hexaacetyl derivative of **37** are also reported. Further work on the reduction products of croconates would be useful.

Scheme 1 summarizes oxidation–reduction relationships among the neutral oxocarbons. It is apparent that the systematics are best developed for the six-membered ring compounds, next for five-membered rings, and least in the four-membered series. Squaric acid is not expected to be stable in reduced form, in that the product would be tetrahydroxycyclobutadiene. The tetraanion of tetrahydroxycyclobutadiene might be isolable, however.

C. Esterification

A number of 'esters', actually enol ethers, are known for squaric and croconic acids (Table 5). The dimethyl and diethyl esters of croconic acid can be prepared by the reaction of methyl or ethyl iodide with silver croconate[58,83]. The esters are orange-red solids stable to 200°, but rapidly hydrolysed by moist air to croconic acid. Unlike croconic acid, they are soluble in most organic solvents.

TABLE 5. 'Esters' (enol ethers) of oxocarbon acids

R^1	R^2	m.p. (°c)	b.p. (°c)/p (torr)	Reference
CH_3	CH_3	58		84
CH_3	H	132–4		84
C_2H_5	C_2H_5		89–91/0·4?	84
n–C_3H_7	n–C_3H_7			44
n–C_4H_9	n–C_4H_9		139/0·5	42, 43
CH_3	CH_3	114–5	250/740	83, 58
C_2H_5	C_2H_5	58	174–5/3	83
C_2H_5	H	150 d.	19/747	83

A colourless diethyl acetal, **39**, is obtained when either croconic acid or silver croconate is treated with alcoholic hydrogen chloride.

Diazomethane converts this acetal to its dimethyl ester (**40**), a stable yellow liquid[83].

When croconic acid itself is treated with diazomethane in ether, an interesting ring expansion takes place yielding trimethoxy-p-benzoquinone (**42**)[58]. The reaction must involve CH_2 insertion into a ring carbon–carbon bond, and presumably takes place through the intermediate **41**.

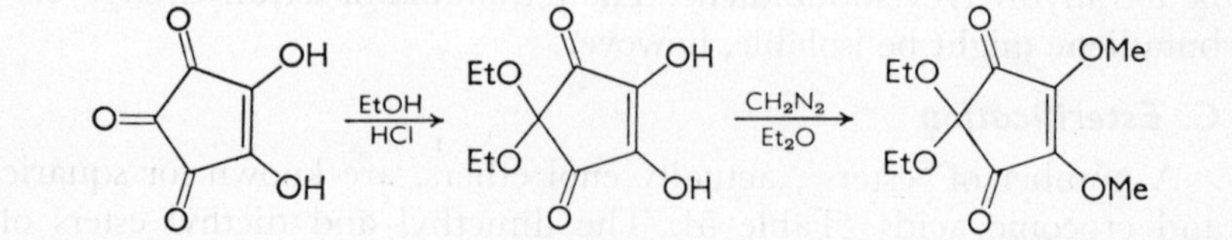

(39)   (40)

Diethyl[43] and dibutyl[42,43] squarate have been made simply by heating the corresponding alcohol with squaric acid, which serves as its own acid catalyst for esterification. Dibutyl squarate has also been made by alcoholysis of perchlorocyclobutenone, and other esters such as di-n-propyl squarate have been obtained by alcoholysis of dichlorocyclobutenedione (section V.D)[44]. These esters are colourless liquids, easily hydrolysed to squaric acid by water.

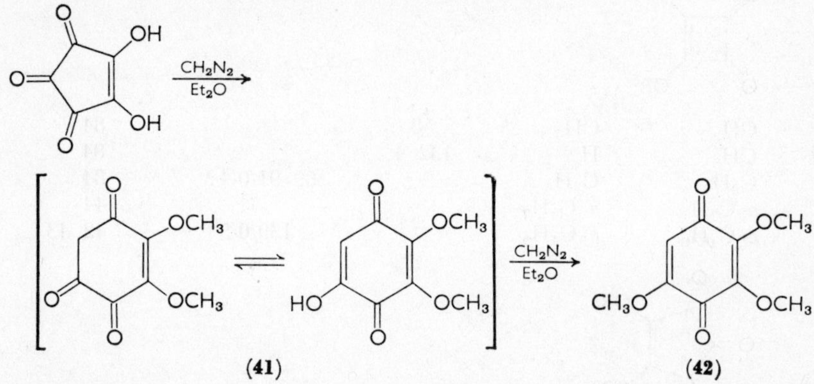

(41)   (42)

When squaric acid is heated with methanol, the half-ester 1-hydroxy-2-methoxycyclobutenedione (**43**), is obtained instead of the dimethyl ester. Compound **43** was also observed in the controlled hydrolysis of dimethyl squarate[84]. Dimethyl squarate itself can be synthesized from silver squarate and methyl iodide, or from squaric acid and diazomethane[84]. In the latter reaction, an orange by-product is also obtained, which may be a ring-enlargement product formed in a reaction similar to that obtained with croconic acid and diazomethane. Unlike the other diesters of squaric acid which are liquid, dimethyl squarate is a low-melting solid. Heating with ethanol in the presence of ethoxide ion converts it to the diethyl ester[84], and other esters could probably be made similarly by transesterification.

Esters of rhodizonic acid do not seem to have been isolated, but should be accessible by the same routes described above.

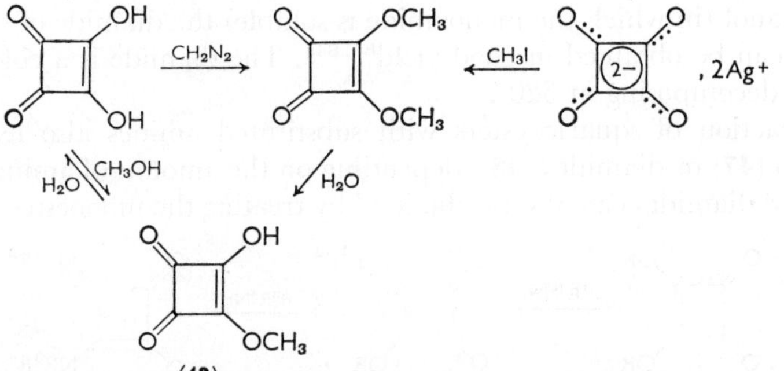

(43)

D. Derivatives of Squaric Acid

It is curious that much more is now known of the chemistry of squaric acid, isolated only in 1959, than about the other oxocarbons which are more than 100 years older. The chemistry of squaric acid derivatives has been reviewed by Maahs[44], and so will be treated only briefly here.

'Squaryl dichloride', dichlorocyclobutenedione (44), can be prepared by the action of SO_3 or oleum on perchlorocyclobutenone, 21 (section II.C.)[43]. It forms yellowish-green crystals melting at 51–52°, and has chemical properties typical for an acyl chloride.

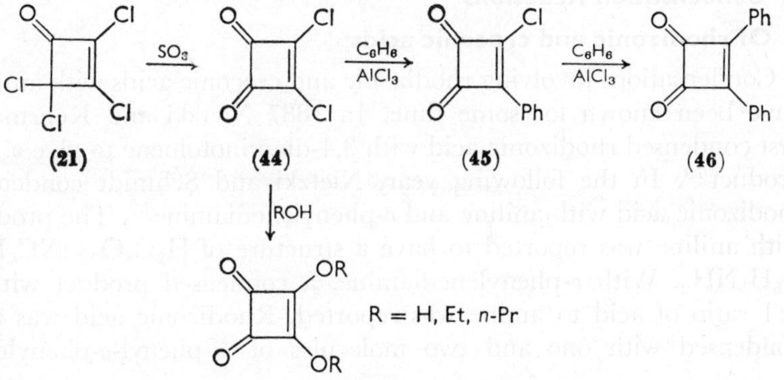

(21)   (44)   (45)   (46)

R = H, Et, n-Pr

Compound **44** is easily solvolysed by water to squaric acid and by alcohols to diesters of squaric acid. With benzene, **44** undergoes Friedel–Crafts reactions to give phenylchloro- (**45**) or diphenylcyclo-butenedione (**46**) with catalytic or stoichiometric amounts of $AlCl_3$, respectively[44].

Ammonolysis of dimethyl squarate with gaseous ammonia in ether yields the monoester monoamide which precipitates out of solution. With aqueous ammonia in methanol or ether, or gaseous ammonia in methanol (in which the monoamide is soluble) the diamide of squaric acid can be obtained in good yield[84, 84a]. The diamide is a colourless solid decomposing at 320°.

Reaction of squaric esters with substituted amines also leads to mono (**47**) or diamides (**48**), depending on the amount of amine used. Mixed diamides can also be obtained by treating the monoester mono-

$$
\underset{(47)}{\overset{R^1R^2NH}{\longrightarrow}} \qquad \underset{(48)}{\overset{R^3R^4NH}{\longrightarrow}}
$$

amides with other amines[44, 86]. Examples of compounds in this series are listed in the review by Maahs and Hegenberg[44].

Squaric acid diacetate (diacetoxycyclobutenedione, m.p. 63–65°) has been synthesized from squaric acid and acetic anhydride. This compound is very reactive and is solvolysed to squaric acid by water or alcohols[85].

Similar acid chlorides, amides, and acyloxy compounds could probably be made from croconic and rhodizonic acids, but up to now they are unknown.

E. Condensation Reactions

I. Of rhodizonic and croconic acids

Condensations involving rhodizonic and croconic acids with amines have been known for some time. In 1887 Nietzki and Kehrmann first condensed rhodizonic acid with 3,4-diaminotoluene to give a 1:1 product[79]. In the following year, Nietzki and Schmidt condensed rhodizonic acid with aniline and *o*-phenylenediamine[87]. The product with aniline was reported to have a structure of $H_2C_6O_5{=}NC_6H_5 \cdot C_6H_5NH_2$. With *o*-phenylenediamine, a condensed product with a 1:1 ratio of acid to amine was reported. Rhodizonic acid was also condensed with one and two molecules of *N*-phenyl-*o*-phenylene-

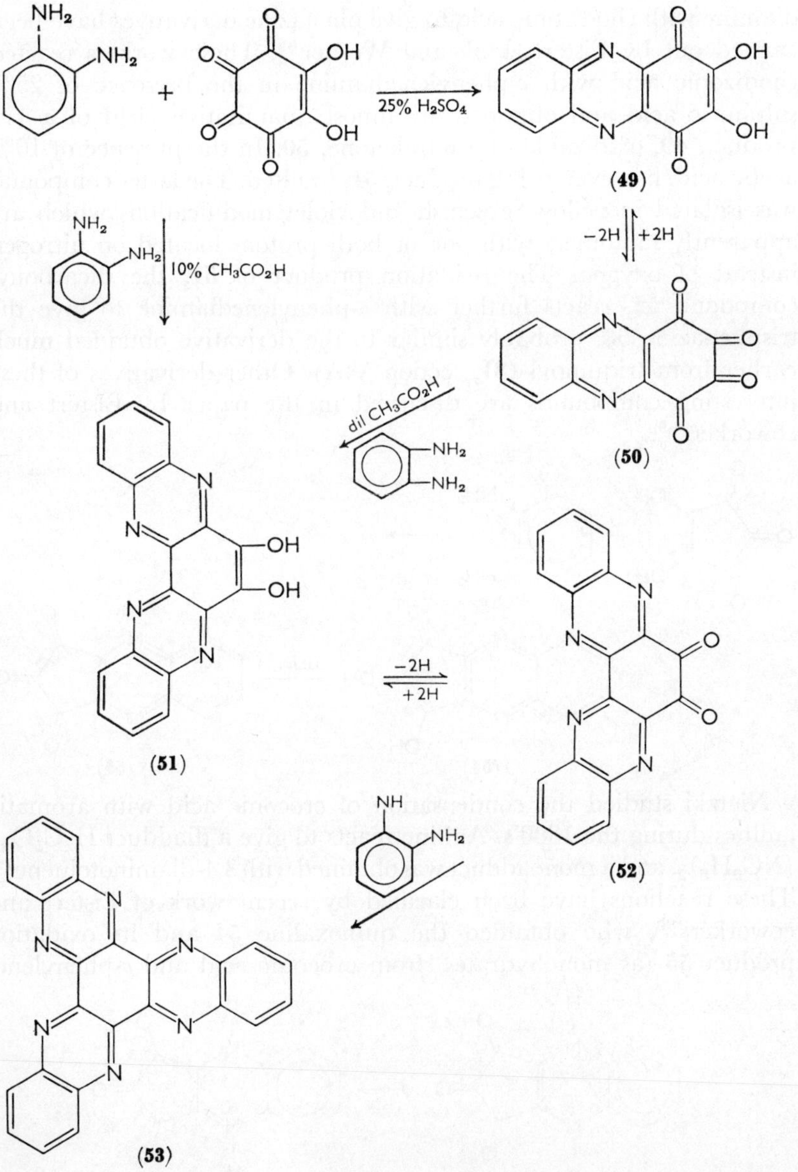

NH2
NH2
+
O
OH
OH
O
O
O
25% H2SO4
N
N
OH
OH
(49)
−2H
+2H
10% CH3CO2H
dil CH3CO2H
(50)
(51)
−2H
+2H
(52)
NH
NH2
(53)

diamine to yield products of empirical formulae $C_{18}H_{12}O_4N_2$ and $C_{30}H_{18}O_2N_4$[88].

More recently, detailed studies on the condensation of o-phenylene-diamine with rhodizonic acid to give phenazine derivatives have been carried out by Eistert, Fink and Werner[89]. These workers reacted rhodizonic acid with o-phenylenediamine in the presence of 25% sulphuric acid and obtained an almost quantitative yield of a 1:1 product, **49**, oxidizable to a tetraketone, **50**. In the presence of 10% acetic acid, however, a 1:2 product, **51**, resulted. The latter compound was isolated in yellow, greenish and violet modifications which are apparently tautomers with one or both protons located on nitrogen instead of oxygen. The oxidation product of **51**, the dicarbonyl compound **52**, reacts further with o-phenylenediamine to give the trisphenazine **53**, probably similar to the derivative obtained much earlier from triquinoyl (**31**, section V.A). Other derivatives of these interesting compounds are discussed in the paper by Eistert and coworkers[89].

(**54**)   (**55**)

Nietzki studied the condensation of croconic acid with aromatic amines during the 1880's. Aniline reacts to give a diadduct $H_2C_5O_3-(NC_6H_5)_2$, and a monoadduct was obtained with 3,4-diaminotoluene[12]. These reactions have been clarified by recent work of Eistert and coworkers[89], who obtained the quinoxaline **54** and its oxidation product **55** (as monohydrates) from croconic acid and o-phenylene-

(**54a**)   (**54b**)

diamine. It is possible that compound **54** exists in the tautomeric form **54a** ↔ **54b** with hydrogen bound to nitrogen[89]. Compound **55**, actually a derivative of leuconic acid, was also obtained by thermal decomposition of **50**, which takes place with disproportionation and with oxidative ring contraction (see section II.B.).

Related dihydropyrazine derivatives in the croconate series were studied by Prebendowski and Malachowski[83], who condensed a diketal of croconic acid (**56**) with ethylenediamine to give both a 2:1 adduct, formulated as **57**, and a 1:1 compound thought to be the dihydropyrazine **58**. On hydrolysis, **58** is converted to the diketo

compound **59**, which reacts further with another mole of ethylene-diamine to give **60**. For all of these products, the structures were deduced only from the elemental analysis. The dimethyl ester of croconic acid (section V.C.) was also condensed with *o*-phenylene-diamine to give **61**, the dimethyl enol ether of **54**.

(**61**)

2. Of squaric acid

Recently, a number of extremely interesting condensation products have been obtained from squaric acid and its esters[90-93]. Some of these are discussed in the review on squaric acid by Maahs and Heyenberg[44], and in a more recent review on 'cyclobutenediylium dyes' by Sprenger and Zeigenbein[92a].

The condensation of squaric acid with activated, 5-substituted pyrroles gives red-violet dyes, formulated as cyanines (**62**) by Triebs and Jacob on the basis of their electron absorption, which corresponds well with those of other cyclotrimethine dyes with betaine structure[94,95]. With excess pyrrole pale yellow tripyrrylcyclobutene derivatives (**63**) are formed. No products are obtained if the pyrrole rings are substituted in both the 2 and 5 positions. If neither position

(**62**)   (**63**)

is substituted, blue-green dyes are formed, apparently with polymeric structure[90].

Many other reactive aromatic and heterocyclic molecules will condense with squaric acid to give highly coloured betaine-like products, 1,3 substituted on the cyclobutene ring. For instance, *N,N*-dialkylanilines[92] condense to give deeply coloured products formulated as **64** (or **65**). Related condensation products are obtained with phloroglucinol[90], α-substituted indoles[91], azulenes[92], barbituric acid[92a] and the betaine bases derived by loss of HI from 2-methyl substituted quinolinium, benzothiazolium and benzoselenazolium iodides[93].

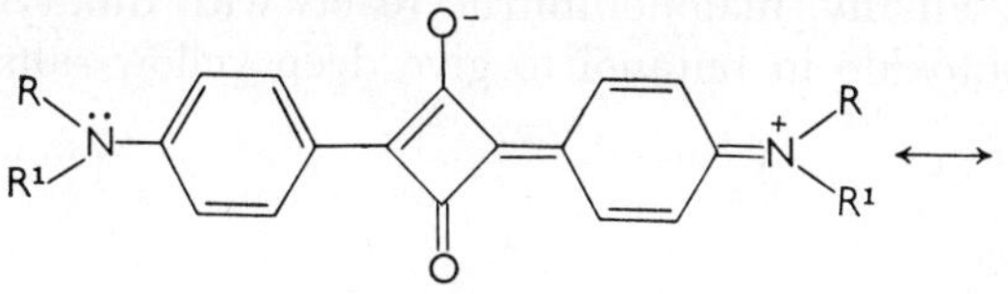

(64)

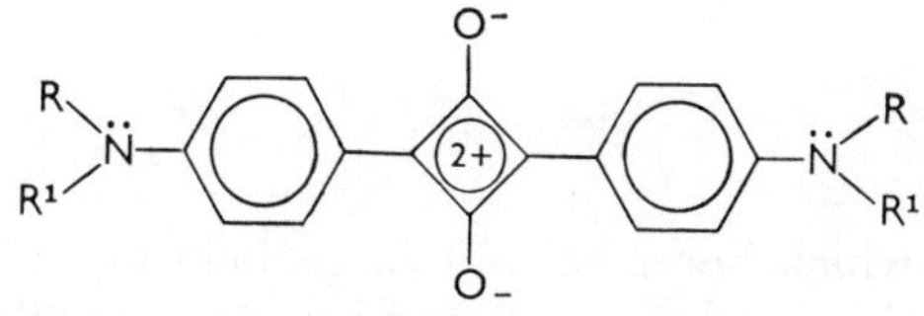

(65)

Sprenger and Ziegenbein[93] have questioned the structures **64** proposed for these dyes by Triebs and Jacob[90,91], and instead suggest that they be formulated as cyclobutenediylium derivatives (**65**). The question however is really one of relative contribution of different limiting structures to the resonance hybrid, which cannot be settled until information about the electron distribution in the dyes is available. To these reviewers, it seems likely that the cyanine form **64** is much more important than the cyclobutenediylium structure **65**.

Squarate esters also undergo condensation with 2- or 4-methyl-pyridinium, quinolinium or benzothiazolium iodides. However, in this case 1,2 rather than 1,3 substitution takes place on the four-membered ring[93]. Sprenger and Ziegenbein formulate the product from *N*-ethyl-4-methylpyridinium iodide as a cyclobutadienylium compound (**66**), but we can write additional and more probable resonance structures such as **67**. Analogous formulae can be written for other members of this series.

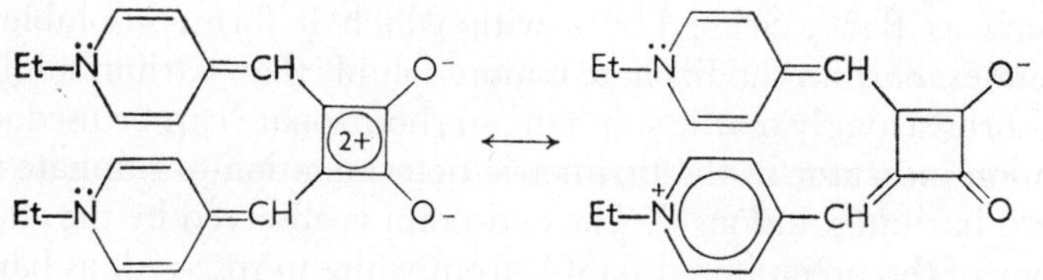

(66)   (67)

Although tertiary aromatic amines and heterocyclic nitrogen compounds condense with squaric acid to form bonds through carbon to the four-membered ring, primary and secondary aromatic amines react quite differently, becoming attached through nitrogen to give yellow-coloured betaine compounds of type **68** (only a single limiting structure is written)[93a]. Finally, malonodinitrile reacts with dibutyl squarate and sodium butoxide in butanol to give deep yellow salts

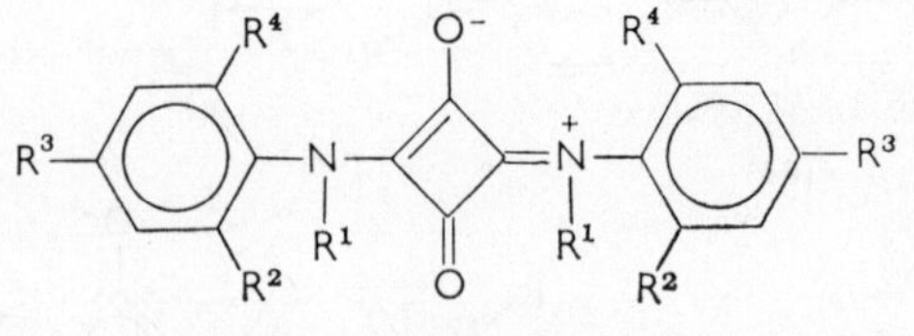

(68)

which Sprenger and Ziegenbein formulate as **69**[93]. Other condensation products with novel structures will probably be made from oxocarbons in the future.

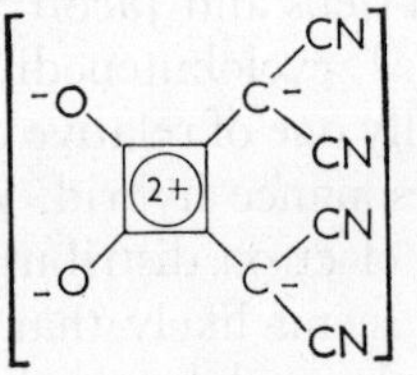

(69)

VI. APPLICATIONS OF THE OXOCARBONS

Rhodizonic acid and the alkali rhodizonates have long been useful in chemical analysis, because of the strongly coloured complexes which rhodizonate forms with many metal ions[94,95]. The analytical uses of rhodizonates have been reviewed by Feigl and Suter[96,96a]. Rhodizonate is useful in spot tests as well as quantitative determinations for metal ions such as Ba^{2+}, Sr^{2+}, Pb^{2+}, with which it forms insoluble red precipitates. Barium rhodizonate is more soluble than barium sulphate, and adsorbs strongly on the sulphate, so rhodizonate can be used as an adsorption indicator in the titrimetric determination of sulphate with standard barium solutions[97]. The end-point is observed by the change in colour of the precipitated $BaSO_4$ from white to rose-red, as barium rhodizonate begins to form. Barium and rhodizonate are also used in turbidimetric determination of small amounts of sulphate[98,98a]. Rhodizonate also forms precipitates with rare earth elements, and can be used to separate them into subgroups[99,99a–b].

Croconate ion has also been suggested[100] as a precipitating reagent in the determination of Ca^{2+}, Sr^{2+} and Ba^{2+}, and the solubility products of the alkaline earth croconates have recently been measured[101].

In biochemistry, rhodizonic acid is also found to have some activity. It serves as a replacement for myoinositol as the substrate for growth of the sugar mould *Saccharomyces carlsbergensis*[102,102a]. Probably because of its capabilities for oxidation and reduction, rhodizonic acid provides some protection to yeasts against damage from radiation[103,103a]. Rhodizonic acid will also prevent the photodecomposition of cystine[104], perhaps because of its ability to reoxidize sulphhydryl groups, i.e. in cysteine, to disulphide linkages[105].

According to patents, rhodizonic acid will increase serum levels of tetracycline antibiotics[106], and rhodizonic and croconic acids will both increase the effectiveness of chlorophyll as an oral deodorant in toothpastes, etc.[107] Several polycarbonyl compounds, including oxocarbons, are reported to be antidiabetic agents[108] and to have *in vitro* activity against influenza virus[109].

In the future, derivatives of squaric acid are likely to find technological application as drugs, dyestuffs, or in other ways. The remarkable new condensation products obtained from squaric acid (section V.E) are likely to lead to a reawakening of interest in derivatives of the other oxocarbons as well.

VII. REFERENCES

1. R. West and D. L. Powell, *J. Amer. Chem. Soc.*, **85**, 2577 (1963).
2. R. West, H.-Y. Niu, D. L. Powell and M. V. Evans, *J. Amer. Chem. Soc.*, **82**, 6204 (1960).
3. C. Brunner, *Schweigger's J.*, **38**, 517 (1823).
4. L. Gmelin, *Ann. Phys.*, **4**, 31 (1825).
5. J. F. Heller, *Ann. Chem.*, **24**, 1 (1837).
6. A. J. Fatiadi, H. S. Isbell and W. F. Sager, *J. Res. Natl. Bur. Std.*, **67A**, 153 (1963).
6a. A. W. Burgstahler and R. C. Barkhurst, *Trans. Kans. Acad. Sci.*, **71**, 150 (1968).
7. A. J. Kluyver, T. Hof and A. G. J. Boezaardt, *Enzymologia*, **7**, 257 (1939); *Chem. Abstr.*, **34**, 6322 (1940).
8. F. Wöhler, *Ann. Phys.*, **12**, 253 (1828).
9. J. Liebig, *Ann. Chem.*, **11**, 182 (1834).
10. B. C. Brodie *Ann. Chem.*, **113**, 358 (1860).
11. J. U. Lerch, *Ann. Chem.*, **124**, 20 (1862).
12. R. Nietzki and T. Benckiser, *Ber.*, **18**, 499, 1833 (1885); **19**, 293 (1886).

13. W. F. Sager, A. Fatiadi, P. C. Parks, D. G. White and T. P. Perros, *J. Inorg. Nuclear Chem.*, **25**, 187 (1963).
14. W. Büchner and E. Weiss, *Helv. Chim. Acta*, **47**, 1415 (1964).
15. R. Nietzki, *Ber.*, **20**, 1617, 2114 (1887).
16. O. Gelormini and N. E. Artz, *J. Amer. Chem. Soc.*, **52**, 2483 (1930).
17. F. Arcamone, C. Prevost and P. Souchay, *Bull. Soc. Chim. France*, 891 (1953).
18. S. Selman and J. F. Eastham, *Quart. Revs.*, **14**, 221 (1960).
19. H. C. Miller, *U.S. Pat.* 2,858,194 (1958).
19a. V. M. Sinclair, R. A. Davies and J. L. Drummond, *Chem. Soc. (London)*, special publication No 22, 260 (1967).
20. V. A. Shushunov, *Zh. Fiz. Khim.*, **23**, 1322 (1949); *Chem. Abstr.*, **44**, 2833 (1950).
21. M. B. Neiman and V. A. Shushunov, *Dokl. Akad. Nauk SSSR*, **60**, 1347 (1950); *Chem. Abstr.*, **45**, 425 (1951).
22. V. A. Shushunov, *Tr. Komissii Anal. Khim., Akad. Nauk SSSR*, **3**, 146 (1951), *Chem. Abstr.*, **47**, 2623 (1953).
23. A. Joannis, *Compt. Rend.*, **116**, 1518 (1893); **158**, 874 (1914).
24. T. G. Pearson, *Nature*, **131**, 166 (1933).
25. L. Hackspill and L. A. van Alten, *Compt. Rend.*, **206**, 818 (1938).
26. E. Weiss and W. Büchner, *Helv. Chim. Acta*, **46**, 1121 (1963).
27. E. Weiss and W. Büchner, *Z. Anorg. Allgem. Chem.*, **330**, 251 (1964).
27a. E. Weiss and W. Büchner, *Chem. Ber.*, **98**, 126 (1965).
28. W. Büchner, *Helv. Chim. Acta*, **46**, 2111 (1963).
29. W. Büchner, *Helv. Chim. Acta*, **48**, 1229 (1965).
30. P. W. Preisler and L. Berger, *J. Amer. Chem. Soc.*, **64**, 67 (1942).
31. H. Y. Niu, *Ph.D Thesis*, University of Wisconsin, *Diss. Abstr.*, **23**, 826 (1962).
32. B. Eistert and G. Bock, *Angew. Chem.*, **70**, 595 (1958).
33. B. Homolka, *Ber.*, **54**, 1393 (1921).
34. R. Kuhn, G. Quadbeck and E. Röhm, *Ann. Chem.*, **565**, 1 (1949).
35. V. G. Brudz, *Russian Pat.* 135,479; *Chem. Abstr.* **55**, 20997, 1961.
35a. E. Ochiai, Y. Kobayashi, T. Haginiwa, S. Takeuchi and M. Fujimoto, *Japan. Pat.* 12,413, *Chem. Abstr.*, **67**, 99732 (1967).
36. P. Moeckel and G. Staerk, *Ger. Pat.* 1,095,823 (1961); *Chem. Abstr.*, **56**, 11500, 1962.
36a. P. Moeckel and G. Staerk, *Z. Chem.*, **7**, 62 (1967).
37. S. Cohen, J. R. Lacher and J. D. Park, *J. Amer. Chem. Soc.*, **81**, 3480 (1959).
38. A. L. Henne and R. P. Ruh, *J. Amer. Chem. Soc.*, **69**, 279 (1947).
39. J. D. Park, C. M. Snow and J. R. Lacher, *J. Amer. Chem. Soc.*, **73**, 2342 (1951).
40. R. West, H. Y. Niu and M. Ito, *J. Amer. Chem. Soc.*, **85**, 2584 (1963).
41. J. D. Park, S. Cohen and J. R. Lacher, *J. Amer. Chem. Soc.*, **84**, 2919 (1962).
42. G. Maahs, *Angew. Chem.*, **75**, 982 (1963); *Int. Ed. Engl.*, 2, 690 (1963).
43. G. Maahs, *Ann. Chem.*, **686**, 55 (1965).
44. G. Maahs and P. Hegenberg, *Angew. Chem.*, **78**, 927 (1966); *Int. Ed. Engl.*, 5, 888 (1966).

45. A. Roedig and P. Bernemann, *Ann. Chem.*, **600**, 1 (1956).
46. G. Carpeni, *Compt. Rend.*, **202**, 1065 (1936); **203**, 75, 1156 (1936); **205**, 273 (1937).
47. G. Carpeni, *J. Chim. Phys.*, **35**, 233 (1938).
48. G. Carpeni, *Compt. Rend.*, **206**, 601 (1938).
49. K. Yamada, N. Mizuno and Y. Hirata, *Bull. Chem. Soc. Japan*, **31**, 543 (1958).
50. M. Washino, K. Yamada and Y. Kurita, *Bull. Chem. Soc. Japan*, **31**, 552 (1958).
51. N. C. Baenziger and D. G. Williams, *J. Amer. Chem. Soc.*, **88**, 689 (1966).
52. Y. Hirata, K. Inukai and T. Tsujiuchi, *J. Chem. Soc. Japan., Pure Chem. Sect.*, **69**, 63 (1948).
52a. S. Skujins, J. Delderfield and G. A. Webb, *Tetrahedron Letters*, **24**, 4805 (1968).
53. D. T. Ireland and H. F. Walton, *J. Phys. Chem.*, **71**, 751 (1967).
53a. D. J. MacDonald, *J. Org. Chem.*, **33**, 4559 (1968).
54. P. Souchay and M. Fleury, *Compt. Rend.* **252**, 737 (1961).
55. B. Carlqvist and D. Dyrssen, *Acta. Chem. Scand.*, **16**, 94 (1962).
56. G. Schwarzenbach and H. Suter, *Helv. Chim. Acta*, **24**, 617 (1941).
56a. E. Patton and R. West, unpublished studies.
57. P. W. Preisler, L. Berger and E. S. Hill, *J. Amer. Chem. Soc.*, **69**, 326 (1947); **70**, 871 (1948).
58. K. Yamada and Y. Hirata, *Bull. Chem. Soc. Japan*, **31**, 550 (1958).
59. M. Ito and R. West, *J. Amer. Chem. Soc.*, **85**, 2580 (1963).
60. W. M. Macintyre and M. S. Werkema, *J. Chem. Phys.*, **42**, 3563 (1964).
61. N. C. Baenziger and J. J. Hegenbarth, *J. Amer. Chem. Soc.*, **86**, 3250 (1964).
62. M. A. Neuman, *Ph.D Thesis*, University of Wisconsin, 1966; *Diss. Abstr.*, **26**, 6394 (1966).
63. J. Kaufman, *J. Chem. Phys.*, **68**, 2648 (1964).
64. M. Cignitti, *Theoret. Chim. Acta*, **5**, 169 (1966).
64a. K. Sakamoto and Y. Ihaya, *Theoret. Chim. Acta*, **13**, 220 (1969).
65. A. Sadô and R. West, unpublished studies.
65a. S. Skujins and G. A. Webb, *Spectrochim. Acta*, Part A, **25**, 917 (1969).
66. R. West and H. Y. Niu, *J. Amer. Chem. Soc.*, **84**, 1324 (1962).
67. W. Büchner and E. Lucken, *Helv. Chim. Acta*, **47**, 2113 (1964).
68. D. Fleury and M. Fleury, *Compt. Rend.*, **258**, 5221 (1964).
69. H. Will, *Ann. Chem.*, **118**, 177 (1858).
70. M. Fleury, *Compt. Rend.*, **261**, 4751 (1965).
71. M. Fleury and P. Souchay, *Compt. Rend.*, **258**, 211 (1964).
71a. M. B. Fleury. P. Souchay, M. Gouzerh and P. Gracian, *Bull. Soc. Chim. France*, 2562 (1968).
72. C. M. Bock, unpublished studies.
73. W. P. Person and D. G. Williams, *J. Phys. Chem.*, **61**, 1017 (1957).
74. P. Souchay and F. Tatibouet, *J. Chim. Phys.*, **49**, C-108 (1952).
75. F. Henle, *Ann. Chem.*, **350**, 330 (1906).
76. F. Bergel, *Ber.*, **62B**, 490 (1929).
77. R. Nietzki and H. Rossmann, *Ber.*, **22**, 916 (1889).
78. R. Nietzki and T. Benckiser, *Ber.*, **19**, 772 (1886).

79. R. Nietzki and F. Kehrmann, *Ber.*, **20**, 322 (1887).

80. H. E. Worne, *U.S. Pat.* 3,227,641 (1966).

80a. J. F. Verchere, M. B. Fleury and P. Souchay, *C. R. Acad. Sci., Paris, Ser. C.*, **267**, 1221 (1968).

81. S. Prebendowski and Z. Rutkowski, *Roczniki Chem.*, **31**, 81 (1957).

82. Z. Rutkowski, *Roczniki Chem.*, **36**, 169 (1962).

83. R. Malachowski and S. Prebendowski, *Ber.*, **71B**, 2241 (1938).

84. S. Cohen and S. G. Cohen, *J. Amer. Chem. Soc.*, **88**, 1533 (1966).

84a. J. E. Thorpe, *J. Chem. Soc.*, B, 435 (1968).

85. A. Triebs and K. Jacob, *Ann. Chem.*, **699**, 153 (1966).

86. G. Maahs and P. Hegenberg, *Ger. Pat. Appl.* c-36340 IVb/120 (1965).

87. R. Nietzki and A. W. Schmidt, *Ber.*, **21**, 1227, 1850 (1888).

88. F. Kehrmann and A. Duret, *Ber.*, **31**, 2437 (1898).

89. B. Eistert, H. Fink and H. K. Werner, *Ann. Chem.*, **657**, 131 (1962).

90. A. Triebs and K. Jacob, *Angew. Chem.*, **77**, 680 (1965); *Int. Ed. Engl.*, **4**, 694 (1965).

90a. S. Skujins and G. A. Webb, *Chem. Commun.*, 598 (1968).

90b. J. Gauger and G. Manecke, *Angew. Chem.*, **81**, 334 (1969).

91. A. Triebs and K. Jacob, *Ann. Chem.*, **712**, 123 (1968).

91a. H. J. Roth and H. Sporleder, *Tetrahedron Letters*, 6223 (1968).

92. W. Ziegenbein and H. E. Sprenger, *Angew. Chem.*, **78**, 937 (1966); *Int. Ed. Engl.*, **5**, 893, 894 (1966).

92a. H. E. Sprenger and W. Ziegenbein, *Angew. Chem., Int. Ed. Engl.*, **7**, 530 (1968).

93. H. E. Sprenger and W. Ziegenbein, *Angew. Chem.*, **79**, 581 (1967); *Int. Ed. Engl.*, **6**, 553 (1967).

93a. G. Manecke and J. Gauger, *Tetrahedron Letters*, 3509 (1967); 1339 (1968).

94. A. Triebs and E. Herrmann, *Ann. Chem.*, **589**, 207 (1954).

95. A. Triebs and R. Zimmer-Galler, *Hoppe-Seyler's Z. Physiol. Chem.*, **318**, 2 (1960).

96. F. Feigl and H. A. Suter, *Ind. Eng. Chem. Anal. Ed.*, **14**, 840 (1942).

96a. R. A. Chalmers and G. M. Telling, *Mikrochim. Acta.* 1126 (1967).

97. E. Pungor and I. Konkoly Thege, *Talanta*, **10**, 1211 (1963).

98. B. E. Andronov, T. I. Sergeeva, T. Y. Gaeva and A. K. Yudina, *Sb. Nauchn. Rabot Inst. Okhrany Truda Vses. Tsentr. Sor. Profsoyuzov*, **1962**, 89; *Chem. Abstr.*, **60**, 15142 (1964).

98a. A. Beszterda, *Chem. Anal.* (Warsaw), **14**, 341 (1969).

99. N. S. Poluektov, *Redkozemel. Elementy, Akad. Nauk SSSR, Inst. Geokhim. i Anal. Khim.*, **190** (1959); *Chem. Abstr.*, **55**, 17335 (1961).

99a. J. P. Pandey and O. C. Saxena, *Mikrochim Acta*, 638 (1968).

99b. O. C. Saxena, *Microchem. J.*, **13**, 222 (1968).

100. J. Faucherre, *Bull. Soc. Chim. France*, **20**, 900 (1953).

101. B. Carlqvist and D. Dyrssen, *Acta. Chem. Scand.*, **19**, 1293 (1965).

102. J. A. Johnston, R. C. Ghadially, R. N. Roberts and B. W. Fuhr, *Arch. Biochem. Biophys.*, **99**, 537 (1962).

102a. P. Dworsky and O. Hoffman-Ostenhof, *Z. Allg. Mikrobiol.*, **7**, 1 (1967).

103. J. L. Seris and A. Bru, *Compt. Rend.*, **255**, 1027 (1962).

103a. A. Bru, J. L. Deris, H. Regis, J. Soubiran and H. Lucot, *J. Radiol., Electro., Med. Nucl.*, **48**, 555 (1967).

104. J. L. Seris, *Compt. Rend.*, **253**, 2152 (1961).
105. J. L. Seris, *Compt. Rend.*, **252**, 3672 (1961).
106. American Cyanamid Co., *Brit. Pat.* 905,092, (1962); *Chem. Abstr.*, **58**, 3277 (1963).
107. W. J. Hale, *U.S. Pat.* 2,815,314 (1957); *Chem. Abstr.*, **52**, 4937 (1958).
108. S. Takeuchi, *Japan J. Pharmeol.*, **17**, 333 (1967).
109. H. Muecke and M. Sproessig, *Arch. Exp. Veterinaermed.*, **21**, 307 (1967).

CHAPTER **5**

Mass spectrometry of carbonyl compounds

J. H. Bowie

The University of Adelaide, South Australia
(*Manuscript received September 1967*)

I. INTRODUCTION

Since the early nineteen sixties, the number of publications in the field of mass spectrometry has grown enormously, and the application of high resolution data, labelling studies, and more recently, computer-aided mass spectrometry, has enabled the interpretation of the fragmentation processes in the mass spectra of many different types of organic molecules. Details of the method, application, and instrumentation of mass spectrometry are available by reference to many excellent books[1-15], reviews[16-22] and compendia of reference data[23-28].

The symbolism used throughout is that developed by Budzikiewicz, Djerassi and Williams[8,9] from initial proposals by McLafferty[4,16] and Shannon[29]. In the text, fragmentation processes of cations are indicated by arrows ($\frown$, heterolytic cleavage—two electron shift), while fragmentations of radical ions (odd-electron species) are indicated by fishhooks ($\frown$, homolytic cleavage—one electron shift). The use of the fishhook is not intended to imply that two-electron shifts may not operate in odd-electron species. Structures are drawn for ions in order to relate the fragmentation processes to the structures of the molecules in the ground state. These structures are intended to be nominal, as there are now many examples[53] where molecular ion rearrangements occur. It is generally accepted (reference 1, pp. 251–262, reference 2, pp. 153–157 and reference 30) that evidence for a one-step decomposition process (e.g. ion A → ion B) in a mass spectrum is given by the presence of a metastable peak, the position of the peak m^* being given by the expression $m^* = (m_{(B)})^2/m_{(A)}$. It has recently been suggested[31] that metastable ions may also arise from two-step or multistep fragmentation processes. Nevertheless, metastable peaks are of considerable value for the interpretation of fragmentation patterns, and where the positions of such peaks are available for any mass spectrum which is illustrated as a figure, both the fragmentation processes and the presence of metastable ions (designated by an asterisk) will be indicated in that figure.

II. ALIPHATIC CARBONYL COMPOUNDS

A. Saturated Aldehydes and Ketones

The energy necessary to remove one of the lone pair electrons from the oxygen of the carbonyl group is approximately 10 ev[32] (1 ev = 23 kcals/mole). The energy of the electron beam is generally maintained at 70 ev, thus producing optimum fragmentation of the molecule. Fragmentation processes may be specific for particular types of organic molecules, and aliphatic ketones and aldehydes fragment by two major processes.

I. Cleavage α to C=O

This fragmentation is the fundamental process for all saturated aliphatic aldehydes and ketones. The relative abundances of **2** and **3** will depend upon both the strength of the two α carbon–carbon bonds, and the relative stabilities of the two cations.

$$
\begin{array}{ccc}
\overset{\cdot\,+}{\underset{\|}{O}} & & \\
R'\!-\!C\!\overset{\curvearrowleft}{-}\!R & \xrightarrow{\;-R\cdot\;} & R'\!-\!C\!\equiv\!\overset{+}{O} \\
(R, R' = \text{H or alkyl}) & & \\
(\mathbf{1}) & & (\mathbf{2})
\end{array}
$$

$$
\begin{array}{ccc}
\overset{\cdot\,+}{\underset{\|}{O}} & & \\
R'\!\overset{\curvearrowleft}{-}\!C\!-\!R & \xrightarrow{\;-R'\cdot\;} & R\!-\!C\!\equiv\!\overset{+}{O} \\
(\mathbf{1}) & & (\mathbf{3})
\end{array}
$$

The spectra of saturated aldehydes have been studied by Gilpin and McLafferty[33] and those of saturated ketones by Sharkey and coworkers[34]. The spectra (Figures 1 and 2) of propionaldehyde (**4**) and butan-2-one (**5**) illustrate the α-cleavage process. The molecular ion of propionaldehyde may either lose a hydrogen atom or an ethyl radical to form **6** and **7** respectively. [18]O labelling[33] shows m/e 29 in Figure 1 to correspond to **7**, while exact mass measurements[35] show

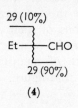

(4)

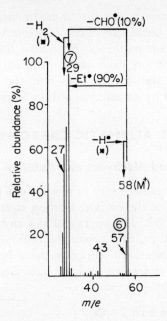

FIGURE 1

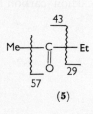

(5)

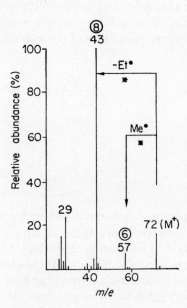

FIGURE 2

it to contain in addition a small portion (10%) of the ethyl cation. α-Cleavage in **5** produces both **6** and **8**. The stable acetyl cation (**8**) constitutes the base peak of the spectrum.

$$Et—C{\equiv}\overset{+}{O} \qquad H—C{\equiv}\overset{+}{O} \qquad Me—C{\equiv}\overset{+}{O}$$

$$\text{(6) } m/e\text{ 57} \qquad\quad \text{(7) } m/e\text{ 29} \qquad\quad \text{(8) } m/e\text{ 43}$$

2. Cleavage β to C=O

When the alkyl substituent attached to the carbonyl group contains three or more carbon atoms with at least one hydrogen attached to the γ-carbon atom, a rearrangement process occurs, which involves a cleavage β to the carbonyl group with associated γ-hydrogen migration. This is known[8,36,37] as the McLafferty rearrangement[38]. Although other plausible mechanisms have been proposed[39,40], the rearrangement can be described as the homocyclic process **9** → **10** + **11**. The charge is retained by the more stable fragment, which for ketones and α-branched aldehydes is generally the enol radical ion **10**. Aliphatic aldehydes which have a suitable side chain (as in **9**),

but no α-branching, show M − CH₂=CH—OH (M − 44) peaks in their spectra[33]. This shows that some of the charge may be retained by the olefin fragment **11**. It is important to note that this site-specific γ-hydrogen rearrangement has been substantiated by deuterium labelling studies[41], and that the process will only occur when the γ-hydrogen–oxygen distance is less than 1.8 Å[42,43], a distance that can be easily attained in aliphatic compounds. Deuterium isotope effects for various McLafferty rearrangements have been studied[44], but as yet no definite conclusions have been reached concerning the mechanistic implication of the results. The spectrum (Figure 3) of

10*

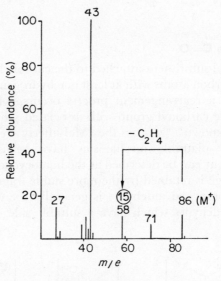

FIGURE 3

pentan-2-one (**12**) illustrates the β-cleavage process. The ion **15** at
m/e 58 arises by loss of ethylene from the molecular ion via a McLafferty rearrangement.

Double rearrangement becomes possible when both alkyl substituents of a ketone contain three or more carbon atoms and a
γ-hydrogen atom. The presence of these rearrangement ions in a
spectrum may be used to determine the size of each side chain.
Djerassi and coworkers[41] have shown by deuterium labelling studies
that the second rearrangement also shows complete γ-hydrogen transfer specificity. It has also been shown[45] that enol–keto tautomerism
is not a prerequisite for the second rearrangement. Both rearrangement
processes can be seen in the mass spectrum (Figure 4) of nonan-5-one
(**13**). Although it is not possible to distinguish between hydrogen
migration to oxygen or carbon in the second rearrangement[46], ion

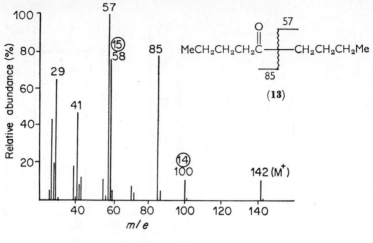

FIGURE 4

14 (formed by loss of propylene from the molecular ion of **13**) is represented here as decomposing by allylic cleavage with the γ-hydrogen atom rearranging to the enolic double bond to yield ion **15**:

(14) m/e 100 → − MeCH=CH₂ → **(15)** m/e 58

B. Unsaturated Aldehydes and Ketones

αβ-Unsaturated ketones and aldehydes also fragment by α-cleavage[47]. The α-cleavage process in αβ-unsaturated ketones occurs β in preference to α with respect to the double bond. This feature can be seen in the spectrum (Figure 5) of mesityl oxide (**17**), where **16** (M − Me·) is the base peak, while the acetyl cation (**8**) has only a relative abundance of 55%.

$$\text{Me}_2\text{C=CH—C}\overset{+}{\equiv}\text{O}$$

(16) m/e 83

J. H. Bowie

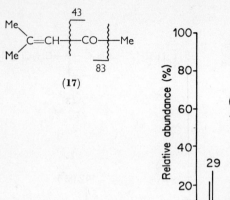

FIGURE 5

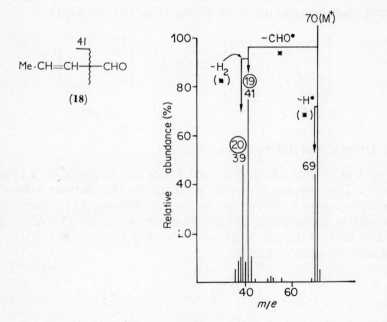

FIGURE 6

$\alpha\beta$-Unsaturated aldehydes may cleave either α (which is an energetically unfavourable process) or β to the double bond. The latter cleavage gives rise to abundant $M - 1$ ions. Stable allylic cations are formed when the substituent is greater than vinyl, e.g. the spectrum (Figure 6) of crotonaldehyde (**18**) shows a pronounced peak at m/e 41, due to the allyl cation **19**. The ion **19** fragments by loss of hydrogen to yield the cyclopropenyl cation **20** (m/e 39).

$$
\overset{\overset{\cdot\,+}{\underset{\|}{O}}}{\text{Me—CH=CH—C—H}} \xrightarrow{-\text{CHO}\cdot} \text{Me—CH=CH}^+ \longrightarrow \text{CH}_2\text{=CH—CH}_2^{\,+} \xrightarrow{-\text{H}_2} \triangle
$$

m/e 70 m/e 41   (**19**) m/e 41   (**20**) m/e 39

The McLafferty rearrangement neither occurs for $\alpha\beta$-unsaturated substituents, nor when the only available γ-hydrogen atom is attached to a double bond. In the latter case β-cleavage and γ-hydrogen rearrangement would cause an allene to be eliminated, e.g. in 4-methylhept-6-en-3-one[48].

C. Aliphatic Diketones

I. α-Diketones

Aliphatic α-diketones fragment by cleavage of the C—C bond between the two carbonyl groups[49,50], producing the base peak of the spectrum, provided that each side chain contains less than five carbon atoms and is not branched. When a straight side chain contains five or more carbon atoms, the base peak of the spectrum is the propyl cation (C_3H_7^+, m/e 43). When the side chain is branched, the base peak of the spectrum is produced by the appropriate alkyl cation, a tertiary cation being more stable and more abundant than a secondary cation. In the spectrum (Figure 7) of decane-5,6-dione (**21**), α-cleavage produces **22** (m/e 85), which decomposes by loss of carbon monoxide to produce the n-butyl cation (**23**, m/e 57). No metastable ion is present to indicate that **23** is formed directly from the molecular ion. The McLafferty rearrangement is not observed in the mass spectra of aliphatic α-diketones.

$$
\text{Me(CH}_2)_3\text{—C} \overset{+}{\equiv} \text{O} \qquad\qquad \text{Me(CH}_2)_2\text{CH}_2^{\,+}
$$

m/e 85 m/e 57
(**22**)   (**23**)

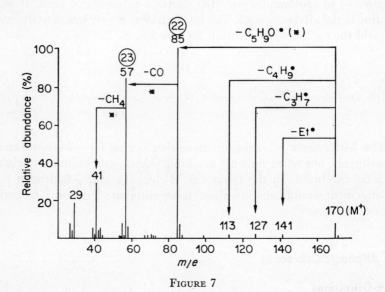

FIGURE 7

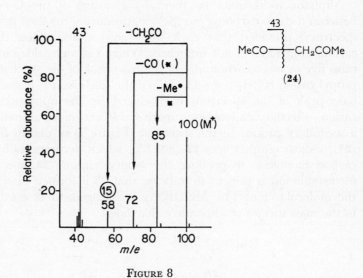

FIGURE 8

2. β-Diketones

The spectra of aliphatic β-diketones have been reported by Williams and coworkers[51] and by Schamp and Vandewalle[52]. The mass spectrum (Figure 8) of acetylacetone (24) shows the normal α-cleavage processes, together with loss of ketene from the molecular ion. The

$$\left[\begin{array}{c} \text{O} \quad \overset{H}{\curvearrowright} \text{CH}_2 \\ \underset{Me}{C} \quad \underset{CH_2}{C} \quad O \end{array}\right]^{\ddagger} \xrightarrow{-CH_2CO} \left[\begin{array}{c} OH \\ \underset{Me}{C} = CH_2 \end{array}\right]^{\ddagger}$$

(25) m/e 100   (15) m/e 58

$$\left[\begin{array}{c} \text{O} \quad \overset{H}{\curvearrowright} \overset{Et}{CH} \\ \underset{Me}{C} \quad \underset{CH}{C} H_2 \\ \quad\quad COMe \end{array}\right]^{\ddagger} \xrightarrow{-EtCH=CH_2} \left[\begin{array}{c} HO \\ \underset{Me}{C} \overset{}{\underset{CH}{}} COMe \end{array}\right]^{\ddagger} \longrightarrow \begin{array}{l} m/e\ 85(25\%), \\ 72, 58, \\ 43(90\%) \end{array}$$

(27) m/e 156   (28) m/e 100

$$\left[\begin{array}{c} \text{O} \quad \overset{H}{\curvearrowright} CH_2 \\ \underset{Me}{C} \quad \underset{CH}{C} \quad O \\ \quad n\text{-}C_4H_9 \end{array}\right]^{\ddagger} \xrightarrow{-CH_2CO} \left[\begin{array}{c} HO \\ \underset{Me}{C} = CH-CH_2 \{ \} n\text{-}C_3H_7 \end{array}\right]^{\ddagger} \xrightarrow{-C_3H_7}$$

(29) m/e 156   (30) m/e 114

$$\begin{array}{c} HO \\ \underset{Me}{C} = CH-CH_2^+ \end{array}$$

(31) m/e 71

M − CH$_2$CO process proceeds by β-cleavage with γ-hydrogen rearrangement to form the enol radical ion (25 → 15)[51]. The third process in this spectrum, M − CO, is a skeletal rearrangement process[53] of the type [ABC]‡ → [AC]‡ + B[54] (see Section VII). The spectra of acetylacetones alkylated at the 3-position[51] retain the features of that of acetylacetone itself. The enol radical ion produced by the M − CH$_2$CO process may now cleave in the allylic position (β to the double bond). In addition, the alkyl side chain may be eliminated by the normal McLafferty rearrangement producing the enol form of the acetylacetone radical ion which decomposes as indicated in Figure 8. These processes can be seen in the spectrum

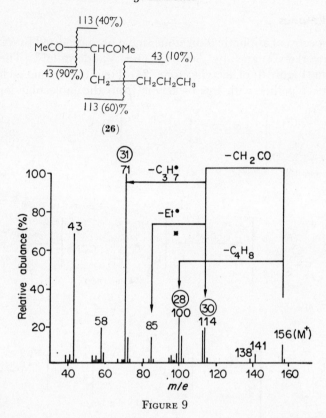

FIGURE 9

(Figure 9) of 3-*n*-butylpentane-2,4-dione (**26**), viz. the McLafferty rearrangement (**27 → 28**), and the loss of ketene followed by allylic
cleavage (**29 → 30 → 31**).

3. γ-Diketones

γ-Diketones, like simple ketones, fragment by α-cleavage. There is
only minimal fragmentation of the central C—C bond between the
two carbonyl groups[55]. α-Cleavage with hydrogen rearrangement is
also a feature of the spectra of γ-diketones, e.g. the spectrum of
hexane-2,5-dione exhibits an M − CH_2CO peak. The spectrum (Figure 10) of 4,4-dimethylheptane-2,5-dione (**32**), as well as showing the
normal α-cleavage processes, exhibits an M − C_2H_4CO ion (*m/e*
100). It is not clear whether the hydrogen migrates to carbon or oxygen,
but the rearrangement must proceed by a non-concerted mechanism
(compare with the concerted process **25 → 15**).

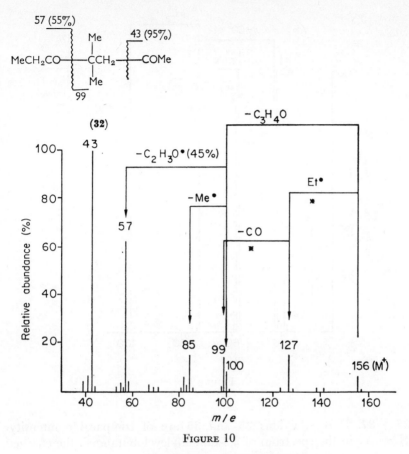

FIGURE 10

III. MONOCYCLIC KETONES

A. Saturated Monocyclic Ketones

I. Cyclobutanones

The base peak in the spectrum of cyclobutanone is the ketene radical ion (m/e 42), produced by elimination of ethylene from the molecular ion[56]. The high resolution mass spectra of a variety of alkylated cyclobutanones have been studied by Fétizon and co-workers[57]. In these spectra, fragmentation depends on the position of the alkyl substituent, and both charged olefins and ketenes may be produced. The spectrum (Figure 11) of 2,2-dimethylcyclobutanone (**33**) shows two competing fragmentations, viz. **34** → **35**, and **34** →

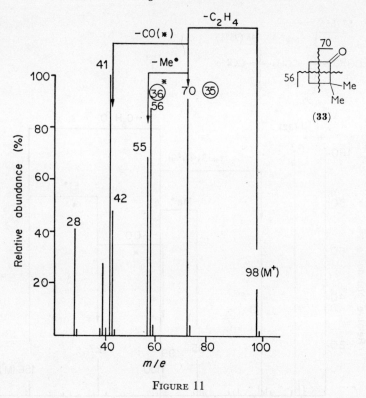

FIGURE 11

36 + 37. Here, the ions **35** and **36** are of comparable intensity. However in the spectrum of 3,3-dimethylcyclobutanone, the charged olefin **36** is produced almost to the exclusion of the ketene radical ion **37**. The McLafferty rearrangement is observed in the spectrum of 2-*t*-butylcyclobutanone.

2. Cyclopentanones

The interpretation of the spectrum (Figure 12) of cyclopentanone (**38**) has been aided by high resolution data[58] and by the spectrum of cyclopentanone-2,2,4,4-d_4[59]. The base peak (*m/e* 55) is produced by the process **39 → 40**, and *m/e* 28 by **39 → 41**. Alkylcyclopentanones behave similarly[60]. Deuterium labelling studies have shown that the M − C_2H_4 process in the spectrum of 2-ethylcyclopentanone is produced by the McLafferty rearrangement[60], and that this re-

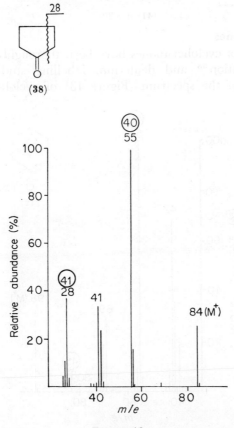

FIGURE 12

arrangement in the spectrum of 2-s-butylcyclopentanone involves specifically a γ-hydrogen of the Me group[70].

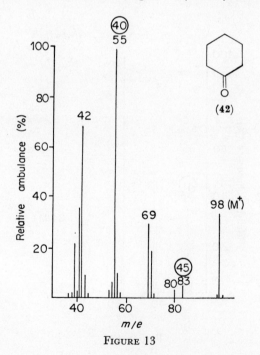

3. Cyclohexanones

The spectra of cyclohexanones have been thoroughly investigated by high resolution[58] and deuterium labelling studies[60-62]. The interpretation of the spectrum (Figure 13) of cyclohexanone (42)

FIGURE 13

has been aided by a consideration of the spectra of cyclohexanone-2,2,4,4-d_4, -3,3,5,5,-d_4, -4,4-d_2, and -4-d[60,62]. The base peak (as in the case of cyclopentanone) is **40** (m/e 55). This may be produced either in a stepwise manner (compare **39 → 40**) or in a concerted

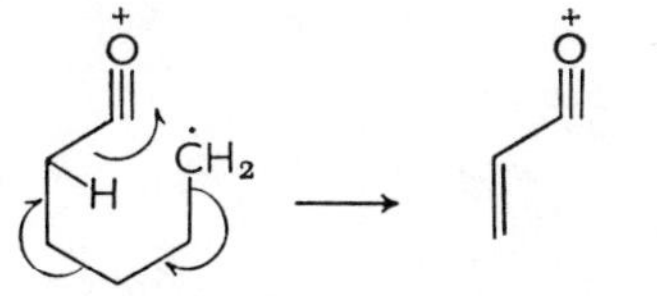

(**43**) m/e 98   (**40**) m/e 55

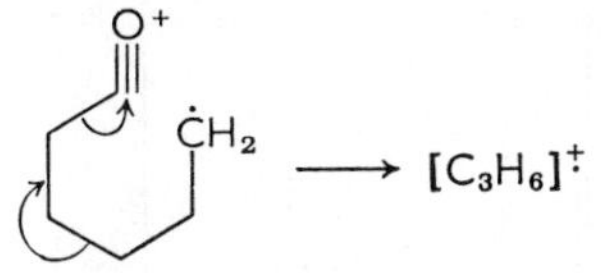

(**43**)   (m/e 42)

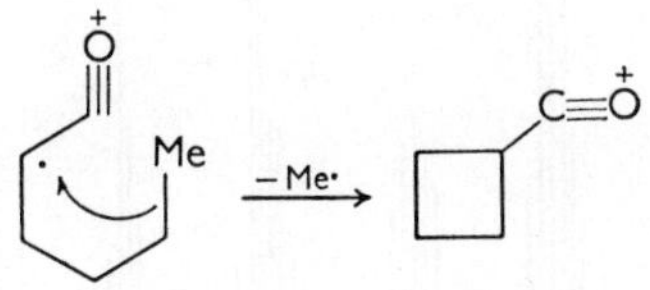

(**44**)   (**45**) m/e 83

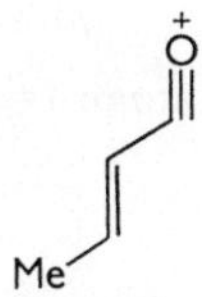

(**46**) m/e 69

manner (**43 → 40**). Other processes observed are **43 → 45** and **43 →** [C_3H_6]‡. The alkylated cyclohexanones behave similarly: e.g. when there are methyl substituents at both $C_{(3)}$ and $C_{(5)}$ the base peak will be **46** (m/e 69)[60,63]. In addition, the base peak in the spectrum of menthone (5-α-methyl-2-β-isopropylcyclohexanone) arises by the process M $-$ C_3H_6, viz. by β-cleavage with γ-hydrogen rearrangement[61,64,65].

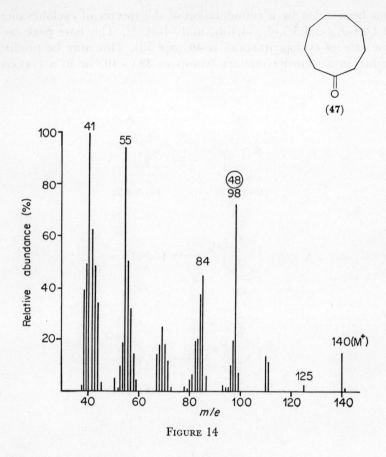

FIGURE 14

4. Higher Cyclic Ketones

Although the carbonyl group largely determines the course of the fragmentation of the C_4 to C_6 monocyclic ketones, this is not the case as the ring size is increased. Hydrocarbon fragments become more prominent in the spectra of the C_7 to C_{10} ketones[58,60,62]. This can be seen in the spectrum (Figure 14) of cyclononanone (47) which is dominated by hydrocarbon fragments; e.g. m/e 55 is mainly due to $C_4H_7^+$ as shown from the spectrum of cyclononanone-2,2,9,9-d_4[62]. Figure 14 also shows a large $M - C_3H_6$ peak (m/e 98), and it has been proposed[62] that this may be represented as the cyclohexanone molecular ion **48** (m/e 140 → **48**). The same ion is present in the spectrum of cyclo-octanone[62].

B. Unsaturated Monocyclic Ketones

I. Cyclopentenones

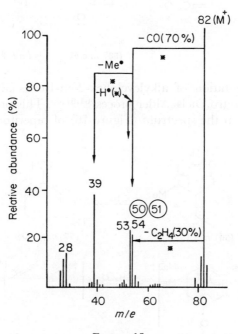

FIGURE 15

Cyclopent-2-en-1-one (**49**), on electron impact, fragments by loss of both ethylene and carbon monoxide from the molecular ion plausibly to give the cyclobutene (**50**) and cyclopropenone (**51**) radical ions[66]. This is summarized in Figure 15. 3-Methylcyclopent-2-en-1-one fragments analogously, although an additional process, M − Me·, is observed.

(**50**) m/e 54 (70%)   (**51**) m/e 54 (30%)

2. Cyclohexenones

Cyclohex-2-en-1-one[67] unlike **49**, fragments specifically by the retro Diels–Alder process[68] as follows:

(**52**) m/e 96   (**53**) m/e 68   (m/e 40)   (**54**) m/e 39

The fragmentation of alkylcyclohex-2-en-1-ones also occurs primarily by the retro Diels–Alder process[64,67,69]. This process produces the base peak in the spectrum (Figure 16) of piperitone (**55**), where

(**56**) m/e 152   (**57**) m/e 110

(**59**) m/e 138   (**60**) m/e 96

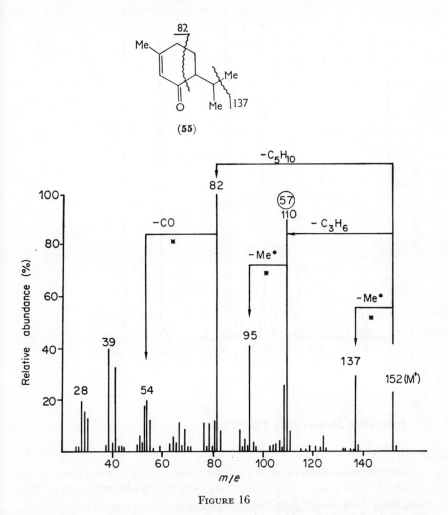

FIGURE 16

the side chain may be eliminated by the McLafferty rearrangement
(**56** → **57**). The interpretation of the spectrum of piperitone has
been aided by high resolution and deuterium labelling studies[69].

When a cyclohexenone is β,γ-unsaturated, the retro Diels–Alder
process produces an entirely different fragmentation[67] as shown in the
spectrum (Figure 17) of 3,5,5-trimethylcyclohex-3-en-1-one (**58**),
where the base peak is produced by the process $M - CH_2CO$
(**59** → **60**). Other processes are summarized in Figure 17.

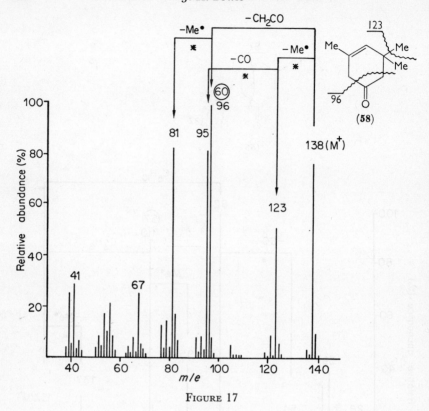

FIGURE 17

C. Saturated Monocyclic Diketones

I. α-Diketones

The spectra of a series of aliphatic monocyclic diketones have been reported[50], and their fragmentations depend considerably upon the ring size. Two major processes can be seen in the spectrum (Figure 18) of 3,3,5,5-tetramethylcyclopentane-1,2-dione (**61**), namely **62** → **64** and **62** → **65**.

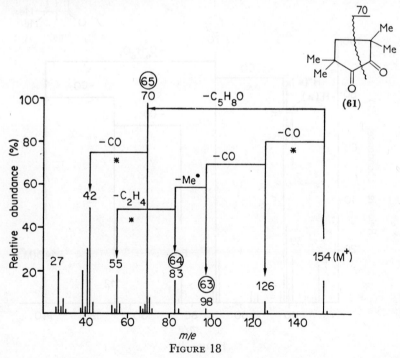

$$(62) \xrightarrow{-C_4H_8O} \left[\underset{Me}{\overset{Me}{>}} C=C=O \right]^{\ddot +}$$

(62)   (65) m/e 70

$$(66) \xrightarrow{-C_6H_8O_2} \left[\underset{Me}{\overset{Me}{>}} C=CH_2 \right]^{+}$$

(66) m/e 168   (67) m/e 56

3,3,6,6-Tetramethylcyclohexane-1,2-dione fragments differently from the above compounds. Although the process M − CO − CO is observed, the base peak is due to the hydrocarbon fragment **67** (*m/e* 56). There is no evidence to suggest that **67** is formed in a one-step process from the molecular ion, but a suggested mechanism is outlined in **66 → 67**.

FIGURE 18

2. β-Diketones

The spectrum (Figure 19) of 2,2,4,4-tetramethylcyclobutane-1,3-dione (**68**) has been discussed by Turro and colleagues[71]. The major process is the formation of the ion **65** (*m/e* 70).

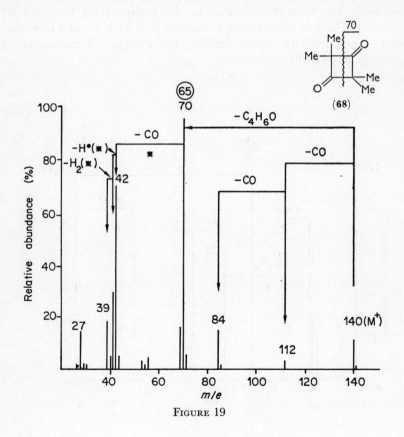

(*m/e* 140)   (**65**) *m/e* 70

The interpretation of the spectra of cyclohexane-1,3-diones[72–75] has been aided by extensive high resolution measurements[73,74] and deuterium labelling studies[74,75]. The fragmentations are largely dependent on the substitution pattern, but the general features can be

FIGURE 19

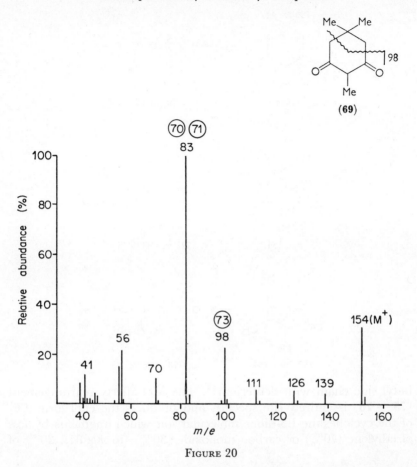

Figure 20

seen in the spectrum (Figure 20) of 2,5,5-trimethylcyclohexane-1,3-dione (**69**)[73,75]. The base peak (m/e 83) is a doublet due to **70** (10%) and **71** (90%). Ions of the type **71** are the base peaks in the spectra of the majority of substituted cyclohexane-1,3-diones, and ions of the type **70** are always present in the spectra of compounds containing the O—C—C(R)—C—O unit[76,124]. A portion of the molecular ion is present in the enol form (**72**), as the M − C_4H_8 ion (m/e 98) must be formed by the retro Diels–Alder process, **72** → **73**.

The complex fragmentation processes in the mass spectrum (Figure 21) of 2-*n*-butylcyclohexane-1,3-dione (**74**) have been elucidated by the application of high-resolution data, and from the spectra of the four deuterated analogues obtained by labelling each carbon unit in the

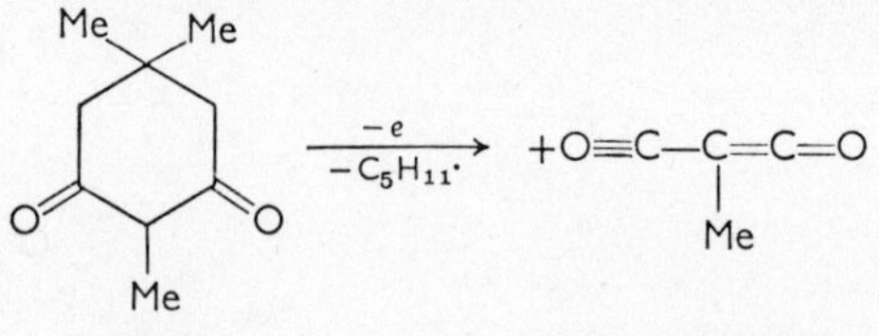

$$\xrightarrow[-C_5H_{11}\cdot]{-e}$$

(69)

(70) m/e 83 (10%)

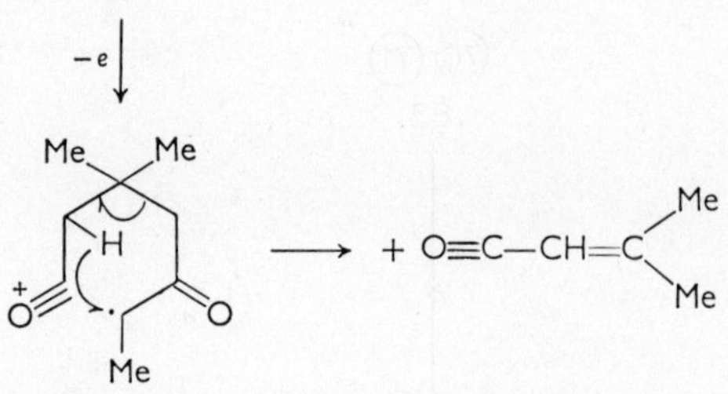

$-e$

m/e 154

(71) m/e 83 (90%)

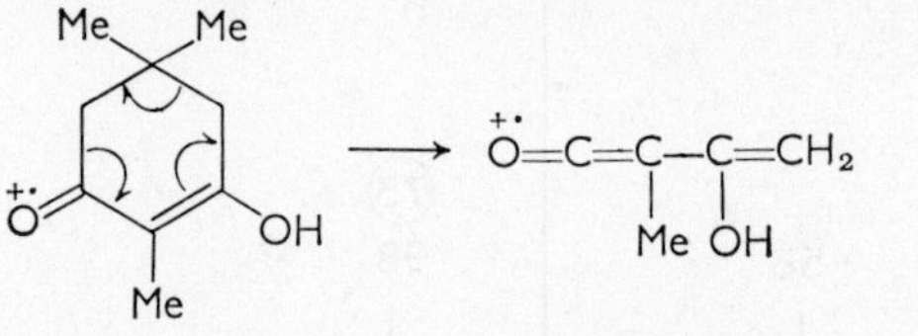

(72) m/e 154

(73) m/e 98

butyl side chain with deuterium[74]. The McLafferty rearrangement (75 → 76) is entirely site specific, and produces the enol form (76) of the cyclohexane-1,3-dione molecular ion which fragments by loss of ethylene (70%) or carbon monoxide (30%) (to m/e 84). 30% of

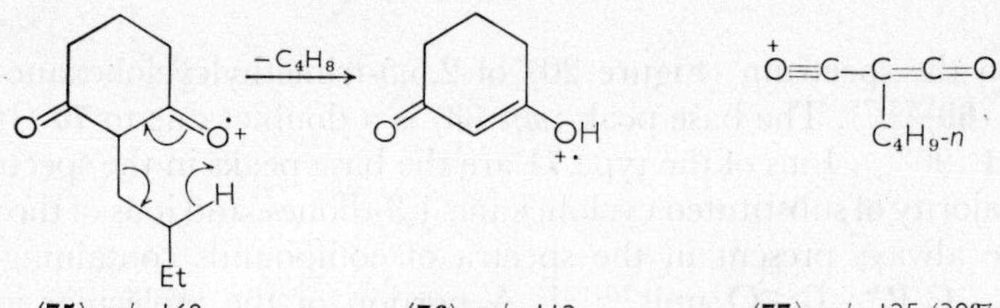

$$\xrightarrow{-C_4H_8}$$

(75) m/e 168

(76) m/e 112

(77) m/e 125 (30%)

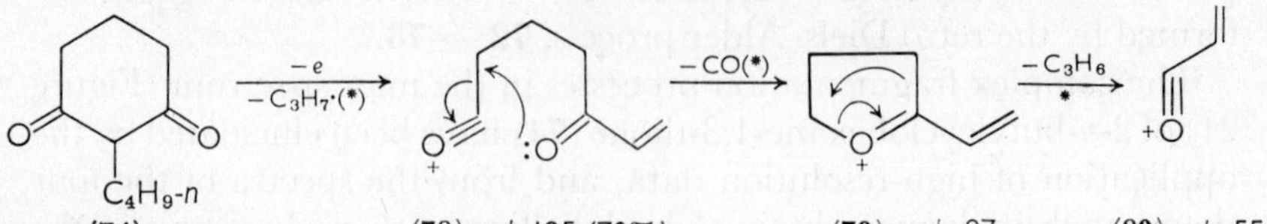

(74)

$$\xrightarrow{-e}_{-C_3H_7\cdot(*)}$$

(78) m/e 125 (70%)

$$\xrightarrow{-CO(*)}$$

(79) m/e 97

$$\xrightarrow{-C_3H_6}_{*}$$

(80) m/e 55

m/e 125 (M − 43) arises by loss of a propyl radical from the ring to produce **77** (compare with **70**), while the complement (70%) is produced by loss of a propyl radical from the side chain to produce **78**, the fragmentation of which ultimately produces **80**.

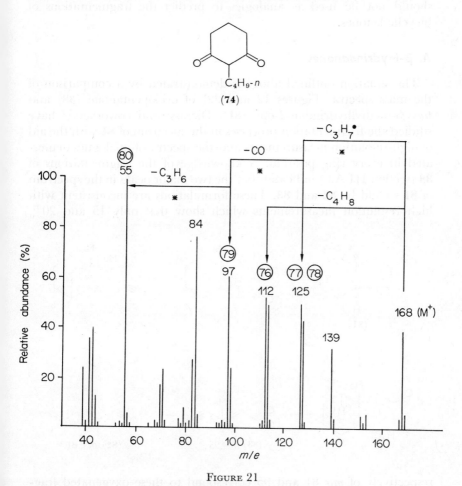

FIGURE 21

IV. BICYCLIC KETONES

The carbonyl group in simple aliphatic and monocyclic ketones is able to localize the positive charge and therefore controls the course of the fragmentation. This is not the case with bicyclic and polycyclic systems, where the increased hydrocarbon portions of the molecules cause the fragmentations to be more random. As the carbonyl group in bicyclic systems no longer directs the fragmentations in a predictable manner, the fragmentations of the simple monocyclic ketones should not be used as analogies to predict the fragmentations of bicyclic ketones.

A. β-Hydrindanones

The situation outlined above is demonstrated by a comparison of the mass spectra (Figures 12 and 22) of cyclopentanone (38) and *trans*-8-methylhydrindan-2-one (81). Djerassi and coworkers[77] have studied the fragmentation processes in the spectrum of 81 with the aid of high-resolution measurements and the spectra of analogues deuterated in every ring position. A knowledge of the fragmentations of 38 (section III.A.2) indicates that the two major peaks in the spectrum of 81 should be 82 and 83. These formulations are inconsistent with high resolution measurements which show that only 15 and 20%

(81)

(83) m/e 95   (82) m/e 81

respectively of m/e 81 and 95 correspond to these oxygenated fragments.

The M − 42 peak (m/e 110) arises by loss of ketene as shown in 84 → 85. The additional hydrogen lost in the formation of 86 (m/e 109), originates from the angular methyl group. The formation of the base peak (m/e 95) involves complete loss of the A ring together with a hydrogen atom coming mainly from $C_{(6)}$ (some loss from $C_{(9)}$ is also

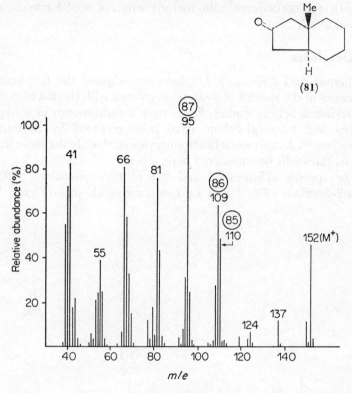

FIGURE 22

11 + c.c.g. II

noted), and consequently the major portion of m/e 95 may be represented as **87**.

B. Decalones

Djerassi and colleagues[78,79] have investigated the fragmentation processes in the spectra of α- and β-decalones with the aid of extensive deuterium labelling studies. Exact mass measurements in the spectra of *cis*- and *trans*-2-decalone have been reported by Beynon and coworkers[80]. As this work has been reviewed[81], only the more important features will be considered here.

The spectra (Figures 23 and 24) of *trans*-1-decalone (**88**) and *trans*-2-decalone (**89**) should be compared with that (Figure 13) of

(88)

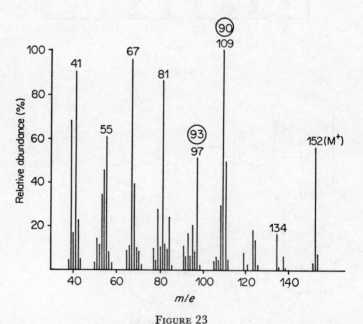

Figure 23

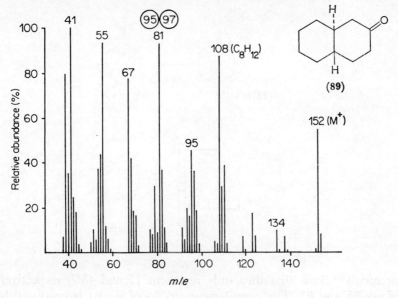

FIGURE 24

cyclohexanone (**42**). Although the 'cyclohexanone' fragmentation (section III.A.3) occurs to some extent, competing fragmentations arise from the bicyclic skeleton.

Trans-1-decalone may fragment by the processes indicated in **88** to give **90** and **91**. Part of the m/e 55 peak is due to the $C_4H_7^+$ species. The ion at m/e 97 (M − 55) is produced by the rearrangement **92** → **93**. Below m/e 90, the spectrum is dominated by hydrocarbon fragments; e.g. the ion m/e 81 ($C_6H_9^+$) is produced by the two fragmentations depicted in **94**.

If the carbonyl group of *trans*-2-decalone (**89**) controls the course of the fragmentation, then the abundances of the ions **91** and **95** should be pronounced (compare section III.A.3). Exact mass meas-

(**90**) m/e 109   (**88**)   (**91**) m/e 55
                                              percentage not specified

(**92**) *m/e* 152   (**93**) *m/e* 97

(**94**)

urements[80] show that they only represent 17 and 45% respectively of *m/e* 55 and 81. The complement (55%) of *m/e* 81 is produced by the rearrangement **96 → 97.**

(**95**) *m/e* 81 (45%)   (**89**)   (**91**) *m/e* 55 (17%)

(**96**) *m/e* 152   (**97**) *m/e* 81 (55%)

It is of interest to note that although the mass spectra of *cis*- and *trans*-2-decalone are almost identical[78,80], when an angular methyl group is present in a 1-decalone, the spectra of the *cis* and *trans* compounds may be significantly different (e.g. those of **98** and **99**)[79].

(98)   (99)

C. Unsaturated Bicyclic Ketones

The spectra of several unsaturated bicyclic ketones have been reported[81] and the fragments in their spectra can be envisaged as arising from a combination of 'decalone' and 'cyclohexene' type fragmentations. The spectrum of **100** shows an $M - C_2H_4$ peak, which arises from a retro Diels–Alder reaction, similar to the fragmentation of cyclohex-2-en-1-ones (section III.B.2). A second process is $M - CH_2CO$ which involves an unfavourable cleavage α to a double bond. A similar loss of ketene is observed in the spectrum of α-tetralone (see section VI.D). The retro Diels–Alder process cannot occur in the spectrum of **101**; instead the spectrum exhibits a pronounced $M - CH_2CO$ peak. In addition to the hydrocarbon fragments in the spectrum of **101′**, a retro Diels–Alder reaction operating in ring B produces a prominent $M - C_4H_6$ peak.

(100)   (101)   (101′)

V. POLYCYCLIC KETONES

A. Saturated Steroidal Ketones

The studies by Djerassi's group on ketosteroids and related compounds have resulted in an outstanding series of papers, giving details of the complex fragmentations of steroidal ketones and providing extensive information on deuterium-labelling techniques. Much of this work has been concerned with androstane derivatives; viz. 1-[82], 2-[83], 3-[84,85], 4-[86], 6-[87], 7-[88], 11-[89,90], 15-[91] and 17-keto-androstane[92]. Cholestan-2-one[83], -6-one[87], -16-one[93,94] and pregnan-12-one[94,95] and -20-one[96] have also been extensively investigated. This work has been reviewed to 1965[97,98]. Other groups have made significant contributions to this field, and references to their work may be found in the later review[98] and in the papers cited above.

Hydrocarbon fragments usually dominate the spectra of steroidal ketones (an exception is the spectrum[89,90] of 5α-androstan-11-one). This confirms the earlier conclusion (section IV) that the carbonyl group does not direct the fragmentation of polycyclic ketones in a predictable manner, hence the effectiveness of mass spectrometry in the determination of the position of the carbonyl group is somewhat limited in these cases. As much of this work has been reviewed[98], recourse is made to a recent example to demonstrate the extreme complexity of the fragmentation processes in the mass spectra of a ketosteroid.

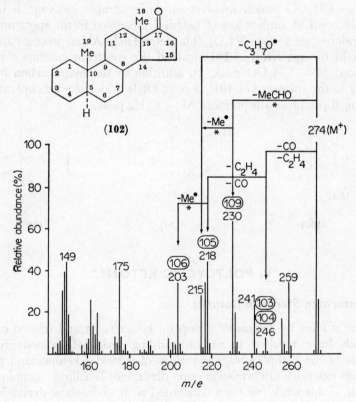

FIGURE 25

$$\left[\text{}\right]^{\ddot{+}} \xleftarrow[-C_2H_4]{-e} \quad \begin{matrix}(102)\\ \text{m.w. 274}\end{matrix} \quad \xrightarrow[-CO]{-e} \left[\text{}\right]^{\ddot{+}}$$

(103) m/e 246   (104) m/e 246

(105) m/e 218   (106) m/e 203

(107) m/e 274   (108) m/e 274   (109) m/e 230

(110)

The decomposition modes in the mass spectrum (Figure 25) of 5α-androstan-17-one (102) have been studied with the aid of high-resolution data, metastable peaks, and the spectra of 5α-androstan-17-one-3α-d_1, -7β-d_1, -8β-d_1, -15α-d_1, -12,12-d_2, -16,16-d_2, and -18,18,18-d_3[92]. This work clearly demonstrates that *a priori* predictions of the fragmentation modes cannot be made, and that extensive labelling studies are necessary for the interpretation of the spectrum.

The peak at *m/e* 246 is a doublet due to 103 and 104, and both these ions may fragment to produce 105 (*m/e* 218). It is also possible that 105 may be formed directly from the molecular ion. The ion at *m/e* 203 is produced by loss of a methyl radical $C_{(19)}$ from 105. The most characteristic fragment in the spectrum is *m/e* 230 which is formed by the process M − CH_3CHO, involving a double hydrogen transfer from $C_{(7)}$, $C_{(8)}$ and $C_{(12)}$ (labelling) and possibly $C_{(14)}$. Transfer of a hydrogen atom from either $C_{(7)}$, $C_{(8)}$ or $C_{(12)}$ produces an ion (e.g. 107 → 108) which may fragment with the loss of the hydrogen at $C_{(14)}$ via a McLafferty rearrangement (108 → 109). Fragmentations producing *m/e* 215 and 175 are shown in 110. In many respects the spectrum of 102 is similar to that of androstane itself[99].

The location of the keto function in a steroid molecule may be determined in favourable cases by mass spectral analysis of a suitable derivative, containing a group which will direct the fragmentation in a specific manner, e.g. ethylene ketals, ethylene thioketals and dimethylamino derivatives. This subject has been reviewed[100]. It is instructive to compare Figure 25 with the spectrum of 5α-androstan-17-one ethylene ketal, which is very simple in that it exhibits only a strong *m/e* 99 fragment. Deuterium labelling studies[100] show that *m/e* 99 is produced by the process 111 → 112.

(111) *m/e* 318   (112) *m/e* 99

A very important feature which emerges from the study of the fragmentations of ketosteroids, concerns the distance effect operative in the McLafferty rearrangement. This rearrangement (section II.B) has been shown to be site-specific for aliphatic ketones[41] and for suitably substituted 16-ketosteroids (e.g. **113**)[101]. As 5α-androstan-11-one (**114**) fragments mainly by cleavage of the $C_{(9)}$—$C_{(10)}$ bond[89], and 5α-androstan-15-one (**115**) by cleavage of the $C_{(8)}$—$C_{(14)}$ bond[91] it would seem reasonable to expect that the cleavages should operate by the rearrangements outlined in **114** and **115**. Deuterium labelling studies[89,91] show that these hydrogens are not involved in the re-arrangements. As the γ-hydrogen oxygen distances in **113–115** are 1·52, 1·84 and 2·28 Å respectively, it follows that the γ-hydrogen rearrangement cannot occur by this mechanism when the distance between the γ-hydrogen and the recipient oxygen atom is greater than 1·8 Å[97].

(113)   (114)   (115)

B. Unsaturated Steroidal Ketones

The spectra of unsaturated ketosteroids are simpler and more readily interpretable than those of saturated steroidal ketones. The major fragmentations of four selected examples are briefly summarized in formulae **116–119**.

Loss of ketene is an important process in the spectra of **116** and **117**[102], as is B ring cleavage with and without hydrogen rearrangement respectively. Cleavage of the B ring is also an important process in the spectra of **118**[103] and **119**[104], with the oxygen-containing fragments retaining the charge. Although the retro Diels–Alder process in **119** produces m/e 218, no similar process is noted in the spectrum of **117**. This is also in conflict with the spectrum of the related bicyclic compound **100**, which shows an appreciable M − C_2H_4 peak (section IV.C).

11*

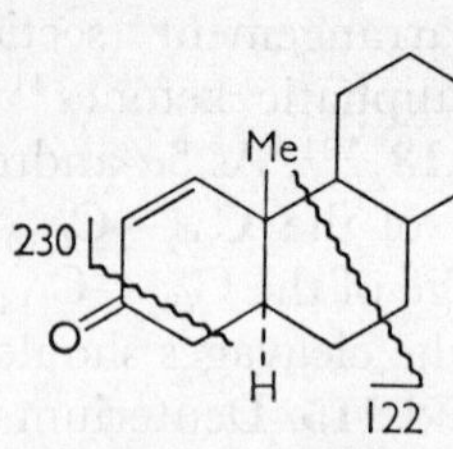

(116)

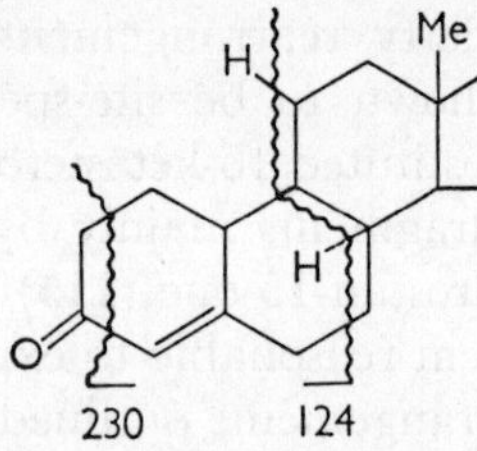

(117)

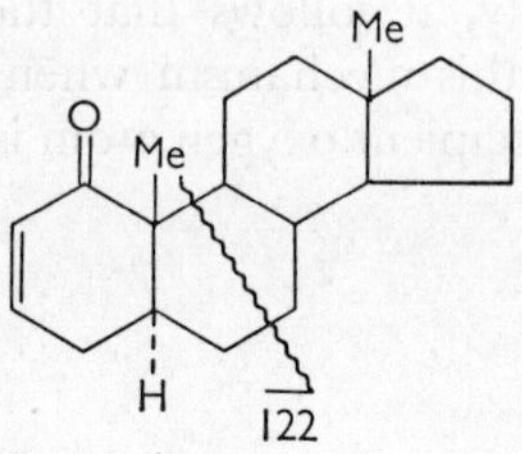

(118)

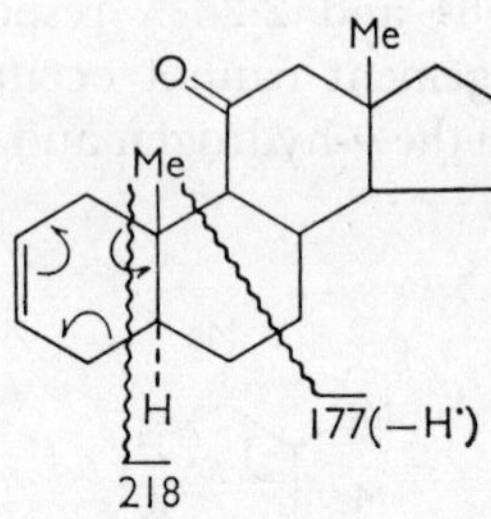

(119)

C. Miscellaneous Polycyclic Ketones

Steroidal ketones containing aromatic A and B rings (e.g. estrogens, dehydroestrogens and equilenin derivatives) do not fragment primarily through the carbonyl group[105-107]. Instead, the major decomposition modes occur by initial cleavage of benzylic bonds.

The complex behaviour of a series of pentacyclic triterpenoid ketones upon electron impact has been reviewed[108].

VI. AROMATIC CARBONYL COMPOUNDS

A. Simple Aldehydes and Ketones

The mass spectra of aromatic aldehydes and ketones are generally simple and amenable to analysis. Prominent molecular ions are a feature of these spectra. The behaviour[109,110] of benzaldehyde derivatives on electron impact is illustrated by the spectrum (Figure 26) of benzaldehyde (120). Loss of a hydrogen atom from the molecular ion produces the stable benzoyl cation (124, m/e 105), which decomposes by loss of carbon monoxide to $C_6H_5^+$ (125, m/e 77). It has been suggested that $C_6H_5^+$ is probably linear and not the phenyl cation[141].

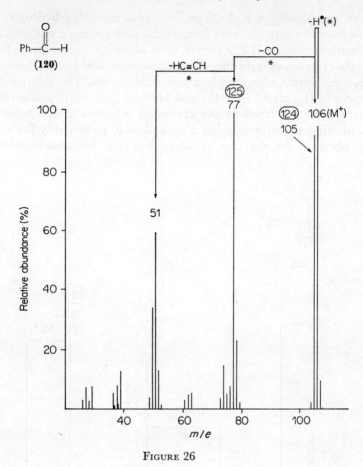

FIGURE 26

Acetophenones and benzophenones also behave simply[111], fragmenting by the processes **122** → **125** and **123** → **125** respectively. α-Cleavage in acetophenone (**122**, Figure 27) also produces the acetyl cation m/e 43. Alkyl aryl ketones containing a γ-hydrogen

$$Ph-\overset{\overset{\cdot+}{\overset{\displaystyle O}{\|}}}{C}-R \xrightarrow[*]{-R\cdot} Ph-C\equiv O^+ \xrightarrow[*]{-CO} C_6H_5^+$$

(121) m/e 106 (R = H)   **(124)** m/e 105   **(125)** m/e 77
(122) m/e 120 (R = Me)
(123) m/e 172 (R = Ph)

atom may undergo β-cleavage with concomitant γ-hydrogen rearrangement to give the enol form of the acetophenone ion radical[111]. (Compare section II.B). It has been elegantly demonstrated[111,112] that α-cleavage of substituted acetophenones and benzophenones on electron impact obeys a Hammett relationship. The relative abundances of the acetyl (m/e 43) and benzoyl (m/e 105) cations may be quantitatively related to the Hammett σ values of the substituent by plotting an expression log z/z_0 against σ. Reasonably linear plots are obtained for the case of acetophenones. Electron-withdrawing

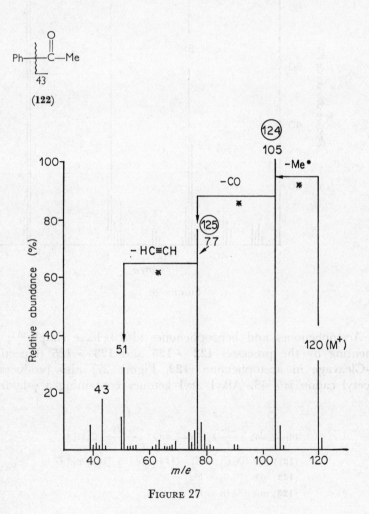

FIGURE 27

$$z = \frac{\text{Relative abundance of MeCO}^+ \text{ in the spectrum of } XC_6H_4COMe}{\text{Relative abundance of } [XC_6H_4COMe]^{+\cdot}}$$
$$(\text{for } X = H, z = z_0)$$

substituents promote α-cleavage, while electron-donating groups reduce α-cleavage. This is a most important observation, as it shows that the extent of these α-cleavage processes in simple aromatic ketones may be predicted by normal ground-state principles. Similar straight line correlations are obtained[113] when either the ionization potentials of substituted acetophenones or the appearance potentials of the substituted benzoyl cations (which occur in the spectra of acetophenones), are plotted against the Hammett σ function.

B. Unsaturated Ketones

Williams and coworkers[114] have studied the fragmentation of benzalacetophenone with the aid of the spectra of analogues deuterated

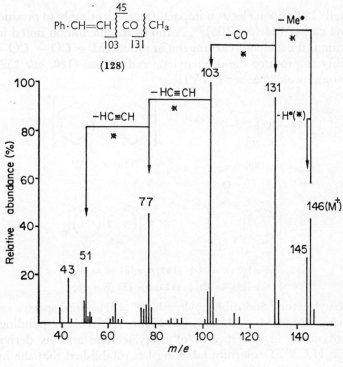

FIGURE 28

at every available position. Apart from exhibiting the expected α-cleavage processes, the spectrum of benzalacetophenone shows a pronounced M − 1 ion. Deuterium labelling studies are consistent with the loss of a hydrogen radical from the A ring to form the stable oxonium ion **127** by an intermolecular aromatic substitution reaction, viz. **126** → **127**. This is a general process for compounds of the general formula PhCH=CH—COR[114], and can be seen in the spectrum (Figure 28) of benzalacetone (**128**). M − 1 ions have also been observed in the spectra of aryl β-diketones (section VI.C) and monothio derivatives of aryl β-diketones[115].

(**126**) m/e 224   (**127**) m/e 223

C. Diketones

Benzil (**129**), upon electron impact, fragments simply to produce the benzoyl cation **124** (m/e 105)[50]. A minor fragmentation noted in this spectrum is the skeletal rearrangement process M − CO − CO − H$_2$ plausibly to produce the diphenylene radical ion (**130**, m/e 152, 1% of the base peak) (see section VII).

(**129**)   (**130**) m/e 152

m/e 162 (R = Me) (**132**) m/e 161 (R = Me)
m/e 224 (R = Ph)) (**133**) m/e 223 (R = Ph)

Benzoylacetone and dibenzoylmethane (**131**) afford spectra exhibiting pronounced M − H· ions (see Figure 29)[51]. Corresponding ions are not noted in the spectra of acetylacetone and its derivatives (section II.C.2). Deuterium labelling has established that the hydrogen lost originated from the aromatic ring. This is consistent with

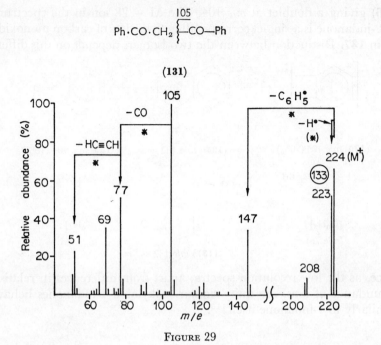

$$Ph \cdot CO \cdot CH_2 \overbrace{}^{105} CO—Ph$$

(131)

FIGURE 29

the formation of the stable oxonium species **132** and **133** (compare section VI.B). In general the presence of a pronounced $M - 1$ peak is either indicative of an aldehyde group or of a species which may form a very stable cation by loss of a hydrogen radical.

The spectra of alkyl substituted benzoylacetone derivatives, apart from containing the benzoyl cation as the base peak, are similar to those of acetylacetone derivatives (section II.C.2). Rearrangements involving losses of ketene and/or an alkene (from the side chain) by six-membered transition states, produce enol fragment ions which then decompose by allylic cleavage[51]. All spectra (cf. Figure 29) contain m/e 69 ($\overset{+}{O}{\equiv}C—CH{=}C{=}O$) peaks (see section III.C.2).

D. Indanones and Tetralones

The behaviour of indanones and tetralones upon electron impact has been studied by labelling and high resolution studies[67] and by computer-aided mass spectrometry[116].

The molecular ion **134** of 1-indanone may fragment by loss of either carbon monoxide (probably to **135**) or ethylene (probably to

136) giving a doublet at *m/e* 104. The M − 28 ion in the spectrum of 2-indanone is a singlet corresponding to the loss of carbon monoxide from **137**. Distinction between the two isomers depends on this differ-

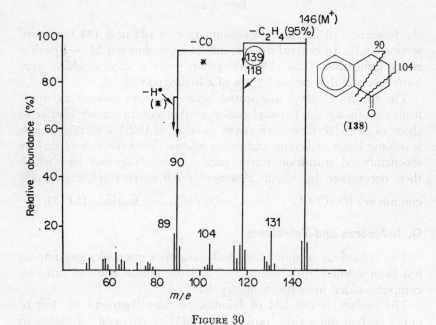

ence, as the low resolution spectra, apart from differences in relative abundance of ions, are quite similar. Alkylated 1-indanones behave similarly to 1-indanone itself[116].

FIGURE 30

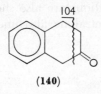

(140)

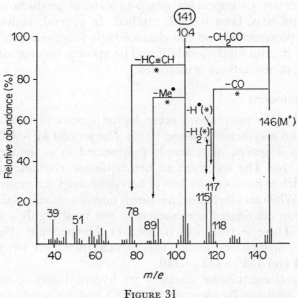

FIGURE 31

The carbonyl group does not control the course of the fragmentation of 1- (138) and 2-tetralone (140) [67]. Their fragmentations arise by the retro Diels–Alder processes outlined below, as was verified by

(138)   **(139)** m/e 118

(140)   **(141)** m/e 104

labelling studies[67]. Other fragmentations are also shown in Figures 30 and 31. Alkyl derivatives of 1-tetralone also undergo the retro Diels–Alder fragmentation[116].

E. Quinones

Quinones are an important group of natural products and their mass spectra have been recently studied. In general, molecular ion peaks are pronounced and the characteristic fragmentation patterns are useful in structural elucidation. The spectra become simpler as the degree of aromaticity is increased.

I. Benzoquinones

Two major fragmentations occur in the spectra of benzoquinone (**142**)[117] and alkylbenzoquinones[76]: (i) The process $M - CO - CO$ produces a fragment most simply represented as an ionized cyclobutadiene. (ii) The spectrum of benzoquinone contains an ion at m/e 82 which is produced by loss of acetylene from the molecular ion (see **142**). When an alkylbenzoquinone is unsymmetrically substituted, the fragment eliminated corresponds to the most highly substituted acetylene. These features can be seen in the spectrum (Figure 32) of 2,3-dimethylbenzoquinone (**143**) which fragments by processes **143** → **144** and **143** → **145** → **146**.

When a benzoquinone contains an hydroxyl group, additional fragments originate by cleavage of the 1,2 and 4,5 bonds (or 3,4 and 1,6 bonds) with accompanying rearrangement of the hydrogen from

(**142**)

(**143**)   (**144**) m/e 80 m/e 77

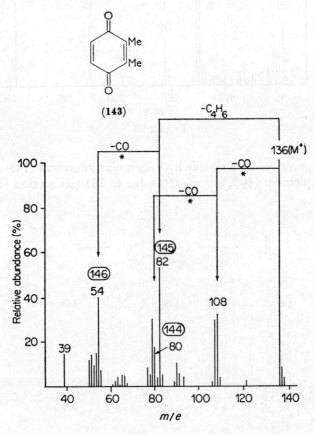

the hydroxyl group[76]. The spectrum (Figure 33) of 2-hydroxy-5-methylbenzoquinone (**147**) contains ions at m/e 69 and 70. Sixty per cent of the ion at m/e 69 is formed by the process **148 → 149** while

FIGURE 32

324　　　　　　　　　　J. H. Bowie

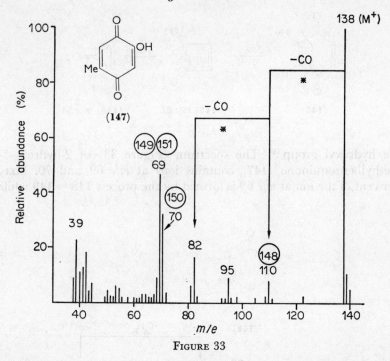

FIGURE 33

m/e 70 is produced by a double hydrogen rearrangement (**148 → 150**). The complement (40%) of *m/e* 69 is due to **151** (see section III.C.2).

Benzoquinones with isoprenoid substituents fragment specifically through the side chain[118-123]. The spectrum[123] of ubiquinone Q3 (**152**) exhibits allylic cleavage of the side chain. The base peak of the spectrum is produced by the process **153** → **154**. The formation of a stable species (e.g. **154**) is a general fragmentation process of isoprenoid benzoquinones and of vitamin K derivatives[125].

(**152**)

(**153**) (**154**) m/e 235

2. Naphthaquinones

Naphthaquinones behave in a well defined manner upon electron impact. 1,4-Naphthaquinone (**155**) fragments initially by the process M − CO to **156**, m/e 130. This ion decomposes by loss of either CO (to **159**) or acetylene (to **157**), thus producing a 'doublet' at M − 54 and M − 56[117,124]. 1,2-Naphthaquinone (**158**) behaves differently, as the process M − CO − CO predominates[50]. The spectra of alkyl 1,4-naphthaquinones retain the features of **155**, but losses of the alkyl radical from various ions are noted[124,125].

The initial fragmentations of hydroxy-1,4-naphthaquinones which have the hydroxyl substituent in the A ring are similar to that of **155**, but when the hydroxyl is in the B ring a specific hydrogen transfer process occurs, analogous to that observed in the spectra of hydroxy-benzoquinones (section VI.E.1)[124]. This rearrangement, which has been substantiated by deuterium labelling, can be seen in the

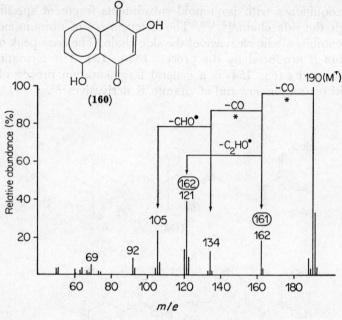

FIGURE 34

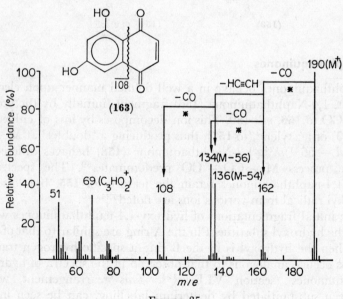

FIGURE 35

(155)   **(156)** m/e 130   **(157)** m/e 104

(158)   **(159)** m/e 102   (m/e 76)

spectrum (Figure 34) of 2,5-dihydroxynaphthaquinone (**160**) (viz. **161** → **162**). This spectrum should be compared with that (Figure 35) of 5,7-dihydroxynaphthaquinone (**163**) where the normal 'doublet' at M − 54 and M − 56 is observed.

Mass spectrometry has aided the structure elucidation of the spinochrome pigments[126], vitamin K analogues[125] and of naphthaquinones containing the 'aphin' side chain (see section VI.E.3)[127].

(160)   **(161)** m/e 162   **(162)** m/e 121

3. Polycyclic Quinones

Anthraquinone and phenanthraquinone both fragment by the process M − CO − CO to form the biphenylene radical ion (m/e 152), this fragmentation being more pronounced in the spectrum of phenanthraquinone[117]. The spectra of substituted anthraquinones[117] are much simpler than those of benzoquinones and naphthaquinones.

The molecular ion is generally the base peak, and the fragmentations follow those of anthraquinone, with additional breakdown through the substituents.

(164)   (165)

(166)   (167)

Mass spectrometry is most useful for structural investigations when the anthraquinone or phenanthraquinone contains an aliphatic side chain. The anthraquinone **164** fragments as indicated to form the 1,6,8-trihydroxy-3-methyl-anthraquinone (emodin) molecular ion[128]. Ochromycinone (**165**) fragments initially as a 1-tetralone system (section VI.D), then by additional loss of carbon monoxide[129]. The fragmentation patterns in the spectra of the rhodomycinone and pyrromycinone pigments are extremely complex[130,131]. This work has been reviewed[132]. The base peak (m/e 314) in the spectrum of α-rhodomycinone (**166**) arises by retro Diels–Alder cleavage of ring D. Other peaks observed are due to losses of water and carbon monoxide from various ions[131]. The side chain of piloquinone (**167**) undergoes a characteristic McLafferty rearrangement[133].

The mass spectra of aphin derivatives are very simple[127]. No fragmentation of the dihydroxyperylene quinone system is noted. Loss of two molecules of acetaldehyde from the molecular ion of erythroaphin fb (**168**) possibly produces **169**, m/e 422. This ion fragments by loss of an electron to form the doubly charged species m/e 211 (base peak). Loss of methyl from the molecular ion is also an observed process.

(168)

(169) *m/e* 422   *(m/e* 211)

F. Tropones and Related Compounds

The mass spectrum of tropone (**170**) exhibits pronounced loss of carbon monoxide (M − 28) to produce the benzene ion radical (*m/e* 78)[134,135]. Benzotropone (**171**) fragments similarly[135] with the production of the naphthalene ion radical (*m/e* 128).

(170)   **(171)**

Substituted tropones may also fragment by initial loss of carbon monoxide[135]. Tropolone (**172**) fragments by loss of carbon monoxide to the phenol radical ion (**173**). Further decomposition of **173**[136]

J. H. Bowie

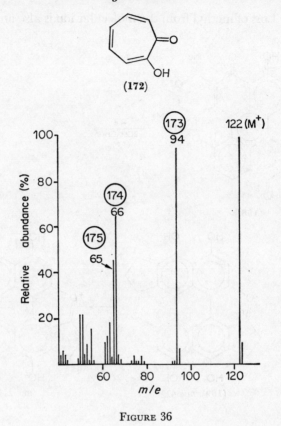

(172)

FIGURE 36

ultimately produces the cyclopentadienyl cation (**175**) (see Figure 36). The M – CO process is even pronounced in the spectrum of

colchicine (**176**) [137]. The base peak (**177**) of this spectrum is produced by loss of acetamide from the M – CO peak.

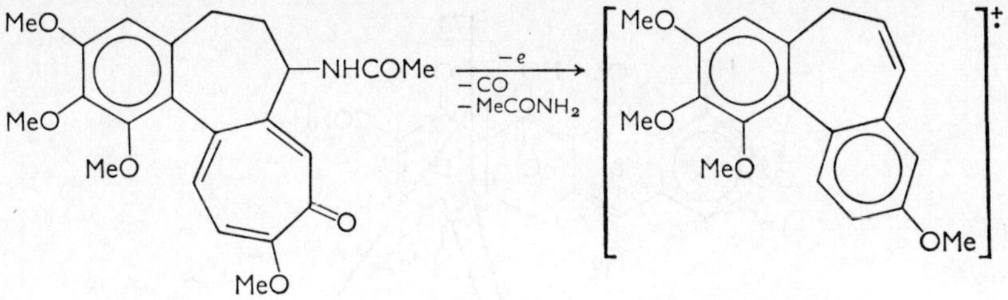

(176)   (177) *m/e* 312

G. Miscellaneous Aromatic Carbonyl Compounds

There are certain classes of aromatic carbonyl compounds that are not discussed in this chapter. Obvious examples are the flavonoid, coumarin and xanthone pigments. These are not included because they are best classified as pyrone derivatives. Alkaloids which contain carbonyl groups have been discussed by Djerassi, Budzikiewicz and Williams[9]. Only colchicine (**176**, section VI.F) and atherospermidine (**178**) are included here.

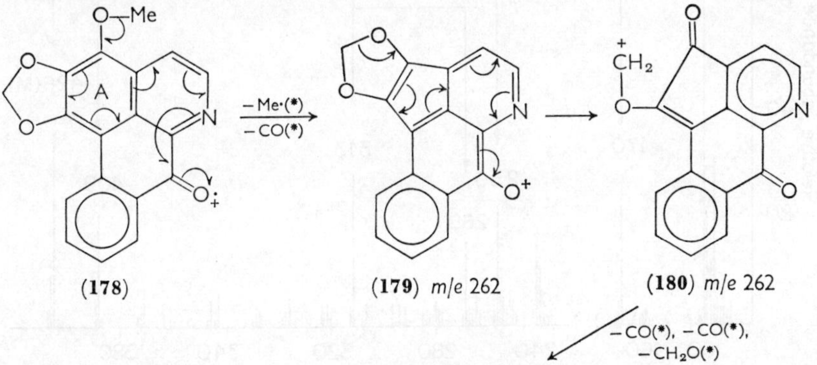

(178)   (179) *m/e* 262   (180) *m/e* 262

−CO(•), −CO(•), −CH₂O(•)

[C₁₃H₆N]‡
(181) *m/e* 176

The fragmentations of alkaloids which contain a carbonyl group are generally not controlled by the carbonyl group[9]. The alkaloids of the spermatheridine series are exceptions to this generalization. When atherospermidine (**178**) is subjected to electron impact, all the substituents on ring A are lost by concerted processes involving the carbonyl group (**178** → **181**)[138].

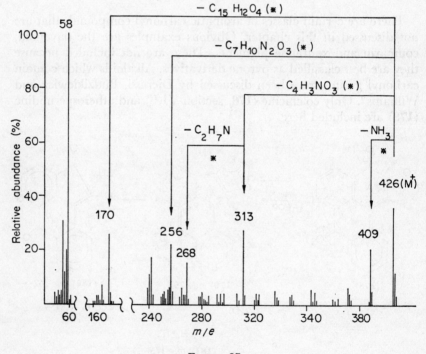

FIGURE 37

The mass spectra of tetracycline derivatives are complex, most of the fragmentation occurring in the D ring[139]. The spectrum of 5α,6-anhydrotetracycline (**182**) is illustrated in Figure 37, and the major fragmentations are outlined below.

VII. SKELETAL REARRANGEMENT PROCESSES

Until several years ago, reports of the occurrence of migration of groups other than hydrogen in the mass spectrometer were rare. Since 1965, the discovery of many such processes (now termed skeletal rearrangements), has shown that *a priori* prediction of the fragmentation modes of certain classes of organic compounds is a difficult undertaking. Even now it is not possible to predict with any certainty when skeletal rearrangements may occur. The presence of these processes at present places restrictions on the use of the extremely elegant 'element-mapping' technique, developed by Biemann from his studies of computer-aided mass spectrometry[140]. This technique assumes that either skeletal rearrangement processes are absent in mass spectra or that their occurrence is predictable. It is to be hoped that this deficiency will be overcome in the future.

The early work on skeletal rearrangement processes has been reviewed by Brown and Djerassi[53]. Although skeletal rearrangement processes are not yet fully understood, they may be divided into two broad classes: (a) Processes of the type ABC → AC + B. These may occur from either an odd or an even electron species. A and C are

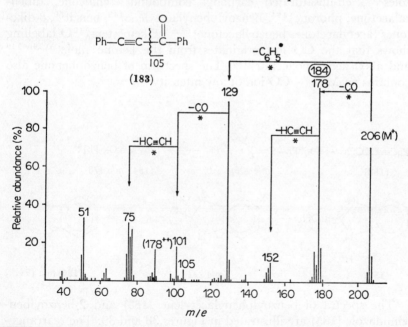

FIGURE 38

initially not joined together and B is a neutral molecule (if this definition is used, the M − CO process in quinones (section VI.E) is not a skeletal rearrangement process[53]). (b) Many skeletal rearrangement processes can only be explained by specific rearrangement of the molecular ion to give a structure which is quite different from the ground state structure. Recognition of such processes is becoming relatively common, and they often involve migration of substituents.

The skeletal rearrangement processes of type (a) observed in the spectra of simple carbonyl compounds generally involve the loss of carbon monoxide from the molecular ion. The relative abundances of the ions produced range from less than 10% (cf. acetylacetone, section II.C.2, and benzil, section VI.C) to the base peak of the spectrum (see Figure 38). A process of this type often requires a double bond (C=C or C=O) to be ionized in order that the incipient radical may migrate to an electron-deficient centre[54]. The migrating group presumably does not become free, but migrates at the same time as the neutral molecule is ejected.

The following carbonyl compounds show M − CO peaks in their mass spectra: benzophenone[141], 2-acylthiophens[142], 2-acylbenzimidazoles[143], $\alpha\beta$-unsaturated carbonyl compounds (chalcone, dibenzalacetone, phorone)[144], benzoylphenylacetylene[145], benzil[50], β-diketones (acetylacetone, benzoylacetone)[51,52], β-ketoesters (^{18}O labelling shows that the CO lost originates from the ketone moiety)[146–149] and alkyl acetothioacetates[148]. The spectrum of benzophenone also contains an M^{2+} − CO ion of low intensity[150].

The spectra of benzoylphenylacetylene (183) and 2-benzoylbenzimidazole (185) are illustrated in Figures 38 and 39. The rearrangement processes involved are 183 → 184 and 185 → 186. The ions

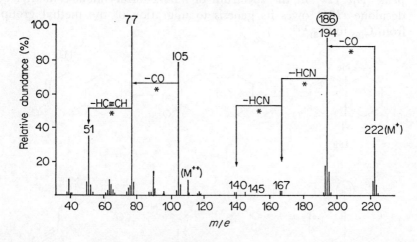

(185)

FIGURE 39

184 and **186** fragment exactly as do the molecular ions of diphenyl-acetylene and 2-phenylbenzimidazole respectively.

Molecular ion rearrangements have been reported for cyclic ketones. The formation of the base peak (m/e 74) in the spectrum of 4-methoxycyclohexanone (**187**) involves a methoxyl migration, plausibly represented by the process **187** → **188**[151]. This process is more pronounced than the McLafferty rearrangement in the spectrum of 2-ethyl-4-methoxycyclohexanone. Analogous processes are observed in the spectrum of 4-hydroxycyclohexanone[151].

(187)   (188) m/e 74

Migration of a methyl group from $C_{(9)}$ to $C_{(1)}$ in the molecular ion of trans-Δ^3-10-methyl-2-octalone (189) is necessary for the formation of an abundant ion at m/e 69 (189 $\rightarrow$ 190)[152]. Methyl migrations also occur when 2-arylidene-9-methyl-1-decalones are subjected to electron impact[153]. Deuterium labelling studies have shown that the base peak (m/e 121) in the spectrum of trans-2-furfurylidene-9-methyl-1-decalone (191) owes its genesis to migration of the methyl group from $C_{(9)}$ to $C_{(2)}$[153].

(189)   (191)   (192) m/e 121   (190) m/e 69

VIII. ACKNOWLEDGEMENTS

Thanks are expressed to Drs. G. E. Lewis and D. H. Williams who read the manuscript and suggested several important additions.

IX. REFERENCES

1. J. H. Beynon, *Mass Spectrometry and Its Applications to Organic Chemistry*, Elsevier, Amsterdam, 1960.
2. K. Biemann, *Mass Spectrometry: Organic Chemical Applications*, McGraw-Hill, New York, 1962.
3. R. I. Reed, *Ion Production by Electron Impact*, Academic Press, London, 1962.
4. F. W. McLafferty (Ed.), *Mass Spectrometry of Organic Ions*, Academic Press, New York, 1963.
5. J. H. Beynon and A. E. Williams, *Mass and Abundance Tables for Use in Mass Spectrometry*, Elsevier, Amsterdam, 1963.
6. C. A. McDowell (Ed.), *Mass Spectrometry*, McGraw-Hill, New York, 1963.
7. C. Brunée and H. Voshage, *Massenspektrometrie*, Verlag Karl Thiemig, Munich, 1964.

8. H. Budzikiewicz, C. Djerassi and D. H. Williams, *Interpretation of Mass Spectra of Organic Compounds*, Holden-Day, San Francisco, 1964.

9. H. Budzikiewicz, C. Djerassi and D. H. Williams, *Structure Elucidation of Natural Products by Mass Spectrometry*, Vols. 1 and 2, Holden-Day, San Francisco, 1964.

10. R. W. Kiser, *Introduction to Mass Spectrometry and Its Applications*, Prentice-Hall, Englewood Cliffs, N.J., 1965.

11. R. I. Reed, *Applications of Mass Spectrometry to Organic Chemistry*, Academic Press, New York, 1966.

12. D. H. Williams and I. Fleming, *Spectroscopic Methods in Organic Chemistry*, Chapter 5, McGraw-Hill, England, 1966.

13. F. W. McLafferty, *Interpretation of Mass Spectra*, Benjamin, New York, 1966.

14. H. C. Hill, *Introduction to Mass Spectroscopy*, Heyden and Sons, London, 1966.

15. H. Budzikiewicz, C. Djerassi and D. H. Williams, *Mass Spectrometry of Organic Compounds*, Holden-Day, San Francisco, 1967.

16. F. W. McLafferty in *Determination of Organic Structures by Physical Methods*, (Eds. F. C. Nachod and W. D. Phillips) Academic Press, New York, 1962, pp. 93–173.

17. J. D. Morrison, *Rev. Pure Appl. Chem.*, **12**, 117 (1962).

18. K. Biemann, *Techniques in Organic Chemistry*, Vol.11 (Ed. A. Weissberger) John Wiley and Sons, New York, 1963, pp. 259–316.

19. K. Biemann, *Ann. Rev. Biochem.*, **32**, 755 (1963).

20. F. W. McLafferty and R. S. Gohlke, *Chem. Eng. News*, **42**, 96 (1964).

21. G. Spiteller and M. Spiteller-Friedmann, *Angew. Chem., Int. Ed. Engl.* **4**, 383 (1965).

22. R. I. Reed, *Quart. Rev.*, **20**, 527 (1966).

23. F. W. McLafferty, *Mass Spectral Correlations*, American Chemical Society, Washington D.C., 1963.

24. R. S. Gohlke, *Uncertified Mass Spectral Data*, The Dow Chemical Co., Michigan, 1963.

25. American Petroleum Institute and Manufacturing Chemists Association, *Catalogue of Mass Spectral Data*, Chemical Thermodynamic Properties Centre, Texas A. and M. University, College Station, Texas.

26. American Society for Testing Materials Committee E-14, *Index of Mass Spectral Data*, A.S.T.M. Philadelphia, 1963.

27. J. Lederberg, *Computation of Molecular Formulas for Mass Spectrometry*, Holden-Day, San Francisco, 1964.

28. D. D. Tunnicliff, P. A. Wadsworth and D. O. Schissler, *Mass and Abundance Tables*, Shell Development Company, California, 1965.

29. J. S. Shannon, *Tetrahedron Letters*, **1963**, 801, also *Proc. Royal Austr. Chem. Inst.* 1964, 328.

30. J. H. Beynon and A. E. Fontaine, *Z. Naturforsch.*, **22a**, 334 (1967) and references therein.

31. J. Seibl, *Helv. Chim. Acta*, **50**, 263 (1967).

32. H. Budzikiewicz, C. Djerassi and D. H. Williams, *Interpretation of Mass Spectra of Organic Compounds*, Holden-Day, San Francisco, 1964, p. 1.

33. J. A. Gilpin and F. W. McLafferty, *Anal. Chem.*, **29**, 990 (1957).

34. A. G. Sharkey, J. L. Schultz and R. A. Friedel, *Anal. Chem.*, **28**, 934 (1956).
35. J. H. Bowie, unpublished observations.
36. J. S. Shannon, *Australian J. Chem.*, **16**, 683 (1963).
37. P. Bommer and K. Biemann, *Ann. Rev. Phys. Chem.*, **16**, 481 (1965).
38. F. W. McLafferty, *Anal. Chem.*, **31**, 82 (1959).
39. F. W. McLafferty, *Interpretation of Mass Spectra*, Benjamin, New York, 1966, p. 131.
40. F. W. McLafferty, *Chem. Commun.*, **1966**, 78.
41. H. Budzikiewicz, C. Fenselau and C. Djerassi, *Tetrahedron*, **22**, 1391 (1966), references therein.
42. C. Djerassi, *Pure Appl. Chem.*, **9**, 159 (1964).
43. C. Djerassi, G. von Mutzenbecher, J. Fajkos, D. H. Williams and H. Budzikiewicz, *J. Amer. Chem. Soc.*, **87**, 817 (1965).
44. J. K. MacLeod and C. Djerassi, *Tetrahedron Letters*, **1966**, 2183.
45. J. K. MacLeod, J. B. Thomson and C. Djerassi, *Tetrahedron*, **23**, 2095 (1967).
46. H. Budzikiewicz, C. Djerassi and D. H. Williams, *Interpretation of Mass Spectra of Organic Compounds*, Holden-Day, San Francisco, 1964, p. 9.
47. J. H. Bowie, R. Grigg and D. H. Williams, unpublished observations.
48. L. Ahlquist, R. Ryhage, E. Stenhagen and E. von Sydow, *Arkiv Kemi*, **14**, 211 (1959).
49. W. Reusch and C. Djerassi, *Tetrahedron*, **23**, 2893 (1967).
50. J. H. Bowie, R. G. Cooks, G. E. Gream and M. H. Laffer, *Australian J. Chem.*, **21**, 1247 (1968).
51. J. H. Bowie, D. H. Williams, S.-O. Lawesson and G. Schroll, *J. Org. Chem.*, **31**, 1384 (1966).
52. N. Schamp and M. Vandewalle, *Bull Soc. Chim. Belges*, **75**, 539, (1966).
53. For a review see P. Brown and C. Djerassi, *Angew Chem. Int. Ed.*, **6**, 477 (1967).
54. J. H. Bowie, D. H. Williams, P. Madsen, G. Schroll and S.-O. Lawesson, *Tetrahedron*, **23**, 305 (1967).
55. S.-O. Lawesson, J. Ø. Madsen, G. Schroll, J. H. Bowie, R. Grigg and D. H. Williams, *Acta Chem. Scand.*, **20**, 1129 (1966).
56. H. J. Hofman, *Tetrahedron Letters*, **1964**, 2329.
57. H. Audier, J. M. Conia, M. Fétizon and J. Goré, *Bull Soc. Chim. France*, **1967**, 787.
58. J. H. Beynon, R. A. Saunders and A. E. Williams, *Appl. Spectr.*, **14**, 95 (1960).
59. P. Natalis, *Bull. Soc. Chim. Belges.*, **67**, 599 (1958).
60. J. Seibl and T. Gäumann, *Z. Anal. Chem.*, **197**, 33 (1963).
61. J. Seibl and T. Gäumann, *Helv. Chim. Acta*, **46**, 2857 (1963).
62. D. H. Williams, H. Budzikiewicz, Z. Pelah and C. Djerassi, *Monatsh. Chem.*, **95**, 166 (1964).
63. H. Budzikiewicz, C. Djerassi and D. H. Williams, *Interpretation of Mass Spectra of Organic Compounds*, Holden-Day, San Francisco, 1964, p. 25.
64. R. I. Reed in *Mass Spectrometry of Organic Ions* (Ed. F. W. McLafferty), Academic Press, New York, 1963, pp. 666–670.
65. B. Willhalm and A. F. Thomas, *J. Chem. Soc.*, **1965**, 6478.

66. J. H. Bowie and S.-O. Lawesson, unpublished observations.
67. J. H. Bowie, *Australian J. Chem.*, **19**, 1619 (1966).
68. K. Biemann, *Mass Spectrometry: Organic Chemical Applications*, McGraw-Hill, New York, 1962, p. 102.
69. A. F. Thomas, B. Willhalm and J. H. Bowie, *J. Chem. Soc. (B)*, **1967**, 392.
70. H. Fritz, H. Budzikiewicz and C. Djerassi, *Chem. Ber.*, **99**, 35 (1966).
71. N. J. Turro, D. C. Neckers, P. A. Leermakers, D. Seldner and P. D. Angelo, *J. Amer. Chem. Soc.*, **87**, 4097 (1965).
72. T. Goto, A. Tatematsu, Y. Nakajima and H. Tsuyama, *Tetrahedron Letters*, **1965**, 757.
73. M. Vandewalle, N. Schamp and H. De Wilde, *Bull Soc. Chim. Belges*, **76**, 111 (1967).
74. M. Vandewalle, N. Schamp and H. De Wilde, *Bull Soc. Chim. Belges*, **76**, 123 (1967).
75. A. Maquestiau and P. Lejeune, *Bull Soc. Chim. Belges*, **76**, 133 (1967).
76. J. H. Bowie, D. W. Cameron, R. G. F. Giles and D. H. Williams, *J. Chem. Soc. (B)*, **1966**, 335.
77. J. Karliner, H. Budzikiewicz and C. Djerassi, *J. Amer. Chem. Soc.*, **87**, 580 (1965).
78. E. Lund, H. Budzikiewicz, J. M. Wilson and C. Djerassi, *J. Amer. Chem. Soc.*, **85**, 941 (1963).
79. E. Lund, H. Budzikiewicz, J. M. Wilson and C. Djerassi, *J. Amer. Chem. Soc.*, **85**, 1528 (1963).
80. J. H. Beynon, R. A. Saunders and A. E. Williams, *Appl. Spectr.*, **14**, 95 (1960).
81. H. Budzikiewicz, C. Djerassi and D. H. Williams, *Interpretation of Mass Spectra of Organic Compounds*, Holden-Day, San Francisco, 1964, Chapter 8.
82. H. Powell, D. H. Williams, H. Budzikiewicz and C. Djerassi, *J. Amer. Chem. Soc.*, **86**, 2623 (1964).
83. J. E. Gurst, and C. Djerassi, *J. Amer. Chem. Soc.*, **86**, 5542 (1964).
84. R. H. Shapiro, D. H. Williams, H. Budzikiewicz and C. Djerassi, *J. Amer. Chem. Soc.*, **86**, 2837 (1964).
85. R. H. Shapiro and C. Djerassi, *Tetrahedron*, **20**, 1987 (1964).
86. J. Gutzwiller and C. Djerassi, *Helv. Chim. Acta*, **49**, 2108 (1966).
87. C. Djerassi, R. H. Shapiro, and M. Vandewalle, *J. Amer. Chem. Soc.*, **87**, 4892 (1965).
88. R. Beugelmans, R. H. Shapiro, L. J. Durham, D. H. Williams, H. Budzikiewicz and C. Djerassi, *J. Amer. Chem. Soc.*, **86**, 2832 (1964).
89. D. H. Williams, J. M. Wilson, H. Budzikiewicz and C. Djerassi, *J. Amer. Chem. Soc.*, **85**, 2091 (1963).
90. D. H. Williams and C. Djerassi, *Steroids*, **3**, 259 (1964).
91. C. Djerassi, G. von Mutzenbecher, J. Fajkos, D. H. Williams and H. Budzikiewicz, *J. Amer. Chem. Soc.*, **87**, 817 (1965).
92. L. Tökés, R. T. La Londe and C. Djerassi, *J. Org. Chem.*, **32**, 1012 (1967).
93. H. Budzikiewicz and C. Djerassi, *J. Amer. Chem. Soc.*, **84**, 1430 (1962).
94. C. Beard, J. M. Wilson, H. Budzikiewicz and C. Djerassi, *J. Amer. Chem. Soc.*, **86**, 269 (1964).
95. C. Djerassi and L. Tökés, *J. Amer. Chem. Soc.*, **88**, 536 (1966).
96. L. Tökés, R. T. La Londe and C. Djerassi, *J. Org. Chem.*, **32**, 1022 (1967).

97. C. Djerassi, *Pure Appl. Chem.*, **9**, 159 (1964).
98. H. Budzikiewicz, C. Djerassi and D. H. Williams, *Structure Elucidation of Natural Products by Mass Spectrometry*, Vol. 2, Holden-Day, San Francisco, 1964, Chapter 20.
99. L. Tökés, *Ph.D Thesis*, Stanford University 1965. Quoted in reference 92.
100. H. Budzikiewicz, C. Djerassi and D. H. Williams, *Structure Elucidation of Natural Products by Mass Spectrometry*, Vol. 2, Holden-Day, San Francisco, 1964, Chapter 18.
101. G. von Mutzenbecher, Z. Pelah, D. H. Williams, H. Budzikiewicz and C. Djerassi, *Steroids*, **2**, 475 (1963).
102. R. H. Shapiro and C. Djerassi, *J. Amer. Chem. Soc.*, **86**, 2825 (1964).
103. D. H. Williams, J. M. Wilson, H. Budzikiewicz and C. Djerassi, *J. Amer. Chem. Soc.*, **85**, 2091 (1963).
104. H. Powell, D. H. Williams, H. Budzikiewicz and C. Djerassi, *J. Amer. Chem. Soc.*, **86**, 2623 (1964).
105. C. Djerassi, J. M. Wilson, H. Budzikiewicz and J. W. Chamberlin, *J. Amer. Chem. Soc.*, **84**, 4544 (1962).
106. J. A. Zderic, H. Carpio, A. Bowers and C. Djerassi, *Steroids*, **1**, 233 (1963).
107. S. N. Ananchenko, V. N. Leonov, V. I. Zaretskii, N. S. Wulfson and I. V. Torgov, *Tetrahedron*, **20**, 1279 (1964).
108. H. Budzikiewicz, C. Djerassi and D. H. Williams, *Structure Elucidation of Natural Products by Mass Spectrometry*, Vol. 2, Holden-Day, San Francisco, 1964, Chapter 23.
109. J. D. McCollum and S. Meyerson, *J. Amer. Chem. Soc.*, **85**, 1739 (1963).
110. T. Aczel and H. E. Lumpkin, *Anal. Chem.*, **33**, 386 (1961).
111. F. W. McLafferty, *Anal. Chem.*, **31**, 477 (1959); F. W. McLafferty in *Advances in Mass Spectrometry* (Ed. J. D. Waldron), Pergamon, London, 1959, pp. 355–364.
112. M. M. Bursey and F. W. McLafferty, *J. Amer. Chem. Soc.*, **88**, 529 (1966).
113. A. Buchs, G. P. Rossetti and B. P. Susz, *Helv. Chim. Acta*, **47**, 1563 (1964).
114. J. Ronayne D. H. Williams, and J. H. Bowie, *J. Amer. Chem. Soc.*, **88**, 4980 (1966).
115. S. H. H. Chaston, S. E. Livingstone, T. N. Lockyer, V. A. Pickles and J. S. Shannon, *Australian J. Chem.*, **18**, 673 (1965).
116. M. C. Hamming and E. J. Eisenbraum, paper presented to the Fourteenth Annual Conference on Mass Spectrometry, Dallas, Texas, May 1966.
117. J. H. Beynon and A. E. Williams, *Appl. Spectr.*, **14**, 156 (1960).
118. D. Misiti, H. W. Moore and K. Folkers, *J. Amer. Chem. Soc.*, **87**, 1402 (1965).
119. B. C. Das, M. Lounasmaa, C. Tendille and E. Lederer, *Biochem. Biophys. Res. Commun.*, **21**, 318 (1965).
120. W. T. Griffiths, *Biochem. Biophys. Res. Commun.*, **25**, 596 (1966).
121. F. Bohlmann and K. M. Kleine, *Chem. Ber.*, **99**, 885 (1966).
122. H. Morimoto, T. Shima, I. Imada, M. Sasaki and A. Ouchida, *Ann. Chem.*, **702**, 137 (1967).
123. R. F. Muraca, J. S. Whittick, C. Doyle Daves, P. Friis and K. Folkers, *J. Amer. Chem. Soc.*, **89**, 1505 (1967).
124. J. H. Bowie, D. W. Cameron and D. H. Williams, *J. Amer. Chem. Soc.*, **87**, 5094 (1965).

125. S. J. Di Mari, J. H. Supple and H. Rapoport, *J. Amer. Chem. Soc.*, **88**, 1226 (1966).
126. D. Becher, C. Djerassi, R. E. Moore, H. Singh and P. J. Scheuer, *J. Org. Chem.*, **31**, 3650 (1966).
127. J. H. Bowie and D. W. Cameron, *J. Chem. Soc. (B)*, **1966**, 684.
128. E. Ritchie, W. C. Taylor and J. S. Shannon, *Tetrahedron Letters*, **1964**, 1437.
129. J. H. Bowie and A. W. Johnson, *Tetrahedron Letters*, **1967**, 1449.
130. H. Brockmann, Jr., H. Budzikiewicz, C. Djerassi, H. Brockmann and J. Niemeyer, *Chem. Ber.*, **98**, 1260 (1965).
131. H. Brockmann, J. Niemeyer, H. Brockmann, Jr. and H. Budzikiewicz, *Chem. Ber.*, **98**, 3785 (1965).
132. H. Budzikiewicz, C. Djerassi and D. H. Williams, *Structure Elucidation of Natural Products by Mass Spectrometry*, Vol. 2, Holden-Day, San Francisco, 1964, Chapter 28.
133. J. Polonsky, B. C. Johnson, P. Cohen and E. Lederer, *Bull Soc. Chim. France*, **1963**, 1909.
134. J. D. McCollum and S. Meyerson, *J. Amer. Chem. Soc.*, **85**, 1739 (1963).
135. J. M. Wilson, M. Ohashi, H. Budzikiewicz, C. Djerassi, Shô Itô and T. Nozoe, *Tetrahedron*, **19**, 2247 (1963).
136. T. Aczel and H. E. Lumpkin, *Anal. Chem.*, **32**, 1819 (1960).
137. J. M. Wilson, M. Ohashi, H. Budzikiewicz, F. Šantavý and C. Djerassi, *Tetrahedron*, **19**, 2225 (1963).
138. I. R. C. Bick, J. H. Bowie and G. K. Douglas, *Australian J. Chem.*, **20**, 1403 (1967).
139. D. R. Hoffman, *J. Org. Chem.*, **31**, 792 (1966).
140. K. Biemann, P. Bommer, D. M. Desiderio and W. J. McMurray in *Advances in Mass Spectrometry* (Ed. W. L. Mead), The Institute of Petroleum, London 1966, Vol. 3, pp. 639–653 and references therein.
141. P. Natalis and J. L. Franklin, *J. Phys. Chem.*, **69**, 2943 (1965).
142. J. H. Bowie, R. G. Cooks, S.-O. Lawesson and C. Nolde, *J. Chem. Soc. (B)* **1967**, 616.
143. S.-O. Lawesson, G. Schroll, J. H. Bowie, and R. G. Cooks, *Tetrahedron*, **24**, 1875 (1968).
144. J. H. Bowie, R. Grigg, D. H. Williams, S.-O. Lawesson and G. Schroll, *Chem. Commun.*, **1965**, 403.
145. S.-O. Lawesson and D. H. Williams, unpublished observations.
146. J. H. Bowie, S.-O. Lawesson, G. Schroll and D. H. Williams, *J. Amer. Chem. Soc.*, **87**, 5742 (1965).
147. J. H. Bowie, R. G. Cooks, S.-O. Lawesson, P. Jakobsen, and G. Schroll, *Chem. Commun.*, **1966**, 539.
148. J. H. Bowie, R. G. Cooks, P. Jakobsen, S.-O. Lawesson, and G. Schroll, *Australian J. Chem.*, **20**, 689 (1967).
149. R. I. Reed and V. V. Takhistov, *Tetrahedron*, **23**, 2807 (1967).
150. F. W. McLafferty and M. M. Bursey, *Chem. Commun.*, **1967**, 533.
151. M. M. Green, D. S. Weinberg and C. Djerassi, *J. Amer. Chem. Soc.*, **88**, 3883 (1966).
152. F. Komitsky Jr., J. E. Gurst and C. Djerassi, *J. Amer. Chem. Soc.*, **87**, 1398 (1965).
153. C. Djerassi, A. M. Duffield, F. Komitsky, Jr., and L. Tökés, *J. Amer. Chem. Soc.*, **88**, 860 (1966).

CHAPTER **6**

Radiation chemistry of ketones and aldehydes

GORDON R. FREEMAN

University of Alberta, Edmonton, Alberta, Canada

I. INTRODUCTION

Absorption of high energy radiation creates ionization and excitation in the absorbing material and the ionized and excited species may undergo chemical changes. In many systems, ionization and excitation can be transferred from one molecule to another, so it is not necessarily the molecule that initially absorbs the energy that ultimately undergoes chemical change. High energy photons (x- and γ-rays) and high energy particles (α- and β-rays, accelerated electrons, accelerated positive ions, neutrons) are used as energy sources. The most common radiations in current use are γ-rays from ^{60}Co and electrons from linear accelerators or Van de Graaff accelerators.

Photochemistry is related to radiation chemistry. Studies in photochemistry have traditionally involved the use of photons that have insufficient energy to produce ionization in most systems. This limitation was imposed mainly for technical reasons. The ionization potentials of many compounds are in the vicinity of 10 electron volts (ev). Energy sources that would emit photons of this energy and reaction vessels that would permit their entrance were not readily available. In recent years these difficulties have been partly overcome and the borderline between radiation chemistry and photochemistry is ill-defined. For example, a hydrogen lamp with a lithium fluoride window emits photons of energies up to 11·8 ev[1]. Many interesting experiments will be possible when photons in the region 20–200 ev can be used.

The photochemistry of ketones and aldehydes, using photons with energies less than 6 ev, was described in an earlier volume of this series[2]. Although photon and electron energies of the order of one million electron volts (1 Mev) were used in most of the radiolysis work reported in the present chapter, it will be seen that there are similarities between part of the radiolysis results and the photolysis results. Photo-excitation leads to a small number of states. Radiation-excitation leads to the initial formation of a relatively large number of states. However, the relative simplicity of the final radiolysis products indicates that either many states react in the same way or that many states settle into a small number of states before chemical reaction occurs.

The radiolysis of ketones and aldehydes has thus far not received as much attention as has that of water, hydrocarbons or alcohols.

II. GENERAL ASPECTS OF RADIATION CHEMISTRY

A. Energy Absorption Processes

Photons and particles with energies much greater than the ionization potential of the absorbing medium will be the main radiations considered in this section.

1. X- and γ-Rays

The number of photons dI adsorbed in a slice of material of thickness dx is proportional to the intensity of the impinging beam, I, and to the thickness of the slice, dx. Hence

$$dI = -\mu I dx \tag{1}$$

where μ is the linear absorption coefficient.

Three photon absorption processes are significant in radiolysis studies[3]: the photoelectric effect, the Compton effect and pair production. The linear absorption coefficient can be expressed as

$$\mu = \tau + \sigma + \pi \tag{2}$$

where τ is the photoelectric absorption coefficient, σ is the Compton absorption coefficient and π is the pair production absorption coefficient.

In the photoelectric effect the photon is completely absorbed by an atom or molecule and an extranuclear electron is ejected (Figure 1). The kinetic energy T of the ejected electron is given by

$$T = h\nu - \omega \tag{3}$$

12*

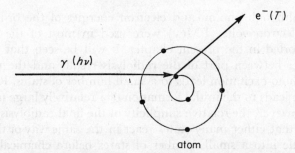

atom

FIGURE 1. Photoelectric effect. $\gamma(h\nu)$ is the incident photon of energy $h\nu$ and $e^-(T)$ is the electron ejected with kinetic energy T.

where $h\nu$ is the incident photon energy and ω is the energy with which the ejected electron was bound in the undisturbed atom or molecule. The photoelectric effect is partly a resonance process, so the electron ejected is usually one from the electron shell that has the binding energy nearest to but below the energy of the photon. Thus, when $h\nu$ is greater than ω of the K electrons in carbon it is usually a K electron that is ejected from the carbon atom. The photoelectric absorption coefficient per atom of absorber, $_a\tau$, increases greatly with increasing atomic number Z of the absorber and decreases rapidly with increasing photon energy:

$$_a\tau \propto Z^m/(h\nu)^n \tag{4}$$

where $m = 4$ to 5 and n decreases from 3.5 to 1 with increasing photon energy. In matter composed of light elements, for example in ketones and aldehydes, τ is negligible for photon energies greater than 0.1 Mev; at energies less than 0.01 Mev, σ and π are negligible compared to τ [3].

The Compton effect corresponds essentially to a 'collision' between the photon and an electron. The photon loses part of its energy to the electron and is deflected from its original direction of propagation (Figure 2). The kinetic energy of the ejected electron is given by

$$T = h\nu - h\nu' - \omega \tag{5}$$

where $h\nu'$ is the energy of the deflected photon. The Compton absorption coefficient per atom of absorber, $_a\sigma$, depends only on the number of electrons in the atom [3]:

$$_a\sigma \propto Z \tag{6}$$

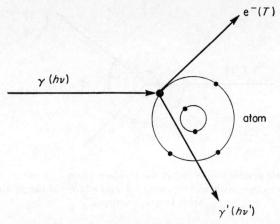

FIGURE 2. Compton effect. $\gamma(h\nu)$ is the incident photon of energy $h\nu$, $\gamma'(h\nu')$ is the deflected photon of energy $h\nu'$ and $e^-(T)$ is the electron ejected with kinetic energy T.

In radiation chemistry, only the energy actually absorbed by the system contributes to the reaction. The energy of the scattered photon, $h\nu'$, is usually not absorbed in the small systems used in radiolysis studies. The total Compton coefficient is therefore divided into two parts, namely, the 'actual' absorption coefficient $_a\sigma_a$ and the 'scatter' coefficient $_a\sigma_s$.

$$_a\sigma = {_a\sigma_a} + {_a\sigma_s} \tag{7}$$

Thus the fraction of the 'total energy absorbed' from the photon beam that is 'actually absorbed' by the system is given by $_a\sigma_a/(_a\sigma_a + _a\sigma_s)$. The Compton effect is important for photon energies in the range from 0·01 to 100 Mev and $_a\sigma_a$ goes through a maximum at about 1 Mev [3]. The γ-rays emitted by ^{60}Co sources are comprised of equal numbers of photons of energies 1·17 and 1·33 Mev. At this energy the photoelectric and pair production processes are negligible by comparison with the Compton process.

Pair production occurs when a photon interacts with the field of a nucleus to produce an electron–positron pair (Figure 3). The sum of the kinetic energies of the electron and positron, T and T', respectively, is given by

$$T + T' = h\nu - 1·02 \text{ Mev} - (\text{nuclear recoil energy}) \tag{8}$$

where 1·02 Mev is the amount of energy that was converted into the 'rest mass' of the electron–positron pair. The recoil energy of the

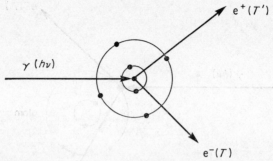

FIGURE 3. Pair production. $\gamma(h\nu)$ is the incident photon of energy $h\nu$, $e^-(T)$ is the product electron with kinetic energy T and $e^+(T')$ is the product positron with kinetic energy T'.

target nucleus is usually negligible by comparison with T or T'. The pair production absorption coefficient per atom of absorber increases with the square of the charge on the absorber nuclei[3]:

$$_a\pi \propto Z^2 \qquad (9)$$

Since two particles with a total rest mass equivalent of 1·02 Mev are created, the process has a threshold energy of 1·02 Mev. In matter composed of light elements, such as ketones and aldehydes, the pair production process is negligible for photon energies below 3 Mev but becomes increasingly important above that value[3].

The net effect of the absorption of a single high energy photon by any of the above processes is that a single molecule is disrupted and a single high kinetic energy electron (or electron–positron pair) is set in motion. Most of the photon energy is transformed into kinetic energy of an electron and is not retained as excitation energy in the target molecule. Overlooking this fact causes uninitiated persons to comment: 'It only takes four or five electron volts to break a bond, so when you hit a molecule with a million electron volts what can you expect besides a mess!'

The high energy electron set in motion by the photon travels through the absorber and loses little bits of its energy to a large number of molecules along its path. For example, a 1·0 Mev photon ionizes a single molecule by the Compton process and gives, on the average, 0·5 Mev to the ejected electron, the other 0·5 Mev being carried off by the scattered photon. The 0·5 Mev electron moves through the absorber and causes the ultimate ionization or excitation of more than 30,000 other molecules before being stopped. The single molecule ionized by the γ-ray interaction is statistically negligible when grouped with those ionized or excited by the high energy electron. Thus it is not surprising that the results of x- or γ-radiolysis are the same as those of high energy electron radiolysis when other reaction conditions, such as the rate of energy absorption in the sample, are the same.

2. High energy electrons

Energetic electrons suffer elastic and inelastic collisions with the electrons and nuclei in molecules. A collision is elastic when the total kinetic energy of the collision partners is the same before and after the collision, otherwise it is inelastic.

Inelastic collisions with bound molecular electrons are usually the predominant mechanism by which a high energy electron loses energy in an absorber[4]. Some of the kinetic energy of the electron is transformed into internal energy in the molecule. If the amount of energy transferred exceeds the ionization potential of the molecule there is a finite probability that it will ionize: the probability of ionization rapidly approaches unity with increasing energy in excess of the ionization potential[5], and much of the excess energy is carried off by the ejected electron. The rate of energy loss of a high energy electron per unit path length in a medium, by inelastic collisions with bound molecular electrons, $-(\mathrm{d}T/\mathrm{d}x)_{\mathrm{coll}}$, may be estimated by the Bethe equation[3,6]:

$$-\left(\frac{\mathrm{d}T}{\mathrm{d}x}\right)_{\mathrm{coll}} = \frac{2\pi e^4 NZ}{m_0 v^2} \left[\ln \frac{m_0 v^2 T}{2I^2(1-\beta^2)} - (2\sqrt{1-\beta^2} - 1 + \beta^2)\ln 2\right.$$
$$\left. + (1 - \beta^2) + 0.125\,(1 - \sqrt{1-\beta^2})^2\right] \qquad (10)$$

where e is the charge on an electron, N is the number of atoms per cm^3, Z is the number of electrons or the total oscillator strength per atom, m_0 is the rest mass of an electron, v is the velocity of the incident electron, T is the relativistic kinetic energy of the incident electron, $\beta = v/c$, c is the velocity of light and I is the logarithmic mean of the excitation energies E_n weighted according to the corresponding oscillator strengths[7],

$$\ln I = \sum_n f_n \ln E_n \qquad (11)$$

where f_n is the fraction of the total optical dipole oscillator strength of the atom for excitation to state n. The Bethe equation (10) involves the assumption that the velocity of the incident electron greatly exceeds that of the atomic orbital electrons (Born approximation), so it applies only when $T \gg I$*. For electrons with $T > 1$ Mev passing through a condensed medium, equation (10) requires a negative 'polarization correction' term on the right hand side[3].

Inelastic collisions of high energy electrons with nuclei produce x-rays (Bremsstrahlung). The rapid deceleration of an electron in the field of a nucleus results in the emission of a quantum of radiation. The electron loses an amount of kinetic energy equal to the energy of the emitted photon. This is an inefficient process for electron energies below 1 Mev and is therefore of marginal interest to radiation chemists. It must be considered during the design of radiation shielding for high energy electron sources.

Incident electrons that have energies greater than about 100 ev do not lose a significant fraction of their energy by elastic collisions with the bound molecular electrons in the absorbing medium[4]. However, this process is important at incident electron energies less than 100 ev.

When the amount of energy transferred by a very high energy electron (e.g. 0·1 Mev) to a molecule, during an inelastic collision, very greatly exceeds the ionization potential of the molecule, the most probable reaction is the ejection of a single electron with a high kinetic energy. If the ionization potential of the molecule is negligible by comparison with the kinetic energy of the ejected electron, the inelastic electron–molecule collision may be treated as an elastic electron–electron collision[4]. This type of collision is also known as a 'knock-on' or 'hard' collision. Inelastic collisions in which smaller amounts of energy are transferred to the molecules are known as 'glancing' or 'soft' collisions, and are much more probable than hard collisions.

In an elastic collision of an electron with a nucleus the electron is deflected but it loses a negligible fraction of its energy[4]. The main

---

* At present f_n versus E_n spectra are not well enough known, so values of I are not obtained directly from equation (11). The Bethe equation for energy loss by fast protons in matter is slightly different from equation (10), but it involves $\ln I$ in essentially the same way. Numerical values of I_i for elements of atomic number Z_i are obtained by fitting $(dT/dx)_{coll}$ data for $\sim 10^8$ ev protons in the elemental material. For a compound that contains N_i atoms of atomic number Z_i and mean excitation potential I_i the value of I is obtained from the equation

$$\ln I = \left(\sum_i N_i Z_i \ln I_i \right) \Big/ \sum_i N_i Z_i$$

effect of this process is to increase the tortuosity of the electron path. Elastic scattering of non-relativistic electrons by nuclei, also known as Rutherford scattering, is important in electron diffraction studies used in the determination of molecular structure.

Nuclear scattering predominates over inelastic scattering by electrons by roughly a factor of Z, the atomic number of the scattering atom[4].

Both positive and negative electrons (positrons and negatrons) lose energy by the above interactions. However, a thermalized positron (energy of the order of 0.1 ev) has a relatively short lifetime; it combines with a negatron to form a positronium atom which annihilates after about 10^{-10} s. The mass of the positronium atom is converted into two gamma rays, each having an energy of 0.51 Mev, travelling in opposite directions[4].

3. Low energy electrons

The transfer of energy from high energy electrons to a medium is quite well described by equation (10) when $T \gg I$. The value of I used in the equation is an average excitation potential of the inner shell and outer shell electrons in the atoms of the absorbing medium. For example, the average excitation potential of the K shell electrons in a carbon atom is about 280 ev [8], that of the L shell electrons is about 11 ev [9], and the value of I for carbon is about 80 ev [7]. However, when the incident electrons have an energy less than about 280 ev in carbon they can not excite the K electrons, so only the L electrons are affected. Under these conditions it becomes necessary to redefine Z and I in equation (10) such that Z is the *effective* total oscillator strength of the atom or molecule with which the incident electron can intereact and I is determined by a suitably truncated summation in equation (11), with f_n being the fraction of the *effective* total oscillator strength of the atom or molecule for excitation to state n of energy $E_n < T$. *Effective* oscillator strengths for low energy electron collisions do not necessarily correspond to optical oscillator strengths because optically forbidden transitions might be strongly excited by low energy electrons[10]. Values of f_n should be obtained from electron scattering data, but unfortunately these are scarce[11-14].

The average excitation potential of the K electrons in oxygen is about 520 ev [8] and the value of I for very high energy incident particles in oxygen is about 100 ev [7]. The value of I for very high energy incident particles in ketones or aldehydes is about 60 ev. For electrons with less than roughly 1000 ev energy in aldehydes or ketones, the values of Z and I in equation (10) should gradually decrease with decreasing T. Since better information is lacking, approximate values of the rates of energy loss by electrons of energies down to about 10 ev can probably be obtained from equation (10) by using suitable values of Z and I. These values could be obtained for a given absorbing medium, with the aid of the oscillator strength versus energy (f_n versus E_n) spectrum of the medium, by the way mentioned in the preceding paragraph.

The theory of energy loss by electrons in the subelectronic excitation energy region of the medium is not yet complete enough to be used with confidence in the estimation of the ranges of the electrons: Equation (10) is not even crudely valid and the rate of energy loss is greatly reduced. In the recent development of a model of non-homogeneous kinetics for ionic reactions in the radiolysis of dielectric liquids[15-17] it was considered preferable to estimate the ranges of low energy electrons by an arbitrary method that has no theoretical justification, rather than to try to justify the use of one of the presently inadequate theories.

4. Charged heavy particles

Energetic protons and other heavy particles are less easily deflected and are decelerated less abruptly than are energetic electrons. Thus, elastic scattering and energy loss by radiation emission (Bremsstrahlung) are much less significant for heavy particles than for electrons of the same energy. In the energy ranges usually used in radiolysis experiments, up to about 30 Mev for helium ions and about 10 Mev for protons, the primary mode of energy loss by heavy particles is through inelastic collisions with the bound electrons of the medium[3,4,7]. Bethe[6] developed the following equation for the rate of collisional energy loss by a heavy particle of charge ze and velocity v passing through matter:

$$-\frac{dT}{dx} = \frac{4\pi z^2 e^4}{m_0 v^2} NZ\left[\ln \frac{2m_0 v^2}{I} - \ln (1 - \beta^2) - \beta^2\right] \qquad (12)$$

where e, m_0, N, Z, I and β mean the same as in equation (10). The derivation of equation (12) involved the Born approximation, which breaks down for protons of energies below about 1 Mev in ketones or aldehydes (because of the K electrons in the oxygen atoms). It is customary to correct equation (12) by adding an 'inner shell correction' term ($-C/Z$) into the square bracket. Values of C/Z have been obtained empirically for protons of various energies in a number of elemental targets and from suitable nomograms for other target materials[18].

The inner shell correction term $(-C/Z)$ obtained for protons of a certain velocity cannot be applied to equation (10) for electrons of the same velocity[19] because an electron has much less energy and is much more readily deflected from its path than is a proton of the same velocity[20].

The density effect correction $(-\delta/2)$ that is added into the square bracket of equation (12) when the value of β approaches unity is negligible for protons with $T < 300$ Mev in ketones and aldehydes[21].

A moving positive particle with charge ze has a high probability of capturing an electron when the particle velocity v is in the vicinity of the Bohr orbital velocity of a K electron in the moving particle, i.e. when[22]

$$v \approx ze^2/\hbar = zc/137 \tag{13}$$

For example, a proton p can capture an electron from the medium through which it is passing:

$$p + \text{medium} \longrightarrow H^* + \text{medium}^+ \tag{14}$$

where H* represents an energetic hydrogen atom and the medium is ionized. The rapidly moving hydrogen atom can lose its electron in a subsequent collision:

$$H^* + \text{medium} \longrightarrow p + e^- \text{(medium)} \tag{15}$$

The net result is that the moving particle is alternately a proton and a hydrogen atom until it has been slowed down so much that reaction (15) can no longer occur. The probabilities of occurrence of reactions (14) and (15) are equal when the particle velocities are $e^2/\hbar = 2\cdot2 \times 10^8$ cm/s, or $T = 25$ kev. The ratio of the reaction probabilities σ_{14}/σ_{15} increases with decreasing energy of the moving particle[22]. The theory of energy transfer from fast heavy particles in this energy region is incomplete and apparently very complex.

Equations (10) and (12) indicate that the rate of energy loss by a high energy electron is similar to that by a heavy charged particle that has the same velocity. Furthermore, since a heavy particle has a much lower velocity than does an electron of the same kinetic energy, the former particle has a much greater rate of energy loss than does the latter in the energy region where equations (10) and (12) are valid. Thus the density of excitation and ionization along the track of a proton or α particle is much greater than that along the track of an electron of the same energy. This is the main cause of the differences in product yields obtained when heavy particles rather than γ-rays or electrons are used in some systems. The name commonly given to $(-dT/dx)$ is 'linear energy transfer' or LET and changes in radiolysis product yields from one type of radiation to another are frequently explained as 'LET effects'[23].

5. Neutrons

Neutrons have no electric charge so they do not interact with the electrons of a system. In systems such as aldehydes and ketones, that have only light elements and relatively low neutron capture cross-sections, high energy neutrons lose their energy mainly through elastic collisions with nuclei (Figure 4). The initial kinetic energy of the neutron is T and that of the nucleus is zero. The amount of energy transferred to the atom (nucleus) is given by[3]

$$\Delta T = \frac{4M}{(M + 1)^2} (T \cos^2 \theta) \tag{16}$$

where M is the ratio of the mass of the atom to that of the neutron and θ is the recoil angle of the atom (see Figure 4). The recoil atom is usually ionized and it loses its kinetic energy in the manner described in section II.A.4.

It is evident from equation (16) that ΔT is greatest when $M = 1$, i.e. for collision with a hydrogen nucleus, and gets progressively smaller as M increases. One may thus calculate that 85% of the energy absorbed by acetone from high energy neutrons is transmitted to the

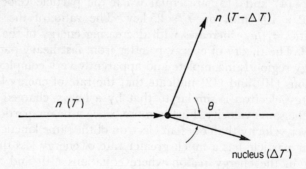

FIGURE 4. Elastic collision of a neutron of energy T with a nucleus, with the transfer of kinetic energy ΔT from the neutron to the nucleus.

medium by recoil protons, 12% is transmitted by recoil carbon ions and 3% by oxygen ions. The results of radiolysis of organic systems by high energy neutrons are therefore expected to be similar to those of proton radiolysis.

High energy charged particles can be generated in a system by the absorption of low energy neutrons by suitable nuclides. For example, if the system contains a boron or a lithium compound, one of the following reactions can occur by the absorption of a thermal neutron [24]:

$$^{10}B + n \longrightarrow {}^{7}Li\ (0\cdot85\ \text{Mev}) + \alpha\ (1\cdot49\ \text{Mev}) + \gamma\ (0\cdot45\ \text{Mev}) \qquad (17)$$

$$^{6}Li + n \longrightarrow {}^{3}H\ (2\cdot73\ \text{Mev}) + \alpha\ (2\cdot05\ \text{Mev}) \qquad (18)$$

The γ-ray produced in reaction (17) would pass out of an ordinary small radiolysis sample without being absorbed. The other product particles of reactions (17) and (18) would be charged and would transfer their energy to the sample in the manner described in section II.A.4.

B. Spacial Distribution of Activated Species

In radiolysis systems the excited and ionized species are not homogeneously distributed in space, but are initially concentrated in the energetic particle 'tracks'. Furthermore, the energy of a high energy particle is transferred in packets of various sizes to the medium. The average amount of energy in a packet is of the order of 10^2 ev, and is sufficient to produce several pairs of ions and several excited molecules. The activated species (ions, electrons, excited molecules and free radicals) stemming from a single packet of energy are formed in close proximity to each other and have a finite probability of reacting among themselves before they can diffuse into the bulk medium. Such a localized grouping of activated species is commonly called a 'spur'. There is a wide distribution of spur populations and population densities; one spur might contain a single ion–electron pair and another might contain several ion–electron pairs plus several excited molecules. Random diffusion forces tend to cause the spur to expand and fade away into the bulk medium surrounding it. The coulombic force between an ion and an electron tends to make them recombine. There is also a finite probability that some of the neutral activated species will encounter and react with other neutral or charged activated species before the spur has faded away.

For a 1 Mev electron in water the average value of $(-\mathrm{d}T/\mathrm{d}x)$ is $0 \cdot 024$ ev/Å [25], and the value is similar for a 1 Mev electron in a liquid ketone or aldehyde. If an average of 10^2 ev is deposited in each spur, the average distance between spurs is several thousand ångstroms. It can be estimated that an average spur in an electron track has an initial diameter of 20–60 Å and that it will be essentially completely dissipated by the time the diameter has increased to 100–300 Å. Thus the electron track is made up of a series of relatively widely separated spurs. A 10 Mev α-particle in water has an average $-\mathrm{d}T/\mathrm{d}x = 9 \cdot 2$ ev/Å [25] and the particle track is a single, long cylindrical spur that has short side branches that are caused by occasional moderately energetic electrons that are set in motion by the harder collisions of the α-particle.

In the gas phase at pressures of one atmosphere and below, the density of the medium is so low that activated species are formed relatively far apart. The concept of spurs has generally been assumed not to apply in gases because it has been assumed that two activated species formed 'near' to each other are sufficiently far apart to cause them to diffuse into the bulk medium rather than react together. The pressures of gases that are required for spur formation with different types of radiation will doubtless be the subject of many future investigations.

Reactive solutes are frequently added to interfere with the 'normal' reactions of the activated species. For example, some of the normal reactions in an alkane during radiolysis might be

$$RH^+ + e^- \longrightarrow RH^* \tag{19}$$

$$RH^* \longrightarrow R^{\cdot} + H \tag{20}$$

$$H + RH \longrightarrow H_2 + R^{\cdot} \tag{21}$$

Addition of nitrous oxide to the alkane causes reaction (22) to compete with reaction (19), which means that reaction (23) partly replaces reactions (20) and (21). Part of the normal product hydrogen is thereby replaced by the new product nitrogen.

$$e^- + N_2O \longrightarrow N_2O^- \tag{22}$$

$$N_2O^- \longrightarrow N_2 + O^- \tag{23}$$

In general, the concentration of activated species in a spur is several orders of magnitude greater than that in the bulk medium. For this reason, much higher concentrations of reactive solutes, commonly called scavengers, are required to interfere with spur reactions than are required to interfere with reactions of activated species in the bulk medium.

Since the activated species are not homogeneously distributed in space, especially in condensed phase systems, non-homogeneous kinetics is required for the quantitative interpretation of radioylsis results in terms of reaction mechanisms. Two models of non-homogeneous kinetics for the radiolysis of liquids have been developed. One model is based upon random diffusion of the activated species and it neglects the coulombic interaction between ions[26-28]. The other model treats the coulombic interaction as the main cause of reaction between activated species in spurs[16,17]. It is significant that the random diffusion model was developed for water, which has a high dielectric constant, and that the coulombic interaction model was developed first for liquid alkanes, which have low dielectric constants. Both models are obviously gross oversimplifications of reality and much more work has to be done to improve them or to develop new ones.

C. General Types of Radiolysis Reactions

The chemically significant activated species initially formed by the absorption of high energy radiation appear to be mainly positive ions, electrons and electronically excited molecules. These can react to form other ions and excited molecules, free radicals and stable product molecules. Ions are ultimately neutralized, excited molecules ultimately decompose or are de-excited and free radicals ultimately react with each other. All of these general types of reactions must ultimately be considered when the overall mechanism of the radiolysis of a system is being devised.

III. REACTIVE INTERMEDIATES

Even in the most extensively studied radiolysis systems, namely, water, alcohols, and hydrocarbons, the exact natures of the reactive intermediates are not known. The radiolysis of ketones has received much less attention and that of aldehydes has thus far been almost ignored. The purpose of the present section is to provide a background against which future studies, as well as the limited ones of the past, may be viewed. The reactive intermediates in radiation chemistry may be classified into positive ions, electrons, negative ions, excited molecules and neutral free radicals.

A. Positive Ions

I. Formation

Primary positive ions are formed by the interactions of energetic charged particles with the molecules in the irradiated medium. When the primary radiation is not charged, e.g. γ-rays or neutrons, the ionizations produced directly by the primary radiation are numerically negligible by comparison with those produced by the recoil electrons or recoil ions (see sections II.A.1 and II.A.5).

It seems likely that some of the positive ions are formed in electronically excited states and that others are formed in the ground electronic state. Some of the ions are vibrationally excited, due to the Franck–Condon effect and to internal conversions from lower vibrational levels of higher electronic states to higher vibrational levels of lower electronic states. The initial ionizing interaction of an energetic particle with a ketone or aldehyde molecule might remove an electron from either an alkyl group or the carbonyl group of the molecule. The ionization potential of an alkyl group is higher than that of a carbonyl group: the ionization potentials of $CH_3\overset{.}{C}O$ and CH_3COCH_3 are 7·9 ev and 9·9 ev, respectively, whereas those of $C_2H_5^{.}$ and C_3H_8 are 8·7 ev and 11·2 ev, respectively[29]. Thus, if an electron is removed from an alkyl group in a ketone or aldehyde the charge will tend to migrate to the carbonyl group. This type of charge migration is included in the meaning of the term internal conversion.

New positive ions are formed by subsequent reactions of other positive ions.

2. Reactions

a. Decomposition and deactivation of excited ions. Ions that have sufficient excitation energy can decompose by fragmentation or rearrangement. The average amount of time required for a decomposition reaction to occur decreases with increasing excitation energy above a certain minimum energy requirement (threshold). If an excited ion undergoes a sufficient number of collisions before decomposition can occur, the excitation energy may be removed from the ion and the decomposition prevented. It is generally noted that molecular fragmentation occurs to a greater extent during radiolysis in gas phase than in condensed phase systems. This is probably because fragmentation of both excited ions and neutral molecules occurs readily in gases at about one atmosphere pressure or below, whereas collisional deactivation and the recombination of radical fragments in cages (see section III.E.2) can occur in liquids and solids.

Presumably the excited ions can undergo much the same types of decomposition and deactivation reactions as can the corresponding excited molecules (see section III.D.2), with the added consideration that the charge centre in the ion might be chemically active.

b. Ion–molecule reactions. In hydrocarbon systems, the reactions of positive ions with neutral molecules can be roughly classified according to the number of electron vacancies in the valence shell of the reactant ion[30]. A hydrocarbon ion that has no electron vacancy in its valence shell can donate a proton to another species that has a higher proton affinity. For example[31],

$$CH_5^+ + C_2H_6 \longrightarrow CH_4 + C_2H_7^+ \qquad (24)$$

A hydrocarbon ion that has one electron vacancy in its valence shell can undergo reactions that have the appearance of those of a mono-radical. For example[32,33],

$$CH_4^{\cdot +} + CH_4 \longrightarrow CH_5^+ + CH_3^{\cdot} \qquad (25)$$

$$CH_4^{\cdot +} + C_2H_4 \longrightarrow C_3H_8^{\cdot +} \qquad (26)$$

However, it is difficult to tell whether reaction (25) is actually a hydrogen atom abstraction or a proton transfer reaction. Reaction (26) is also more complex than a simple free-radical addition to a double bond; once the initial addition complex of the $CH_4^{\cdot +}$ and C_2H_4 has formed, proton migration occurs[33]. A carbonium ion lacks two electrons and has a completely empty orbital in the valence shell, and it tends to abstract H^- from other species[34]:

$$CH_3^+ + C_2H_6 \longrightarrow CH_4 + C_2H_5^+ \qquad (27)$$

Ion–molecule reactions in ketones and aldehydes have so far received little attention. The proton transfer reaction (28) probably

$$CH_5^+ + CH_3CHO \longrightarrow CH_4 + CH_3\overset{+}{C}HOH \qquad (28)$$

can occur because the proton affinity of acetaldehyde (180 kcal/mole[35]) is greater than that of methane (118 kcal/mole[31]). The carbonyl molecular ion is a free radical so hydrogen atom abstraction might occur (reaction 29). However, it is not possible to distinguish

$$(CH_3)_2CO^{\cdot +} + CD_3COCD_3 \longrightarrow (CH_3)_2COD + {}^{\cdot}CD_2COCD_3 \qquad (29)$$

reaction (29) from the proton transfer reaction (30) on the basis of present knowledge. Polarization forces might keep the $(CH_3COCH_3$—

$CD_3COCD_3)\overset{+}{\cdot}$ activated complexes together long enough to allow internal rearrangements to cause products of reaction (29) to be formed from some of the complexes and products of reaction (30) to be formed from others.

$$(CH_3COCH_3)\overset{+}{\cdot} + CD_3COCD_3 \longrightarrow CH_3CO\dot{C}H_2 + (CD_3)_2\overset{+}{C}OH \qquad (30)$$

More speculation about ion molecule reactions in carbonyl compounds would not be worthwhile here. There is obviously a great need for experimental data in this field.

c. Charge exchange. Charge exchange has been observed to occur between simple ions and molecules in mass spectrometers[36]. For example[36],

$$Xe\overset{+}{\cdot}(^2P^o_{\frac{3}{2}}) + C_2H_4 \longrightarrow Xe + (C_2H_4^+\cdot)^* \qquad (31)$$

The recombination energy of the $Xe\overset{+}{\cdot}(^2P^o_{\frac{3}{2}})$ ion is 12·1 ev[36] and the ionization potential of ethylene is 10·5 ev[9], so reaction (31) is 1·6 ev (37 kcal/mole) exothermic. The energy liberated in reaction (31) is probably mainly localized in the polyatomic product, so it is shown to be excited, $(C_2H_4^+\cdot)^*$.

The charge exchange reaction between $Xe\overset{+}{\cdot}(^2P^o_{\frac{3}{2}})$ and acetone is 3·7 ev exothermic. The activation energy for the decomposition of an unexcited acetone ion into an acetyl ion and a methyl radical is probably nearly 3 ev, because the reaction is 2·4 ev endothermic and the activation energy of its reversal can be assumed to be a few tenths of an electron volt. Thus reaction (32) in the gas phase is

$$Xe\overset{+}{\cdot}(^2P^o_{\frac{3}{2}}) + CH_3COCH_3 \longrightarrow Xe + (CH_3COCH_3^+\cdot)^* \qquad (32)$$

$$(CH_3COCH_3^+\cdot)^* \longrightarrow CH_3CO^+ + CH_3^- \qquad (33)$$

probably frequently followed by reaction (33). In the liquid or solid phase, substantial deactivation of the acetone ion might occur before reaction (33) could occur. However, if acetone exchanged charge with an ion that had a larger recombination energy, such as $Ar^+(^2P^o_{\frac{3}{2}})$ (recombination energy 15·8 ev[36]), reaction (33) might occur to an appreciable extent even in the condensed phases.

Charge exchange with atomic ions is used as a method of producing molecular ions of known energies.

Exchange of charge between complex molecules and complex molecular ions is relatively difficult to study because of the probable simultaneous occurrence of ion–molecule reactions. However, information on this topic is needed for the elucidation of the overall mechanisms of the radiolysis of complex molecular systems.

d. Neutralization. Ion neutralization reactions are difficult to study directly in radiolysis systems, so the literature contains little more than reasonable speculation on the subject. The products formed from the neutralization of a given positive ion should depend strongly on the nature of the negative species that takes part in the reaction. Discussion of ion neutralization will be deferred to section III.C.2.e.

B. Electrons

1. Possible states of extra electrons in different media

It is necessary to distinguish between several possible states of electrons that are not bound to individual molecules, namely the free, quasi-free, trapped and solvated states. The free and quasi-free states are non-localized states, whereas the trapped and solvated states are localized states of electrons.

An electron can be truly free only in a vacuum. A free electron in an applied electric field is continuously accelerated as long as it is in the field, so the conventional term 'mobility' has no meaning with regard to free electrons.

If the electron is in a medium that hinders its motion, the medium can remove, by means of collisions, the energy that the electron acquires by acceleration in an electric field. In an electric field of constant strength, the electron acquires a constant 'drift velocity' in the direction of the field. The electron mobility is then defined as the quantity (drift velocity/field strength). This situation applies to quasi-free and solvated electrons.

A quasi-free electron is one that interacts only slightly with the surrounding medium. If an extra electron is in a low pressure gas that does not capture electrons, such as helium or nitrogen, the time-averaged distance between the electron and the molecules of the gas is so large that the time-averaged interaction between them is very small. The electron is deflected and exchanges energy each time it collides with a molecule, but it moves freely between collisions for distances that are long compared to a molecular diameter. The mobilities of electrons in gases such as the rare gases, hydrogen and nitrogen are dependent upon the gas density and the applied electric field strength, but at atmospheric pressure, room temperature and very low field strengths they are 10^3–10^4 cm^2/v s [37,38].

The net interactions of nearly-thermal energy (~ 0.1 ev) electrons with argon, krypton and xenon are so low[39,40] that extra electrons are quasi-free even in the liquids of these substances. The mobilities of extra electrons is 470 cm^2/v s in liquid argon at 90°K, and is 1300 cm^2/v s in liquid krypton at 120°K.

However, the repulsive interaction between electrons and helium atoms is sufficiently great[40] that a thermal electron in liquid helium is localized in a bubble[41], so it becomes a solvated electron. Contrary to a recent prediction[42], it is quite possible that a thermal electron in liquid neon is also localized in a bubble, because the energy depend-ence of the electron scattering cross-section of neon[40] and the possible magnitude of the zero-point energy of the localized electron was neglected.

The quasi-free state of electrons occurs in media where the potential energy barriers to electron translation are not sufficiently large to confine an electron in a small space for a finite time.

In cases where the electron–medium interactions provide suffi-ciently great potential energy barriers to electron translation through the medium, a low energy electron becomes localized in a small space. A localized electron diffuses through the medium at a rate similar to that of a molecular ion. The mobilities of localized electrons[43] and of positive helium ions[44] in liquid helium at 4·2°K are 2·2 × 10^{-2} and 5 × 10^{-2} cm^2/v s, respectively; those of electrons[45] and sodium ions[45] in liquid ammonia at 240°K are 9·1 × 10^{-3} and 1·5 × 10^{-3} cm^2/v s, respectively; those of electrons[46,47] and sodium ions[48] in water at 298°K are, respectively, 1·8 × 10^{-3} and 0·52 × 10^{-3} cm^2/v s. It seems very likely that electrons form localized states also in some liquid hydrocarbons[15,17].

No completely satisfactory model for the structure of solvated electrons has yet been proposed[49], with the exception of the structure of electrons solvated in liquid helium[50]. The continuous dielectric models[49] are consistent only with electron localization in polar liquids. A very crude general model for electron localization in polar and non-polar dielectric liquids has been proposed[30]: present experi-mental results seem to be consistent with the qualitative picture of the solvated electron occupying an interstitial cavity in the liquid, with a fraction of the electron density extending into surrounding cavities. The medium around the cavity is polarized and it is contracted some-what by the long range polarization attractions (electrostriction). The central cavity might be larger than a normal interstitial position in the liquid, due to short range repulsions and the zero-point energy requirement of an electron confined in a small space[30,51].

The trapped state of an electron is not in thermal equilibrium with the surrounding medium: it is a transient state that is transformed into the solvated state over a period approximately equal to the dielectric relaxation time of the medium. The solvated state is in thermal equilibrium with the medium.

2. Reactions

a. Attachment to neutral molecules. Electrons can attach themselves to suitable molecules to form negative ions,

$$e^- + AB \longrightarrow (AB^-)^* \tag{34}$$

If the reactant molecule has a positive electron affinity, the attachment reaction is exothermic and the heat of reaction is contained as excitation energy in the product ion. The excited ion has four possible fates: it might release the electron, reforming the reactant (reaction 35); it might dissociate into two fragments (reaction 36); it might be

$$(AB^-)^* \longrightarrow AB + e^- \tag{35}$$

$$(AB^-)^* \longrightarrow A + B^- \tag{36}$$

$$(AB^-)^* + M \longrightarrow AB^- + M \tag{37}$$

$$(AB^-)^* + C \longrightarrow AB + C^- \tag{38}$$

stabilized by a collision with another molecule (reaction 37); or it might transfer the electron to another molecule during a collision (reaction 38). The excitation energy of the ion is usually vibrational and it can be dissipated amongst the various vibrational modes in the molecule. The larger the number of modes, the less likely is reaction (35) and the more likely is one of reactions (36)–(38) to occur. Reaction (36) is favoured by a large amount of excitation energy per vibrational mode in $(AB^-)^*$, by a weak A—B^- bond and by a low concentration of collision partners M or C in the system (these include AB molecules). When reaction (36) occurs, the overall process of reactions (35) and (36) is known as dissociative attachment. Reaction (37) is favoured by the opposites of the above-mentioned conditions that favour reaction (36). Reaction (38) is a possible method of ion stabilization, although it would be difficult to distinguish experimentally from reaction (37) if C is AB or has a lower electron affinity than AB. In the latter case the electron would probably be transferred back to an AB molecule in a later collision (see section III.C.2.c). If C has a greater electron affinity than AB, reaction (38) could only be distinguished on a reaction time basis from the corresponding reaction of the stabilized ion (see section III.C.2.c),

$$AB^- + C \longrightarrow AB + C^- \tag{39}$$

Negative carbonyl ion formation in the gas phase has not yet received much attention, but the few available results[52,53] are readily explainable in terms of reactions (34)–(37). For electrons with energies in the thermal region, the efficiency of electron attachment, i.e. the electron attachment coefficient, appears to be inversely proportional to the electron speed[52]. The average electron speed varies from one experimental system to another, so the quantity that is most readily compared from one compound to another is the product of the attachment coefficient and the electron velocity, called the attachment frequency ν of the compound[52]. Values of ν were determined for several compounds mixed with diluent gases at total pressures of a few tens of torr[52]. The values of ν depended somewhat on the diluent gas (ethylene, carbon dioxide or methanol). The average values obtained for a few carbonyl compounds are listed in Table 1.

The attachment frequencies can be related to the rate constants of reactions (34)–(37) as follows: in the experimental apparatus[52], the electrons could be captured by molecules as the electrons drifted in an electric field from their source (a photocathode) to their point of measurement (a grid that collected the electrons but not the negative ions). The decay of the electron concentration during the time of passage between the source and the collecting grid can be represented by expression (40), for electrons of a given velocity distribution (e.g. for thermal electrons at a given temperature).

$$-\frac{d[e^-]}{dt} \propto \nu[e^-][AB] \tag{40}$$

A steady state kinetics treatment of reactions (34)–(37) yields equation (41).

$$-\frac{d[e^-]}{dt} = k_{34}\left(1 - \frac{k_{35}}{k_{35} + k_{36} + k_{37}[M]}\right)[e^-][AB] \tag{41}$$

By comparing expressions (40) and (41) for electrons of the same velocity distribution it can be seen that

$$\nu \propto k_{34}\left(1 - \frac{k_{35}}{k_{35} + k_{36} + k_{37}[M]}\right) \tag{42}$$

TABLE 1. Thermal electron attachment frequencies.

Compound	$\nu(s^{-1}\ torr^{-1})^{a,b}$
1. $CH_3COCOCH_3$	1×10^9
2. $CH_3COCOC_2H_5$	2×10^9
3. $CH_3COCH_2COCH_3$	3×10^4 p
4. $CH_3COCH_2CH_2COCH_3$	2×10^3 p
5. $CH_2{=}CHCHO$	8×10^5 p
6. $CH_2{=}CHCOCH_3$	6×10^5 p
7. $(CH_3)_2C{=}CHCOCH_3$	5×10^5 p
8. $C_6H_5COCH_3$	6×10^6 p
9. $C_6H_5CH_2COCH_3$	2×10^5 p
10. CH_3COCH_3	400^c
11. $CHOCHO$	10^{8d}

^{a} L. Bouby, F. Fiquet-Fayard and H. Abgrall, *Compt. Rend.*, **261**, 4059 (1965). Where more than one diluent gas was used, the average value of ν is reported.
^{b} p is the gas pressure in torr.
^{c} This is an upper limit because the observed effect might have been due to impurities.
^{d} R. N. Compton and L. Bouby, *Compt. Rend.*, **264**, 1153 (1967).

Under conditions where $(k_{35} + k_{36}) \ll k_{37}[M]$, expression (42) reduces to

$$\nu \propto k_{34} \qquad (43)$$

and the observed attachment frequency should have a relatively high value and should be independent of the gas pressure. This was observed for the conjugated diketones biacetyl and pentane-2,3-dione (compounds 1 and 2, Table 1).

Under conditions where $k_{35} \gg (k_{36} + k_{37}[M])$ and $k_{36} < k_{37}[M]$, expression (42) reduces to

$$\nu \propto k_{34}k_{37}[M]/k_{35} \qquad (44)$$

and the attachment frequency should have a relatively low value and be proportional to the pressure. This was observed for compounds 3–9 in Table 1.

The attachment frequency of acetone (compound 10, Table 1) was very low and should have been proportional to the pressure, contrary to observation. For this reason it seems very probable that the observed attachment was due to the presence in the system of a trace (< 1 p.p.m.) of an impurity such as biacetyl.

It was pointed out[52] that conjugated double bonds are important in the attachment process.

The value of k_{35} is very small for the biacetyl negative ion. The average lifetime for autoionization of the ion, when formed from thermal electrons, is $1 \cdot 2 \times 10^{-5} \, s^{53}$ (i.e. $k_{35} = 8 \times 10^4 \, s^{-1}$). The average time between collisions for an ion in the systems used to obtain the values in Table 1 was 10^{-8}–10^{-9} s, so $k_{37}[M] \gg k_{35}$ and it is clear why the value of ν for biacetyl was observed to be independent of pressure. It can also be deduced from the results of Compton and coworkers[53] that $k_{36} \ll k_{35}$ in biacetyl. Since the 2,3 bond of biacetyl is relatively weak $(70 \, \text{kcal/mole}^{54})$ by comparison with the bonds in most of the other compounds in Table 1, it follows that the value of k_{36} is probably negligible for all of these compounds when they attach thermal electrons. Radiolysis systems usually have pressures greater than 1 torr and reaction (37) apparently dominates reaction (36) at these concentrations of carbonyl compounds.

Comparison of the electron attachment frequencies of compounds 5 and 6 in Table 1 indicates that the value of ν for acetaldehyde would be approximately as small as that of acetone in the vapour phase. On the other hand, acetaldehyde reacts rapidly with solvated electrons in liquid ethanol[55,56] and acetone reacts rapidly with solvated electrons in liquid water[57]. The difference between the attachment coefficient of the thermal quasi-free electrons in the gas phase and that of the solvated electrons in the liquid solutions probably reflects differences in the rate constants of reactions (34)–(36) as well as the difference in the concentration of M for reaction (37) in the two phases. For example, if the gas phase reaction (45) is more exothermic than is the aqueous phase reaction (46), the reverse of reaction (45) probably has a greater rate constant than has the reverse of reaction (46). This would mean that k_{35} has a greater value in the gas than in the liquid phase.

$$e^- + CH_3COCH_3 \longrightarrow (CH_3COCH_3^-)^* \qquad (45)$$

$$e_{aq}^- + (CH_3COCH_3)_{aq} \longrightarrow (CH_3COCH_3^-)_{aq}^* \qquad (46)$$

Electron attachment probably occurs readily in the pure liquids acetone and acetaldehyde. Attachment of thermal electrons would be extremely rapid in the condensed states of the other compounds listed in Table 1.

Negative ions in liquid and solid ketones and aldehydes have been studied by various spectroscopic techniques[58,60]. The carbonyl anion can be represented as follows:

In aliphatic aldehydes and ketones structure **1** predominates because of the electronegativity of the oxygen atom. In aromatic aldehydes and ketones, especially if the aromatic groups contain electronegative substituents[58] or fused rings, some of the negative charge is delocalized over the aromatic groups.

One of the many questions that remain to be answered is whether, in liquid phase carbonyl compounds, electrons have time to reach the solvated state before they are captured by the carbonyl groups. The electrons might be captured when they have been greatly slowed down but before they have become trapped. On the other hand, they might be captured after being trapped but before they have settled completely into the solvated state. If the capture process is relatively slow, the electron becomes fully solvated before being captured. For example, if the most stable geometry of the negative ion is significantly different from the most stable geometry of the uncharged carbonyl compound, the electron might not be captured until it meets a molecule with a favourable geometry. This might require sufficient time that the electron becomes trapped, or even solvated, before the capture process occurs.

The time required for electron solvation in many simple organic liquids that have viscosities of 1–10 millipoise should be 10^{-12}–10^{-11} s.

b. Neutralization. It is probable that most electrons generated in most pure dielectric liquids have sufficient time to become solvated before being neutralized[15]. Whether or not most of the electrons generated in the radiolysis of carbonyl compounds are captured by the carbonyl groups before being neutralized is not known. The discussion at the end of the previous section is relevant to this question.

For the moment we will assume that nearly all of the electrons generated in ketones and aldehydes are captured before being neutralized, and discussion of the neutralization reaction will be deferred to section III.C.2.e.

C. Negative Ions

I. Formation

Essentially all of the initial negative ions formed in irradiated systems apparently result from the attachment of electrons to molecules (see section III.B.2.a). There is no evidence that the heterolytic dissociation reaction (47) occurs in any radiolysis system studied:

$$AB^* \longrightarrow A^+ + B^- \tag{47}$$

where AB* is an excited molecule. In discussions of negative ion formation one frequently encounters reaction (48)[61,62]. However,

$$e^- + AB \longrightarrow e^- + A^+ + B^- \tag{48}$$

reaction (48) is really an overall process composed of the elementary steps (47) and (49), and it has a very low probability of occurrence in most systems.

$$e^- + AB \longrightarrow e^- + AB^* \tag{49}$$

Other negative ions may be formed by the reactions of the initial negative ions. It should be stressed that dissociative electron attachment is not a single step process, but is merely reaction (34) followed by reaction (36).

2. Reactions

a. Autoionization, decomposition and deactivation of excited ions. See section III.B.2.a.

b. Thermal decomposition. It is interesting to note that the rate constant for the decomposition of the N_2O^- ion to $N_2 + O^-$ at $110°$ ($k = 3 \times 10^3 \, s^{-1}$ for thermal activation[63]) is greater than that for the decomposition of the neutral N_2O molecule to $N_2 + O$ at the same temperature[64] by a factor of 10^{27}. A factor of about 10^4 of this enormous difference is probably explained by the fact that a forbidden singlet $\rightarrow$ triplet transition is involved in the decomposition of molecular N_2O[65], whereas no spin change is involved in the decomposition of the ion (doublet $\rightarrow$ doublet). The electron affinity of the $O(^3P)$ atom is probably significantly greater than that of the N_2O molecule, so the endothermicity and hence the activation energy of the ion decomposition is probably lower than that of the molecule, which would explain much of the rest of the difference in the decomposition rates.

Similarly, the decomposition of a ketone or aldehyde ion [e.g. reaction (50)] should have a lower activation energy than has the

$$CH_3COCH_3^- \longrightarrow CH_3CO^- + CH_3^· \tag{50}$$

decomposition of the corresponding neutral molecule [e.g. reaction (51)] because the electron affinity of the radical should be greater than

$$CH_3COCH_3 \longrightarrow CH_3CO^· + CH_3^· \tag{51}$$

that of the molecule. Thus the rate constants for the decomposition of carbonyl negative molecular ions should be greater than those for the decomposition of the corresponding neutral molecules under the same conditions of temperature and pressure.

c. Electron transfer (charge exchange). An electron can be transferred from a negative ion to a suitable molecule during a collision[66]:

$$AB^- + C \longrightarrow AB + C^- \qquad (52)$$

For reaction (52) to occur at an appreciable rate at temperatures in the vicinity of $300°K$, C must have a greater electron affinity than has AB. Furthermore, AB^- must be sufficiently long lived to be able to encounter a C and react with it.

There is a great need for accurate values of the electron affinities of many types of compounds and radicals. Apart from the crude value ($\gtrsim 25$ kcal/mole[53]) for biacetyl, no values of electron affinities for carbonyl compounds and radicals appear to be available.

Solvation effects doubtless influence the values of electron affinities, so values are required for compounds in solvents of various polarities as well as in the gas phase.

d. Brønsted base reaction. Carbonyl negative ions are Brønsted bases and can accept protons from Brønsted acids. For example, reaction

$$(CH_3)_2CO^- + ROH \longrightarrow (CH_3)_2\dot{C}OH + RO^- \qquad (53)$$

(53) occurs during the radiolysis of solid phase solutions of acetone

$$CH_3CHO^- + C_2H_5OH_2^+ \longrightarrow CH_3\dot{C}HOH + C_2H_5OH \qquad (54)$$

in alcohols at $-196°$[60]. Reaction (54) occurs during the radiolysis of liquid ethanolic solutions of acetaldehyde[55,56].

The ion $(CH_3)_2\overset{+}{C}OH$ is probably formed during the radiolysis of pure acetone (section III.A.2.b), so one of the neutralization reactions will be

$$(CH_3)_2CO^- + (CH_3)_2\overset{+}{C}OH \longrightarrow (CH_3)_2CO^* + (CH_3)_2\dot{C}OH^\ddagger \qquad (55)$$

where one or both of the neutral products are formed in excited states $(*,\ddagger)$.

e. Neutralization. Charge neutralization commonly occurs by the transfer of either an electron or a proton. Neutralization reactions that involve solvated electrons obviously occur by electron transfer. The above ion–ion neutralization reaction (55) probably also occurs mainly by electron transfer, although proton transfer might make a contribution. Reaction (54) occurs by proton transfer.

The products of a neutralization reaction depend on the natures of the reactant ions. Positive ions that can act as Brønsted acids are associated, or 'clustered', with several of their parent molecules in the gas phase at pressures in the vicinity of and greater than 1 torr[67]. In acetone at 0·4 torr, 50% of the positive ions observed with a mass spectrometer were $H^+(CH_3COCH_3)_2$[67]. Clustering also occurs about positive ions in acetaldehyde vapour[67]. Negative ions have also been observed in acetone vapour by high pressure mass spectrometry[62], but the vast majority of the ions observed at ion-chamber pressures greater than 0·01 torr might have been due to impurities. The main negative ion that was clearly formed from acetone was $CH_3COCH_2^-$ and its concentration decreased from 39% to 3% of the total observed negative ion concentration as the acetone pressure increased from 0·001 to 0·05 torr. It is therefore not yet worthwhile to speculate about the products of neutralization reactions in the vapour phase radiolysis of simple carbonyl compounds.

The clustering of molecules about ions in the gas phase corresponds to partial solvation, but most of the ions that undergo neutralization in the liquid state are completely solvated. The reaction of a given pair of ions is less exothermic the more highly they are solvated. This may be one of the reasons that the yields of fragmentation products are much greater in gas- than in liquid-phase radiolysis.

In the liquid phase, the ions are initially scattered along the high energy particle tracks. They are not homogeneously distributed in the system, so their neutralization reactions have to be described by non-homogeneous kinetics. The kinetically significant electrons are the secondary electrons set in motion by the high energy primary charged particles (see sections II.A.1 and II.B). The formation and neutralization of ions during the radiolysis of pure dielectric liquids can be represented by the following mechanism[15]. The square brackets around the reactants and products indicate that the entities are inside a spur. The wiggly arrow in reaction (56) represents absorption of energy from the radiation.

$$A \rightsquigarrow [A^+ + e^-] \qquad (56)$$
$$[e^- + mA] \longrightarrow [N^-] \qquad (57)$$
$$[A^+ + A] \longrightarrow [M^+] \qquad (58)$$
$$[M^+ + N^-] \begin{cases} \longrightarrow \text{[geminate neutralization]} & (59) \\ \searrow M^+ + N^- \text{ (free ions)} & (60) \end{cases}$$
$$M^+ + N^- \longrightarrow \text{random neutralization} \qquad (61)$$

Most of the secondary electrons lose their energy so rapidly that they form a localized state while still in the vicinity of their parent ions. Electron localization is represented in reaction (57) and the precise nature of N^- probably goes through several changes. Whether N^- is ultimately a solvated electron or a molecular negative ion depends on the nature of A. In some systems reaction (58) might not occur, in which case M^+ would be A^+. In pure acetone N^- might be $(CH_3)_2CO^-$ and M^+ might be $(CH_3)_2COH^+$, neglecting the solvating molecules associated with these ions.

Reaction (59), also referred to as geminate recombination or initial neutralization, is driven by the Coulombic energy of attraction between the ions. Reaction (60) is simply a diffusion process and is driven by thermal energy (kT). The competition between reactions (59) and (60) at a given temperature depends mainly on the strength of the Coulombic attraction between the pair of ions, which depends on their separation distance and on the dielectric constant of the intervening medium[15], but does not depend on the identities of M^+ and N^-. For this reason the yield of free ions in a given liquid cannot be increased by adding a reactive solute to it, as has sometimes been supposed (unless, of course, the solute increases the dielectric constant of the liquid, which is sometimes the case[17]). The result of this competition is illustrated by the fraction of the total ions generated by reaction (56) that become free ions. Calculated values of this fraction for the γ-radiolysis of several liquids are listed in Table 2.

The free ions diffuse at random and, in the absence of other ions, ultimately meet free ions from other spurs and react with them [reaction (61)].

TABLE 2.   Yields of free ions in the γ-radiolysis of liquids, expressed as fractions of the total ions.

Liquid	Dielectric Constant[a] (20°)	Density[b] (20°, g/cm³)	Free Ions[c] / Total Ions
Cyclohexane	2.0	0.78	0.027
Diethyl ether	4·3	0·71	0·054
Cyclopentanone	13·5	0·95	0·15
Acetaldehyde	~15–20	0·78	0·2–0·3
Acetone	21·4	0·79	0·33
Ethanol	25	0·79	0·37
Methanol	33	0·79	0·48
Water	80	1·00	0·68

[a] Landolt-Börnstein, *Zahlenwerte und Funktionen aus Physik. Chemie. Astronomie. Geophysik. Technik*, Vol. 2, Part 6, Springer, Berlin, 1959, pp. 633–638.
[b] *Handbook of Chemistry and Physics*, 33rd Ed., Chemical Rubber Publishing, Cleveland, 1951.
[c] Calculated by the method of G. R. Freeman and J. M. Fayadh, *J. Chem. Phys.*, **43**, 86 (1965). Recent results indicate that G(total pairs of ions) ≈ 4 in these liquids.

Liquid ketones and aldehydes have moderate values of dielectric constants, so less than half of the ions initially formed during their γ-radiolysis become free ions (Table 2). In a liquid irradiated by a high LET radiation (heavy particles) the fractional yield of free ions is lower than that in the liquid irradiated by a low LET radiation (x- and γ-rays and high energy electrons). This is because the ions are generated closer together by a high LET radiation, so each electron is attracted back towards the region of its origin by the combined fields of several positive ions.

3. Lifetimes

To the best of present knowledge ions generated in the radiolysis of gases at or below atmospheric pressure react essentially according to homogeneous kinetics. The average lifetime τ of ions with respect to neutralization in the absence of an applied electric field can be estimated from the rate of production of ions and the value of the neutralization rate constant ($k \approx 10^{-6}$ cm³/ion s in the gas phase[68]). The rate of production I of ions is given by the quotient of the absorbed dose rate R and the amount of energy W required to form an ion pair:

$$I = R/W \text{ (pairs of ions/cm}^3\text{ s)} \qquad (62)$$

A steady state homogeneous kinetics treatment of the formation and neutralization of ions gives equation (63) for their average lifetime

$$\tau = (kI)^{-\frac{1}{2}} \text{ (s)} \qquad (63)$$

and a typical value of τ is 1×10^{-3} s in a gas at atmospheric pressure with $R = 3 \times 10^{13}$ ev/cm³ s and $W = 25$ ev.

In liquid systems the free ions are neutralized according to homogeneous kinetics and their average lifetimes are also described by equation (63). It may be shown that, to within a factor of two or three, the homogeneous neutralization rate constant for radiolytic ions in dielectric liquids is given by[69]

$$k \approx 2 \times 10^{10}/\epsilon\eta \qquad \text{(1/mole s)} \qquad (64)$$

where ϵ is the dielectric constant of the liquid and η is its viscosity in poise. In liquids that have viscosities in the vicinity of one centipoise (e.g. acetone, acetaldehyde, cyclohexane) the average lifetime of the free ions at a typical γ-ray absorbed dose rate of 10^{16} ev/cm³ s is 10^{-3} s. On the other hand, the ions that undergo geminate neutralization in each of these liquids have a continuous spectrum of lifetimes between about 10^{-7} and 10^{-11} s[70]. The lifetime distributions

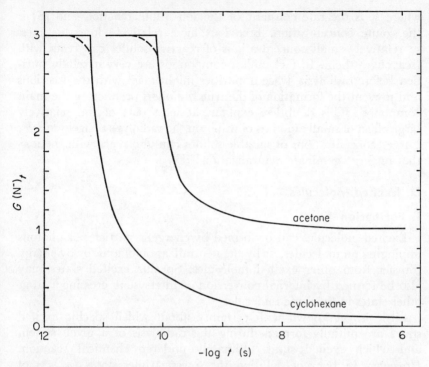

FIGURE 5. Lifetimes of ions in the γ radiolysis of liquid acetone and cyclohexane. $G(N^-)_t$ is the yield of N^- per 100 ev that would be observed after a time t, assuming that 3·0 pairs of ions were formed per 100 ev absorbed in an instantaneous pulse of radiation. The corners at the tops of the curves were rounded intuitively because the model [G. R. Freeman, *J. Chem. Phys.*, **46**, 2822 (1967)] is weak in this region. If $G(N^-)_0 = 4$ instead of 3, the G values in Figure 5 should be multiplied by $\frac{4}{3}$.

are independent of the dose rate and are skewed towards the low lifetimes, with a peak in the vicinity of $10^{-11\pm1}$ s. The relative numbers of ions that have lifetimes greater than or equal to t in acetone and in cyclohexane, for t values between 10^{-12} and 10^{-6} s are shown in Figure 5. The model that was used in the calculations of these spectra is very crude[17], but it fits a number of aspects of the radiolysis of liquids moderately well.

A reactive solute that is in a liquid under irradiation must react with an ion before it is neutralized, or not at all. The average amount of time t_s required for the solute reaction is given by equation (65).

$$t_s = (k_s[S])^{-1} \tag{65}$$

where k_s is the rate constant of the ion–solute reaction and [S] is
the solute concentration. Free ions have relatively long lifetimes,
so relatively small concentrations of reactive solutes can react with
(scavenge) them. In fact, unless compounds are very carefully purified before radiolysis, trace impurities might react with the free ions
and prevent the formation of the true radiolysis products of the main
compound. This doubtless explains at least part of the relatively
large effect of small amounts of water on the radiolysis of acetone[71,72].
Larger concentrations of suitable solutes can also react with the ions
that undergo geminate neutralization[16,17].

D. Excited Molecules

I. Formation

Excited molecules can be formed by energetic particles or photons
impinging on molecules, or by the neutralization of ions, or by energy
transfer from other excited molecules. Specific excited states may
also be formed by internal conversion or intersystem crossing[73] from
other states in the same molecule.

Little is yet known with certainty about which specific excited
states are initially formed during the radiolysis of a given system
and which excited states ultimately undergo chemical reaction.
However, in benzene solution the lowest triplet states (n, π^*) of
cyclohexanone[74] and cyclopentanone[75], which isomerize to open
chain olefinic aldehydes, are formed mainly by energy transfer from
excited benzene molecules, which are probably in their lowest
triplet state.

2. Reactions

Energetic molecules can dissipate their excess energy by chemical
reaction, photon emission or energy transfer. Chemical reactions
of excited ketone and aldehyde molecules were extensively discussed
in the chapter on photochemistry[2]. Photon emission and energy
transfer are described in the recent book by Calvert and Pitts[76].
Various aspects of energy transfer are discussed in detail in such books
as that of Birks[77] and the published reports of various conferences[78-80].
The present discussion will be limited to a brief outline of the general
types of chemical and physical reactions.

The main mode of initial excitation during radiolysis is electronic,
combined with a smaller amount of vibrational excitation due to
the Franck–Condon effect. The electronic energy can in many

instances be transformed into vibrational excitation by internal conversion and intersystem crossing.

a. Chemical reactions These are of five main types: dissociation into free radicals, dissociation into molecules, isomerization, abstraction of a hydrogen atom and addition to a double bond.

The n,π^* excited carbonyl compound that dissociates into a pair of free radicals can be in either a singlet or triplet state. The dissociation has several possible reaction paths, some of which are: (a) predissociation; (b) transition to a high vibrational level of the ground electronic state of the carbonyl group, followed by the rapid surge of part of the vibrational energy into adjacent bonds which breaks an α–C—C bond with a jolt; (c) transition to a high vibrational level of the ground electronic state of the carbonyl group followed by more or less equipartition of the energy in the molecule and the ultimate breakage of a bond; (d) in some instances the Franck–Condon vibrational excitation energy might be sufficiently large for dissociation to occur in the electronically excited state. Although all four of these processes, and even others, might occur, for the commonly occurring dissociation of simple ketones from their lowest triplet state, processes (a) and (b) seem the more likely. Process (a) seems at first sight to be more reasonable than process (b), but the group that contains the electronic energy (C=O) is not the group that dissociates. For example,

$$(CH_3)_2CO^* \longrightarrow CH_3CO^{\cdot} + CH_3^{\cdot} \qquad (66)$$

An interesting correlation of the photo-decomposition of formaldehyde with the geometry and bonding properties of several of its possible excited states has recently been presented[81]. It was suggested that the initially formed singlet n,π^* state undergoes radiationless transitions, facilitated by its non-planar geometry, to single electron population of singlet antibonding states of the CH_2 group ($\sigma^*_{CH_2}$ and $\sigma^*_{CH_2^*}$). The $\sigma^*_{CH_2}$ state has partial bonding between the two hydrogen atoms and decomposes to $H_2 + CO$; the $\sigma^*_{CH_2^*}$ state is H—H antibonding and decomposes to $H + H\dot{C}O$[81].

Remarks similar to the above apply also to the dissociation of excited molecules into smaller molecules. Several dissociation processes occur during the photolysis of glyoxal with 4358 Å light[82]. It has been suggested that Norrish type II decomposition, to form an olefin and an enol, occurs exclusively through the triplet ketone[83].

Several types of isomerization are possible, depending on the nature of the excited molecule. Cycloalkanones can undergo ring

13*

opening to form open chain olefinic aldehydes (reaction 67)[75,84].

$$\text{(cyclopentanone)}^* \longrightarrow CH_2=CHCH_2CH_2CHO \tag{67}$$

Cyclopropyl alkyl ketones can isomerize to propenyl alkyl ketones (reaction 68)[85].

$$CH_3COCH(\overset{CH_2^*}{\underset{CH_2}{\diagup}}) \longrightarrow CH_3COCH=CHCH_3 \tag{68}$$

Alkyl ketones with γ-hydrogens can form cyclobutanols (reaction 69)[86,87].

$$CH_3COCH_2CH_2CH_3^* \longrightarrow CH_3\underset{\underset{CH_2-CH_2}{|}}{\overset{\overset{OH}{|}}{C}}-CH_2 \tag{69}$$

Olefinic carbonyl compounds can undergo cis–trans isomerization (reaction 70)[88].

$$\textit{trans-} \text{ or } \textit{cis-}CH_3COCH=CHCH^* \longrightarrow \textit{cis-} \text{ or } \textit{trans-}CH_3COCH=CHCH_3 \tag{70}$$

Reactions (67) to (70) can presumably have either singlet or triplet state precursors, but the triplet state is greatly favoured in most instances. Reactions (67) to (69) might also have diradical intermediates other than the simple triplet state molecules, but the controversy over diradical and concerted mechanisms will be discussed in section III.E.3.

Triplet carbonyl groups can abstract hydrogen atoms from suitable donors[89,90] and they have reactivities similar to that of t-butoxy radicals[90].

$$R_2CO^* + R'H \longrightarrow R_2\dot{C}OH + R'^. \tag{71}$$

Norrish type II decompositions and reaction (69) involve the intramolecular abstraction of a hydrogen atom from the γ-carbon by the triplet state carbonyl group. Reaction (71) has apparently not been observed when R'H is an R_2CO molecule. When an excited carbonyl molecule meets a ground state carbonyl molecule, perhaps energy transfer occurs instead of reaction(71).

The cycloaddition of triplet carbonyls to olefins, forming oxetanes, can occur when the triplet energy level of the olefin is higher than that of the carbonyl (reaction 72)[91]. When the triplet energy of the

$$R_2CO^* + R_2'C{=}CR_2' \longrightarrow \begin{bmatrix} R_2'C{-}CR_2' \\ \vdots \\ R_2C{-}O \end{bmatrix} \longrightarrow \begin{matrix} R_2'C{-}CR_2' \\ | \quad | \\ R_2C{-}O \end{matrix} \qquad (72)$$

$(n,\pi^* \text{ triplet})$

olefin is less than that of the carbonyl compound, triplet–triplet energy transfer may occur to the exclusion of oxetane formation[91].

b. Photon emission. This physical reaction only occurs in excited states that are sufficiently long lived for the emission process to occur. The amount of time required for emission is greater than about 10^{-9} s for an allowed electronic transition and greater than about 10^{-7} s for a forbidden transition; the lowest triplet states of molecules that have singlet ground states not uncommonly have emission lifetimes of 10^{-3} s in the fluid phases and even longer in the rigid phases[76]. Luminescence of certain aromatic molecules has been used to study electronic energy transfer processes in multicomponent systems[92–94]. However, most excited species generated during radiolysis decompose or are deactivated before they can luminesce[95,96]. Photon emission is not, therefore, a major process in most radiolysis systems.

The emission of infrared radiation during radiolysis has apparently not yet been studied, but one can say intuitively that the importance of infrared emission is negligible from the point of view of energy loss.

c. Energy transfer. The term 'energy transfer', when used by radiation chemists, usually refers to the transfer of electronic excitation energy from one molecule to another. It applies equally well to the transfer of vibrational, rotational and translational energy, but these have so far received little explicit consideration in radiation chemistry.

Electronic energy transfer is of great importance in radiolysis systems because the molecules that initially absorb energy from the energetic charged particles are frequently not the molecules that undergo chemical reaction. This is especially true in the condensed phases. Energy transfer therefore has a great influence in determining the ultimate radiolysis products.

Energy transfer is difficult to study in a single component system. It is readily observed in many multicomponent systems and, by extrapolation, it seems very likely that it also occurs in many single component systems. The following are a few examples of electronic energy transfer involving carbonyl compounds.

Triplet n,π^* carbonyl compounds readily transfer energy to olefins that have lower energy π,π^* states, and the olefins can then undergo cis–trans isomerization[97]. The energy transfer can also occur if the energy of the π,π^* state of the olefin is higher than that of the n,π^* state of the carbonyl, but the reaction is less efficient[97].

During photolysis of a solution of biacetyl in pentan-2-one with 3130 Å light, much more light is emitted by the biacetyl than can be explained by the direct excitation of biacetyl by the light beam. At the same time, the amount of dissociation of the pentan-2-one into ethylene and acetone is much less than that in pure pentan-2-one similarly photolysed[98]. This probably demonstrates transfer of energy from the lowest triplet state (rather than from an excited singlet state as suggested by the original authors) of pentan-2-one to biacetyl.

Triplet state acetone can transfer energy to aldehydes, thereby inducing their decomposition[99]. These reactions were observed in the gas phase, but they would probably also occur in the liquid because triplet acetone can sensitize the cis–trans isomerization of pent-2-ene in the liquid phase[100]. The radiationless-decay lifetime of triplet acetone in liquid hexane at room temperature is 5×10^{-6} s[101], which is long enough to allow the excited molecule to encounter many other molecules by diffusion in the liquid.

The lowest triplet state of benzene can transfer energy to cyclohexanone[74] and to cyclopentanone[75], thereby inducing their isomerization to open chain olefinic aldehydes.

Three conditions are required for efficient transfer of energy from one molecule to another: (a) the reaction must be thermoneutral or exothermic, i.e. the acceptor molecule must have an energy level at the same energy or below that of the excited level of the donor; (b) spin must be conserved in the transfer reaction; (c) the excited donor must have a long enough lifetime to be able to encounter and react with an acceptor molecule.

Four mechanisms have been proposed to explain energy transfer[77,102] and the mechanism is not the same in all systems. The mechanisms differ in the manner of energy migration from the donor D* to the acceptor A.

(i) D* diffuses molecularly through the system until it meets an A molecule. Energy transfer occurs during a physical, molecular encounter.

(ii) The excitation energy migrates through the system as an exciton (an excitation wave, or an energy packet). The exciton can be trapped by an irregularity in the system that has an energy level lower than the energy of the exciton.

(iii) The energy is transferred by a long range interaction between D* and A, over a distance of several molecular diameters. The presence of several inert molecules (solvent) between D* and A is required because when the concentrations of D and A are large this mechanism reduces to the exciton mechanism (ii) [103]. It is sometimes said that in mechanism (iii) a 'virtual photon' passes from D* to A. A real photon is not emitted and re-absorbed in this mechanism.

(iv) The energy migrates by photon emission and reabsorption until it is absorbed by the trapping molecule. The energy of the photon emitted by the 'trap' is too low to be reabsorbed by the medium. This mechanism has received very little experimental support.

For mechanisms (ii)–(iv) the above required condition (a) for efficient energy transfer must be restricted somewhat by adding that the emission spectrum of the donor must overlap the absorption spectrum of the acceptor. This is because the de-excitation of the donor and the excitation of the acceptor do not involve simultaneous distortions of the molecules, i.e. the transitions are Franck–Condon transitions. In mechanism (i) the transfer might involve the intermediate formation of an activated complex, or excimer, that has an appreciable lifetime. The donor and acceptor molecules in the excimer could be distorted from their normal geometries, so the emission and absorption spectra of the participants would not be the same as those of the isolated molecules. Mechanism (i) does not violate the Franck–Condon principle; the principle simply does not apply because the donor and acceptor are in contact for a sufficiently long time for molecular distortion to have time to occur to assist in matching the donor and acceptor energy levels.

Triplet state energy transfer can occur in the solid, liquid and gas phases. In crystals transfer occurs by the exciton mechanism (ii) [104–107]. When the donor and acceptor molecules are in dilute solution in an inert rigid glass, transfer occurs by the virtual photon mechanism (iii) [108,109]. In the liquid phase the molecular diffusion mechanism (i) seems to be favoured [110]. In the gas phase the exciton and virtual photon mechanisms are highly improbable because the molecules are so far apart and it is assumed that triplet state energy transfer occurs by the molecular diffusion mechanism.

The triplet states involved in the above discussion are the lowest triplets of molecules that have singlet ground states. The transfer of energy from excited singlet states of these molecules is less well known. Presumably the above remarks apply also to transfer from singlet states, but their much shorter lifetimes greatly restrict the situations in which energy transfer can occur from them.

E. Neutral Free Radicals

A free radical can be defined as a substance that has one or more unpaired electrons in its valence shell. From this point of view extra electrons (solvated, trapped, etc.) and many positive and negative ions are also free radicals. The present section concerns neutral free radicals.

A free monoradical contains one unpaired electron and a diradical can be defined as an atom or molecule that reacts as though it has two monoradical functions[111]. Many excited molecules can therefore be classified as diradicals. Excited molecules were discussed in section III.D, so the diradicals considered here (section III.E.3) will be mainly those that behave more as a pair-of-doublets[111] state than as a singlet or triplet state species.

I. Formation

Neutral free radicals can be formed by the decomposition of excited molecules and ions, by ion–molecule and ion neutralization reactions and by reactions of other free radicals with molecules. For example:

$$(CH_3)_2CO^* \longrightarrow CH_3CO^\cdot + CH_3^\cdot \tag{66}$$

$$(CH_3COCH_3^{+\cdot})^* \longrightarrow CH_3CO^+ + CH_3^\cdot \tag{33}$$

$$(CH_3)_2CO^{+\cdot} + CH_3COCH_3 \longrightarrow (CH_3)_2\overset{+}{C}OH + \dot{C}H_2COCH_3 \tag{73}$$

$$(CH_3)_2CO^{-\cdot} + ROH \longrightarrow (CH_3)_2\dot{C}OH + RO^- \tag{53}$$

$$(CH_3)_2CO^{-\cdot} + (CH_3)_2\overset{+}{C}OH \longrightarrow (CH_3)_2CO^* + (CH_3)_2\dot{C}OH^\ddagger \tag{55}$$

$$CH_3^\cdot + CH_3COCH_3 \longrightarrow CH_4 + \dot{C}H_2COCH_3 \tag{74}$$

2. Reactions of monoradicals

Many kinds of free radical reactions are described in a new series of books[112], and the information available on the subject is extensive. The detailed discussion of monoradical reactions in carbonyl compounds would require a separate chapter in the present book. The discussion given here will be brief and the greatest emphasis will be given to reactions that involve the carbonyl group itself.

Typical reactions of thermal monoradicals in irradiated saturated carbonyl compounds at room temperature are hydrogen atom abstraction and radical–radical combination:

$$CH_3^{\cdot} + CH_3CHO \longrightarrow CH_4 + CH_3\overset{\cdot}{C}O \tag{75}$$

$$2\,CH_3\overset{\cdot}{C}O \longrightarrow (CH_3CO)_2 \tag{76}$$

The aldehydic hydrogen is easier to abstract than is a hydrogen atom from the alkyl group[113]. In a ketone, the hydrogen atoms on the α-carbons are more readily abstracted than are those on other carbon atoms[113].

In the condensed phases molecular motion is greatly hindered because there is only a small amount of free space between the molecules. Thus, when two radicals are formed by a reaction such as (66) in the liquid or solid phase, they are held next to each other for a finite time until one of the radicals escapes from its twin by a diffusive jump. During the time that they are held next to each other by the solvent 'cage' there is a finite probability that they will react together: for example,

$$(CH_3CO^{\cdot} + CH_3^{\cdot}) \longrightarrow CH_3COCH_3 \tag{77}$$

where the round brackets indicate that the radicals are in the same cage.

Disproportionation apparently does not occur to an appreciable extent between acyl ($R\overset{\cdot}{C}O$) radicals. Radicals that have the unpaired electron on the α-carbon might disproportionate if the β-carbon contains a hydrogen atom, but the disproportionation to combination

$$2CH_3\overset{\cdot}{C}HCOCH_3 \begin{cases} \longrightarrow CH_2{=}CHCOCH_3 + CH_3CH_2COCH_3 & (78) \\[2mm] \longrightarrow \begin{array}{l} CH_3CHCOCH_3 \\ | \\ CH_3CHCOCH_3 \end{array} & (79) \end{cases}$$

ratio, $k_{(78)}/k_{(79)}$ for these radicals is apparently not known.

Radicals do not add to the carbonyl π bond to an appreciable extent. Attachment of a radical to the oxygen of the carbonyl has too high an activation energy; for example reaction (80) would have an activation energy of 14 kcal/mole[114], so the more facile abstraction reaction (81), with activation energy of only 7–8 kcal/mole, occurs instead[113]. Attachment of a radical to the carbon of the carbonyl

$$CH_3CHO + C_2H_5 \begin{array}{c} \nearrow CH_3\dot{C}HOC_2H_5 \qquad (80) \\ \searrow CH_3\dot{C}O + C_2H_6 \qquad (81) \end{array}$$

has not been reported either, but possibly for a different reason; for example the addition reaction (82)[114] and the abstraction reaction (83)[113] have the same activation energy of about 8 kcal/mole, so

$$CH_3^* + CH_2O \begin{array}{c} \nearrow CH_3CH_2O^\cdot \qquad (82) \\ \searrow CH_4 + \dot{C}HO \qquad (83) \end{array}$$

perhaps the frequency factor of reaction (82) is much smaller than that of reaction (83); $A_{(83)} \approx 4 \times 10^7$ l/mole s[113].

If a carbonyl molecule contains much more energy than it needs to decompose, for example if it absorbed a relatively large quantum of energy, the radicals formed by its decomposition might have sufficient energy to decompose further:

$$CH_3COCH_3^{**} \longrightarrow CH_3CO^* + CH_3^* \qquad (84)$$
$$CH_3CO^* \longrightarrow CH_3^* + CO \qquad (85)$$

When acetone is irradiated at an elevated temperature, reaction (85) also occurs by thermal activation of the acyl radicals produced by reaction (66). The CH_3—CO dissociation energy is only 11 kcal/mole[115], so reaction (85) occurs readily at 100–200°c.

In summary, the most probable reactions of monoradicals in saturated carbonyl systems appear to be abstraction, combination and decomposition. In the presence of suitable alkyl radicals or olefinic compounds, the usual disproportionation or addition reactions, respectively, presumably occur as well.

3. Diradical versus concerted mechanism in isomerization reactions

Reactions such as (67)–(69) have more than one possible path. This will be illustrated by reaction (67). A concerted mechanism is possible in which the intermediate never has a significant amount of pair-of-doublets character: A C—C σ-bond would break, a C—C π-bond would form and a hydrogen atom would shift from one carbon to another, all at the same time (reaction 67a).

$$(67a)$$

On the other hand, the excited cyclopentanone molecule might decompose by cleavage of a C—CO bond, producing a diradical that would attain an appreciable amount of pair-of-doublets character as the two free-radical ends separated from each other[111]. The free-radical ends could then either disproportionate or recombine intra-molecularly (reaction (67b), or (86), respectively).

$$(67b)$$

$$(86)$$

There has been a lot of discussion about the relative merits of concerted and diradical mechanisms. Most of the experimental results can be equally well interpreted in terms of either path. Some results, to be mentioned later, seem to require that a diradical inter-mediate be formed. Many results have been presented as evidence in favour of concerted mechanisms[116-118], but it was not realized that the properties of a diradical are different in some respects from those of a mono-radical; most, if not all, of the results presented[116-118] can also be explained by diradical mechanisms[74,119]. There is not room in this article for a detailed critique of the reported results and conclu-sions. Space will only be taken to mention some of the properties of the systems that were neglected in the interpretation of the results. First, the period of rotation of the end groups in a diradical would probably be of the order of 10^{-12}–10^{-11} s, so intramolecular dispro-portionation or recombination could occur in 10^{-11} s or less. Thus, a very high concentration of an extremely reactive radical scavenger, e.g. 100 atm of oxygen, would have to be added to interfere with the intramolecular reactions of diradicals. The excited state precursors of the diradicals probably have longer lifetimes than the diradicals themselves, so the additive would probably attack the excited molecule more readily than it would the diradical (this is true for both excited singlet and triplet carbonyl molecules). Secondly, the small amount of free volume in the liquid phase reduces the rates of molecular motion and distortion in comparison with the rates in the gas phase. It can be shown with the aid of molecular models that the isomeric distribution of hept-5-enals formed from 2-methylcyclopentanone should be very similar from both diradical and concerted mechanisms in the liquid phase. The determination of the *cis/trans* ratio of isomers in the products of the vapour phase reaction would be much more significant. Thirdly, the only type of bond broken in the initial formation of the diradical would be a C—CO bond. And so on.

The following results require the postulation of alkyl acyl diradical intermediates. The vapour phase photolysis of *cis*- and *trans*-2,6-dimethylcyclohexanone each gave the same ratios of *cis*-and *trans*-1,2-dimethylcyclopentane and of unsaturated aldehydes[120]. The photolysis of androsterone caused it to invert to lumiandrosterone[121].

It may be noted that the results of the pyrolysis of cyclopropanes[122] seem to require the postulation of diradical intermediates and there is strong evidence for diradical formation in the pyrolysis of pyrazol-1-ines[123,124].

In some instances, due to molecular structure or lack of free space in the medium, the two radical groups in a diradical are held in close proximity to each other. When this happens the 'unpaired' electrons interact with each other and the diradical has some singlet or triplet character mixed with its pair-of-doublets character. Crude calculations indicate that the amount of interaction is appreciable between thermal radicals separated by less than about 3 Å (the exact interaction distance depends on the shapes and relative orientation of the unfilled orbitals). The lifetime of a diradical is expected to be longer if it has triplet character than if it has singlet character. These suggestions are strongly supported by recent results of various workers[125,126].

The diradicals formed from small-ring cyclic ketones or from any cyclic ketone in the confines of the liquid or solid phase should have considerable singlet or triplet character. If the dissociation of radiation-excited ketones occurs by predissociation (section III.D.2.a), then the diradicals formed from triplet molecules have triplet character, at least initially; those formed from singlet molecules initially have singlet character. The triplet character of a diradical must be inverted to singlet character before internal disproportionation or recombination can occur: disproportionation in the triplet state is highly endothermic because a triplet π,π^* state would have to be formed, and recombination is spin forbidden. This is why diradicals with triplet character should have longer lifetimes than those with singlet character.

The reactions that have the greatest appearance of occurring by concerted mechanisms might ultimately be found to have singlet state precursors that may or may not acquire sufficient pair-of-doublets character to classify them as diradicals.

IV. RADIATION CHEMISTRY OF SPECIFIC COMPOUNDS

It has been general practice in radiation chemistry to interpret results in terms of neutral free radical reactions as long as these reactions were consistent with the results. Radiolysis systems were tested for the presence of free radicals by adding solutes that are very reactive towards radicals. However, recent developments in radiation chemistry, especially concerning the reactions of ions and of excited molecules, have shown that much of the older work has several possible interpretations. Many of the solutes used, such as iodine and conjugated diolefins, react with ions and excited molecules as well as with neutral free radicals, so more selective solutes and other techniques will have to be used.

In this section, the radiation chemistry of carbonyl compounds will be illustrated by briefly discussing the radiolyses of acetone and cyclopentanone.

The yield of a product P is reported as G, which is the number of molecules of P formed per 100 ev of energy absorbed by the system.

A. Acetone

The yields of products from the γ-radiolysis of liquid acetone are sensitive to the water content of the liquid[71,72] and an extensive study of the radiolysis of dry acetone has not been made. However, the yields of the major products from wet acetone, containing about 0.1% water, will give a general idea of the nature of the radiolysis (see Table 3).

The yield of methane was greatly reduced by the addition of 10^{-2} molar iodine or diphenylpicrylhydrazyl (DPPH) to the acetone before radiolysis[127]. It was suggested that 85% of the methane from pure acetone was formed by the reaction of thermal methyl radicals[127]:

$$CH_3^{\cdot} + CH_3COCH_3 \longrightarrow CH_4 + \dot{C}H_2COCH_3 \qquad (74)$$

TABLE 3. Yields of major products from the
γ-radiolysis of liquid acetone[a].

Product	G
H_2	$0·87^b$
CO	$0·83^b$
CH_4	$2·6^b$
C_2H_6	$0·48^b$
$(CH_3COCH_2)_2$	$1·0^c$
CH_3COOH	$0·9^c$
CH_3COOR^d	$0·7^c$
$CH_3COCH_2COCH_3$	$0·41^c$
$(CH_3)_2CHOH$	$0·33^c$
$((CH_3)_2COH)_2$	$0·31^c$

   [a] Dose 10^{20} ev/ml; dose rate 10^{17} ev/ml min; temperature $21 \pm 6°c$; the acetone contained about $0·1\%$ of water.

   [b] P. Ausloos and J. F. Paulson, *J. Amer. Chem. Soc.*, **80**, 5117 (1958).

   [c] J. Kučera, *Coll. Czech. Chem. Commun.*, **30**, 3080 (1965).

   [d] Unresolved acetates.

The presence of iodine would cause reaction (87) to compete with

$$CH_3^* + I_2 \longrightarrow CH_3I + I \qquad (87)$$

reaction (74), thereby reducing the methane yield. The yields of acetonylacetone and acetylacetone were greatly reduced by the presence of oxygen in the system[128]. This is consistent with their formation by the combination of radicals, reactions (88) and (89), and

$$2\ CH_3CO\dot{C}H_2 \longrightarrow (CH_3COCH_2)_2 \qquad (88)$$

$$CH_3CO^· + \dot{C}H_2COCH_3 \longrightarrow CH_3COCH_2COCH_3 \qquad (89)$$

with their reduction in yield by the reaction of the radicals with oxygen, reaction (90).

$$R^· + O_2 \longrightarrow RO_2^· \qquad (90)$$

However, recent advances indicate that oxygen[74,75], iodine and DPPH might have attacked triplet state or ionic precursors of the radicals instead of the radicals themselves. Pulse radiolysis studies of liquid acetone have shown that roughly $1–2\ G$ units of triplet states and roughly $1–2\ G$ units of negative ions are scavengeable by 10^{-2} molar concentrations of solutes such as anthracene[129]. The yield of free ions of each sign (see section III.C.2.e) is also about

$1\ G$ unit[130], and the negative ions observed by pulse radiolysis[129] were probably free ions.

Much more work must be done to resolve the ambiguities of the proposed mechanisms of product formation.

The yields of hydrogen, carbon monoxide and ethane were only slightly affected by the presence of 10^{-2} molar iodine or DPPH in the acetone[127]. These results are consistent with the following suggestions. (a) The acetone species that is a precursor to hydrogen is in a higher excited state than is the species that is a precursor to methane[131]. The more highly excited species presumably has a shorter lifetime and is therefore less subject to attack by a solute. (b) Carbon monoxide is formed by reaction (85). (c) Ethane is formed mainly by the combination of methyl radicals in spurs (see section II.B) or by ion–molecule reactions. The combination of methyl radicals in spurs is further indicated by the increase in ethane yield with increasing LET of the radiation[132] (see section II.A.4).

The yield of acetic acid was reduced by drying the acetone more thoroughly[72]. It was therefore suggested that the acetic acid was formed by the following mechanism[72]:

$$CH_3COCH_3^* \longrightarrow CH_4 + CH_2CO \tag{91}$$

$$CH_2CO + H_2O \longrightarrow CH_3COOH \tag{92}$$

The isopropanol and pinacol may have been formed by reactions (93) and (94), respectively. The $(CH_3)_2\dot{C}OH$ radicals could be

$$2(CH_3)_2\dot{C}OH \begin{cases} \longrightarrow CH_3CHOHCH_3 + CH_3COCH_3 & (93) \\ \longrightarrow ((CH_3)_2COH)_2 & (94) \end{cases}$$

formed in several ways, for example,

$$(CH_3)_2CO^- + (CH_3)_2\overset{+}{C}OH \longrightarrow (CH_3)_2CO^* + (CH_3)_2\dot{C}OH^{\ddagger} \tag{55}$$

$$(CH_3)_2\dot{C}OH^{\ddagger} + M \longrightarrow (CH_3)_2\dot{C}OH + M \tag{95}$$

It is unfortunate that liquid products are so much more difficult to analyse than gaseous products, and that the techniques are so different. These are doubtless the reasons that most workers report only gaseous products, and that most of those who report liquid products neglect gaseous products.

B. Cyclopentanone

The yields of the major products of the radiolysis of liquid cyclopentanone are listed in Table 4. Many of the yields are similar to those of the analogous products from acetone (Table 3). Hydrogen and carbon monoxide are produced in nearly the same amounts in the two systems. The total dimer yield in cyclopentanone ($G = 1\cdot1$) is similar to that of the total dimer (acetonylacetone plus pinacol) in acetone ($G = 1\cdot3$). Cyclopentanol in cyclopentanone ($G = 0\cdot8$) is nearly equivalent to the sum of isopropanol and acetate in acetone ($G = 1\cdot0$). The primary radiolytic reactions are evidently quite similar in the two compounds.

The effects of solutes on the radiolysis of cyclopentanone[75,133,134] indicate that cyclopentanol, pent-4-enal and the dimers each have at least one long-lived precursor and that hydrogen, carbon monoxide and ethylene do not have long-lived precursors. Benzene is a triplet-state sensitizer and a relatively inefficient radical scavenger, whereas

TABLE 4. Yields of major products from the γ-radiolysis of liquid cyclopentanone[a,b].

Product	G
H_2	0·78
CO	0·67
C_2H_4	0·80
C_4H_8 isomers	0·11
$CH_2{=}CHCH_2CH_2CHO$	0·8
	0·8
	0·10
	0·67
total 'dimer'	1·1
polymer[c], $(C_5H_8O)_{4.6}$	0·4

[a] D. L. Dugle and G. R. Freeman, *Trans. Faraday Soc.*, **61**, 1166, 1174 (1965).

[b] Yields extrapolated to zero dose from 10^{19}–10^{20} ev/ml; dose rate 10^{18} ev/ml min; temperature $25 \pm 2°$c; the cyclopentanone contained 0·02% of water.

[c] Excluding dimer; 1·9 G units of monomer reacted to form polymer that had an average degree of polymerization of 4·6.

penta-1,3-diene is an efficient scavenger of both carbonyl triplet-state energy and free monoradicals. A comparison of the solute effects of benzene and penta-1,3-diene indicates that the long-lived precursors of cyclopentanol and of the dimers are monoradicals and that that of pent-4-enal is the lowest triplet cyclopentanone molecule[75,133].

Only about 70 % of the pent-4-enal has triplet cyclopentanone as a precursor. The rest of this product has a shorter-lived precursor, perhaps an excited singlet state[75].

The controversy over diradical and concerted mechanisms, presented for isomerization reactions in section III.E.3, also enters the discussion of mechanisms of formation of carbon monoxide, ethylene and the isomers of C_4H_8 (cyclobutane, but-1-ene, and methylcyclopropane) from cyclopentanone. However, both mechanisms require that the molecules that decompose to ethylene have more vibrational energy than do those that yield C_4H_8, and that the latter have more vibrational energy than do those that form pent-4-enal. The triplet-state molecules have such long lifetimes ($\geq 10^{-7}$ s[75]) that they are probably in their lowest vibrational level before they decompose. It is therefore probable that carbon monoxide, ethylene and C_4H_8 have mainly excited singlet state precursors. These precursors have very short lifetimes and solutes at usual concentrations have only slight effects on them and therefore on their product yields[75].

The formation of pent-4-enal is a ring-opening reaction that is expected to have a positive volume of activation, i.e. the molar volume of the activated complex is expected to be greater than that of the reactant. In agreement with this, the rate of formation of pent-4-enal decreased when the density of the liquid cyclopentanone was increased either by applying pressure to the liquid or by cooling it[119]. The isomerization was nearly completely suppressed by freezing the ketone either by the application of 5000 atm of pressure at 25° or by cooling at atmospheric pressure to $-80°$[119]. Unfortunately, it was not possible to determine the volumes of activation of the individual steps of the radiation-induced isomerization, but the overall value at 1 atm was $\Delta V^{\ddagger} = 2\cdot4$ ml/mole and it increased with increasing pressure.

A detailed investigation of the ionic reactions in irradiated ketones remains to be done. If G (total ionization) $\approx 4\cdot0$ in liquid cyclopentanone, then G (geminate neutralization) $\approx 3\cdot4$ and G (free ions) $\approx 0\cdot6$ (see Table 2). The free ions apparently do not produce a major portion of the hydrogen[133].

V. REFERENCES

1. H. Hurzeler, M. G. Inghram and J. D. Morrison, *J. Chem. Phys.* **28**, 76 (1958).
2. J. N. Pitts, Jr., and J. K. S. Wan in *The Chemistry of the Carbonyl Group*, Vol. 1 (Ed. S. Patai), Interscience, London, 1966, pp. 823–916.
3. H. E. Johns and J. S. Laughlin in *Radiation Dosimetry* (Eds. G. J. Hine and G. L. Brownell), Academic Press, New York, 1956, pp. 49–124.
4. R. D. Evans, *The Atomic Nucleus*, McGraw-Hill, New York, 1955, Chapters 18–21.
5. R. L. Platzman, *Vortex*, **23** (8), 1 (1962).
6. H. A. Bethe and J. Ashkin in *Experimental Nuclear Physics*, Vol. 1, (Ed. E. Segré), John Wiley and Sons, New York, 1953, pp. 166–357.
7. U. Fano in *Studies in Penetration of Charged Particles in Matter*, Natl. Acad. Sci. - NRC Publication 1133, 1964, pp. 281–352, see especially pp. 290–315.
8. Calculated from the wavelengths of the spectroscopic K_α lines, obtained from a table compiled by P. W. Zingaro, North American Philips Co., Mount Vernon, New York.
9. The ionization potential of a carbon atom is 11.3 ev, K. Watanabe, T. Nakayama and J. Mottl, *J. Quantum Spectr. Radiation Transfer*, **2**, 369 (1962).
10. R. L. Platzman, *Radiation Res.*, **2**, 1 (1955).
11. A. Kupperman and L. M. Raff, *J. Chem. Phys.*, **37**, 2497 (1962).
12. J. A. Simpson and S. R. Mielczarek, *J. Chem. Phys.*, **39**, 1606 (1963).
13. E. N. Lasseter and S. A. Francis, *J. Chem. Phys.*, **40**, 1208 (1964).
14. J. P. Doering, *J. Chem. Phys.*, **46**, 1194 (1967).
15. G. R. Freeman and J. M. Fayadh, *J. Chem. Phys.*, **43**, 86 (1965).
16. G. R. Freeman, *J. Chem. Phys.*, **43**, 93 (1965).
17. G. R. Freeman, *J. Chem. Phys.*, **46**, 2822 (1967).
18. For example, U. Fano in *Studies in Penetration of Charged Particles in Matter*, Natl. Acad. Sci - NRC Publication 1133, 1964, p. 324.
19. U. Fano in *Studies in Penetration of Charged Particles in Matter*, Natl. Acad. Sci. - NRC Publication 1133, 1964, p. 282.
20. M. S. Livingston and H. A. Bethe, *Revs. Mod. Phys.*, **9**, 245 (1937), see especially pp. 262–265.
21. Calculated from data on p. 345 of U. Fano in *Studies in Penetration of Charged Particles in Matter*, Natl. Acad. Sci. - NRC Publication 1133, 1964.
22. R. D. Evans, *The Atomic Nucleus*, McGraw-Hill, New York, 1955, pp. 633–637.
23. J. W. T. Spinks and R. J. Woods, *An Introduction to Radiation Chemistry*, John Wiley and Sons, New York, 1964. See especially the items under 'Linear energy transfer' in the index.
24. R. H. Schuler and N. F. Barr, *J. Amer. Chem. Soc.*, **78**, 5756 (1956).
25. J. W. T. Spinks and R. J. Woods, *An Introduction to Radiation Chemistry*, John Wiley and Sons, New York, 1964, p. 36.
26. A. H. Samuel and J. L. Magee, *J. Chem. Phys.*, **21**, 1080 (1953).
27. A. Kupperman in *The Chemical and Biological Action of Radiations*, Vol. 5 (Ed. M. Haissinski), Academic Press, New York, 1961, p. 85.

28. A. Mozumder and J. L. Magee, *Radiation Res.*, **28**, 215 (1966).
29. F. H. Field and J. L. Franklin, *Electron Impact Phenomena*, Academic Press, New York, 1957, pp. 106–111.
30. G. R. Freeman in *Radiation Research*, (Ed. G. Silini), North Holland Publishing Co., Amsterdam, 1967, p. 113.
31. M. S. B. Munson and F. H. Field, *J. Amer. Chem. Soc.*, **87**, 3294 (1965).
32. V. L. Tal'rose and A. K. Lyubimova, *Dokl. Akad. Nauk SSSR.*, **86**, 909 (1952).
33. F. W. Lampe and F. H. Field, *Tetrahedron*, **7**, 189 (1959).
34. F. W. Lampe, J. L. Franklin and F. H. Field in *Progress in Reaction Kinetics*, Vol. 1 (Ed. G. Porter), Pergamon Press, London, 1961, p. 68.
35. A. G. Harrison, A. Ivko and D. van Raalte, *Can. J. Chem.*, **44**, 1625 (1966).
36. For a brief review see E. Lindholm, *Adv. Chem. Ser.*, (**58**) 1 (1966).
37. J. J. Thomson and G. P. Thomson, *Conduction of Electricity Through Gases*, 3rd Ed., Cambridge University Press, 1928, pp. 133–135.
38. A. von Engel, *Ionized Gases*, 2nd Ed., Clarendon Press, Oxford, 1965, pp. 122–127.
39. H. Schnyders, S. A. Rice and L. Meyer, *Phys. Rev.*, **150**, 127 (1966).
40. T. F. O'Malley, *Phys. Rev.*, **130**, 1020 (1963).
41. J. Jortner, N. R. Kestner, S. A. Rice and M. H. Cohen, *J. Chem. Phys.*, **43**, 2614 (1965).
42. J. Jortner, S. A. Rice and E. G. Wilson in *Metal Ammonia Solutions*, W. A. Benjamin, New York, 1964, pp. 222–276, see especially p. 262.
43. L. Meyer, H. T. Davis, S. A. Rice and R. J. Donnelly, *Phys. Rev.*, **126**, 1927 (1962).
44. A. J. Dahm and T. M. Sanders, *Phys. Rev. Letters*, **17**, 126 (1966).
45. Calculated from data of D. S. Burns, *Adv. Chem. Ser.*, (**50**), 82 (1965).
46. Calculated from data of E. J. Hart, quoted by M. Anbar, *Adv. Chem. Ser.*, (**50**), 59 (1965).
47. K. Schmidt and W. Buck, *Science*, **151**, 70 (1966).
48. S. Glasstone, *Textbook of Physical Chemistry*, 2nd Ed., Van Nostrand, Toronto, 1946, p. 909.
49. A good summary of most of the proposed models of electrons in polar liquids is included in reference 42.
50. K. Hiroike, N. R. Kestner, S. A. Rice and J. Jortner, *J. Chem. Phys.*, **43**, 2625 (1965).
51. W. Kauzmann, *Quantum Chemistry*, Academic Press, New York, 1957, p. 188.
52. L. Bouby, F. Fiquet-Fayard and H. Abgrall, *Compt. Rend.*, **261**, 4059 (1965).
53. R. N. Compton, L. G. Christophorou, G. S. Hurst and P. W. Reinhardt, *J. Chem. Phys.*, **45**, 4634 (1966).
54. S. W. Benson, *J. Chem. Educ.*, **42**, 502 (1965).
55. G. E. Adams and R. D. Sedgwick, *Trans. Faraday Soc.*, **60**, 865 (1964).
56. J. J. J. Myron and G. R. Freeman, *Can. J. Chem.*, **43**, 381 (1965).
57. L. M. Dorfman and M. S. Matheson in *Progress in Reaction Kinetics*, Vol. 3 (Ed. G. Porter), Pergamon Press, London, 1965, p. 237.
58. P. H. Rieger and G. K. Fraenkel, *J. Chem. Phys.*, **37**, 2811 (1962).

59. F. S. Dainton, T. J. Kemp, G. A. Salmon and J. P. Keene, *Nature*, **203**, 1050 (1964).

60. T. Shida and W. H. Hamill, *J. Amer. Chem. Soc.*, **88**, 3683 (1966).

61. For example, E. W. McDaniel, *Collision Phenomena in Ionized Gases*, John Wiley and Sons, New York, 1964, p. 382.

62. B. C. de Souza and J. H. Green, *J. Chem. Phys.*, **46**, 1421 (1967).

63. W. J. Holtslander and G. R. Freeman, *Can. J. Chem.*, **45**, 1661 (1967).

64. Calculated for a pressure of 400 torr from data given by F. J. Lindars and C. N. Hinshelwood, *Proc. Roy. Soc. (London)*, **A231**, 178 (1955).

65. A. E. Stearn and H. E. Eyring, *J. Chem. Phys.*, **3**, 778 (1935).

66. G. R. Freeman and W. J. Holtslander, *Chem. Commun.*, 205 (1967).

67. M. S. B. Munson, *J. Amer. Chem. Soc.*, **87**, 5313 (1965).

68. E. W. McDaniel, *Collision phenomena in Ionized Gases*, John Wiley and Sons, New York, 1964, Chapter 12.

69. Calculated from information in G. R. Freeman, *J. Chem. Phys.*, **39**, 988 (1963) and *J. Chem. Phys.*, **41**, 901 (1964).

70. Calculated by the method in reference 17; G. R. Freeman, *Adv. Chem. Ser.* **82**, 339 (1968).

71. R. Barker, *Trans. Faraday Soc.*, **59**, 375 (1963).

72. J. Kučera, *Collection Czech. Chem. Commun.*, **31**, 355 (1966).

73. J. N. Pitts, Jr., F. Wilkinson and G. S. Hammond in *Advances in Photochemistry*, Vol. 1 (Eds. W. A. Noyes, Jr., G. S. Hammond and J. N. Pitts, Jr.), Interscience, New York, 1963, p. 1.

74. A. Singh and G. R. Freeman, *J. Phys. Chem.*, **69**, 666 (1965).

75. D. L. Dugle and G. R. Freeman, *Trans. Faraday Soc.*, **61**, 1174 (1965).

76. J. G. Calvert and J. N. Pitts, Jr., *Photochemistry*, John Wiley and Sons, New York, 1966, Chapter 4.

77. J. B. Birks, *Scintillation Counters*, McGraw-Hill, New York, 1953.

78. *Comparative Effects of Radiation* (Eds. M. Burton, J. S. Kirby-Smith and J. L. Magee), John Wiley and Sons, New York, 1960.

79. *Exciton Symposium* in *Radiation Res.*, **20**, 53–158 (1963).

80. H. Sponer, *Radiation Res.*, *Suppl.*, **1**, 558 (1959).

81. E. W. Abrahamson, J. G. F. Littler and K. Vo, *J. Chem. Phys.*, **44**, 4082 (1966).

82. C. S. Parmenter, *J. Chem. Phys.*, **41**, 658 (1964).

83. R. B. Cundall and A. S. Davies, *Trans. Faraday Soc.*, **62**, 2444 (1966).

84. R. Srinivasan, *J. Amer. Chem. Soc.*, **83**, 4344 (1961).

85(a). J. N. Pitts, Jr. and I. Norman, *J. Amer. Chem. Soc.*, **76**, 4815 (1954).
    (b) L. D. Hess, J. L. Jacobson, K. Schaffner and J. N. Pitts, Jr., *J. Amer. Chem. Soc.*, **89**, 3684 (1967).

86. P. Ausloos and R. E. Rebbert, *J. Amer. Chem. Soc.*, **86**, 4512 (1964).

87. I. Orban, K. Schaffner and O. Jeger, *J. Amer. Chem. Soc.*, **85**, 3033 (1963).

88. R. S. Tolberg and J. N. Pitts, Jr., *J. Amer. Chem. Soc.*, **80**, 1304 (1958).

89. W. M. Moore, G. S. Hammond and R. P. Foss, *J. Amer. Chem. Soc.*, **83**, 2789 (1961).

90. C. Walling and M. J. Gibain, *J. Amer. Chem. Soc.*, **87**, 3361 (1965).

91. D. R. Arnold, R. L. Hinman and A. H. Glick, *Tetrahedron Letters*, 1425 (1964).

92. J. L. Kropp and M. Burton, *J. Chem. Phys.*, **37**, 1742 (1962).

93. J. Yguerabide, N. A. Dillon and M. Burton, *J. Chem. Phys.*, **40**, 3040 (1964).
94. V. A. Krongauz and I. N. Vasil'ev, *Kinetics Catalysis USSR* (Engl. Transl.), **4**, 55 (1963).
95. W. Van Dusen and W. H. Hamill, *J. Amer. Chem. Soc.*, **84**, 3648 (1962).
96. E. A. Rojo and R. R. Hentz, *J. Phys. Chem.*, **69**, 3024 (1965).
97. G. S. Hammond, N. J. Turro and P. A. Leermakers, *J. Phys. Chem.*, **66**, 1144 (1962).
98. J. L. Michael and W. A. Noyes, Jr., *J. Amer. Chem. Soc.*, **85**, 1027 (1963).
99. R. E. Rebbert and P. Ausloos, *J. Amer. Chem. Soc.*, **86**, 4803 (1964).
100. R. F. Borkman and D. R. Kearns, *J. Amer. Chem. Soc.*, **88**, 3467 (1966).
101. P. J. Wagner, *J. Amer. Chem. Soc.*, **88**, 5672 (1966).
102. S. C. Curran, *Luminescence and the Scintillation Counter*, Butterworths, London, 1953.
103. T. Förster, *Ann. Physik*, **2**, 55 (1948).
104. H. Sternlicht and H. M. McConnell, *J. Chem. Phys.*, **35**, 1793 (1961).
105. M. Kasha, *Radiation Res.*, **20**, 55 (1963).
106. P. Avakian and R. E. Merrifield, *Phys. Rev. Letters*, **13**, 541 (1964).
107. R. Voltz, H. Dupont and T. A. King, *Nature*, **211**, 405 (1966).
108. S. Siegel and H. Judeikis, *J. Chem. Phys.*, **41**, 648 (1964).
109. R. E. Kellogg, *J. Chem. Phys.*, **41**, 3046 (1964).
110. G. Porter and F. Wilkinson, *Proc. Roy. Soc. (London)*, **A264**, 1 (1961).
111. G. R. Freeman, *Can. J. Chem.*, **44**, 245 (1966).
112. G. H. Williams (Ed.), *Advances in Free-Radical Chemistry*, Vol. 1, Academic Press, New York, 1965, and subsequent volumes.
113. Derived from information in Table 10 of A. F. Trotman-Dickenson, in *Advances in Free-Radical Chemistry*, Vol. 1 (Ed. G. H. Williams), Academic Press, New York, 1965, pp. 1–38.
114. K. M. Bansal and G. R. Freeman, *J. Amer. Chem. Soc.*, **88**, 4326 (1966).
115. E. Murad and M. G. Inghram, *J. Chem. Phys.*, **41**, 404 (1964).
116. R. Srinivasan in *Advances in Photochemistry*, Vol. 1 (Eds. W. A. Noyes, Jr., G. S. Hammond and J. N. Pitts, Jr.) Interscience, New York, 1963, pp. 83–114, and references therein.
117. R. Srinivasan and S. E. Cremer, *J. Amer. Chem. Soc.*, **87**, 1647 (1965).
118. R. Srinivasan and S. E. Cremer, *J. Phys. Chem.*, **69**, 3145 (1965).
119. D. L. Dugle and G. R. Freeman, *J. Phys. Chem.*, **70**, 1256 (1966).
120. B. Rickborn, R. L. Alumbaugh and G. O. Pritchard, *Chem. Ind. (London)*, 1951 (1964).
121. P. de Mayo in *Advances in Organic Chemistry; Methods and Results*, Vol. 2 (Eds. R. A. Raphael, E. C. Taylor and H. Wynberg) Interscience, New York, 1960, pp. 367–425, see p. 382.
122. B. S. Rabinovitch, E. W. Schlag and K. B. Wiberg, *J. Chem. Phys.*, **28**, 504 (1958) and later work of S. W. Benson and coworkers [*J. Amer. Chem. Soc.*, **84**, 3411 (1962)], H. M. Frey and coworkers [*J. Chem. Soc.*, **1965**, 191], and others.
123. R. J. Crawford and A. Mishra, *J. Amer. Chem. Soc.*, **88**, 3963 (1966).
124. R. J. Crawford and D. M. Cameron, *Can. J. Chem.*, **45**, 691 (1967).
125. J. R. Fox and G. S. Hammond, *J. Amer. Chem. Soc.*, **86**, 4031 (1964).
126. P. Scheiner, *J. Amer. Chem. Soc.*, **88**, 4759 (1966).

127. P. Ausloos and J. F. Paulson, *J. Amer. Chem. Soc.*, **80**, 5117 (1958).
128. P. Drienovsky, *Collection Czech. Chem. Commun.*, **27**, 1450 (1962).
129. S. Arai and L. M. Dorfman, *J. Phys. Chem.*, **69**, 2239 (1965).
130. Calculated by the method in reference 15, using G(total ionization) $= 4$.
131. A. G. Leiga and H. A. Taylor, *J. Chem. Phys.*, **41**, 1247 (1964).
132. R. Barker, *Nature*, **192**, 62 (1961).
133. D. L. Dugle and G. R. Freeman, *Trans. Faraday Soc.*, **61**, 1166 (1965).
134. W. W. Bristowe, M. Katayama and C. N. Trumbore, *J. Phys. Chem.*, **69**, 807 (1965).

Author index

This author index is designed to enable the reader to locate an author's name and work with the aid of the reference numbers appearing in the text. The page numbers are printed in normal type in ascending numerical order, followed by the reference numbers in brackets. The numbers in *italics* refer to the pages on which the references are actually listed. Bold face numbers indicate pages where references are given in the body of the chapter.

If reference is made to the work of the same author in different chapters, the above arrangement is repeated separately for each chapter.

Subject index